RESEARCH TRACKS IN URBANISM

Research Tracks in Urbanism: Dynamics, Planning and Design in Contemporary Urban Territories

Edited by

Alessia Allegri

Lisbon School of Architecture, Universidade de Lisboa, Lisbon, Portugal

Angélica Benatti Alvim

Universidade Presbiteriana Mackenzie, São Paulo, Brazil

Eunice Helena Abascal

Universidade Presbiteriana Mackenzie, São Paulo, Brazil

Joaquin Sabaté

UPC_ETSAB Barcelona School of Architecture, Barcelona, Spain

João Pedro Costa

Lisbon School of Architecture, Universidade de Lisboa, Lisbon, Portugal

Maria Cristina Schicchi

Pontifícia Universidade Católica de Campinas, São Paulo, Brazil

CRC Press
Taylor & Francis Group
Boca Raton London New York

CRC Press is an imprint of the
Taylor & Francis Group, an **informa** business
A BALKEMA BOOK

CRC Press/Balkema is an imprint of the Taylor & Francis Group, an informa business

Typeset by Integra Software Services Pvt. Ltd., Pondicherry, India

Library of Congress Cataloging-in-Publication Data

A catalog record has been requested for this book

Published by: CRC Press/Balkema
Schipholweg 107C, 2316 XC Leiden, The Netherlands
e-mail: enquiries@taylorandfrancis.com
www.routledge.com – www.taylorandfrancis.com

ISBN: 978-1-032-10788-2 (Hbk)
ISBN: 978-1-032-11625-9 (Pbk)
ISBN: 978-1-003-22085-5 (eBook)
DOI: 10.1201/9781003220855

Table of contents

Preface

The first International Research Seminar on Urbanism (SIIU - *Seminario Internacional de Investigación en Urbanismo*) was held 14 years ago in Barcelona. It was organised by the Department of Urbanism and Spatial Planning of the Polytechnic University of Cataluña (UPC - *Universidad Politécnica de Cataluña*) and aimed to bring together young researchers who were conducting their master's and doctoral studies. In a short period of time, the Seminar raised considerable interest beyond the select UPC context, and the number of people interested or wanting to participate increased considerably. The participation of Latin American researchers widely exceeded initial expectations and so in 2013 the V Seminar was, for the first time, organised in two locations at the same time, organised in Barcelona and at a Latin American university, establishing a common platform for discussion. From that point onwards, there was a consolidation of this shared space, allowing for reflection on developments in urbanism and transformation of cities and territories in different Latin American countries.

The Portuguese language has been present ever since the first seminar, and has become increasingly important among participants over time. In 2020, the Seminar was held by the University of Lisbon and by the University of Mackenzie in São Paulo, drawing greater attention to lusophone research. Undoubtedly, these studies focused on issues that are particular to Portugal and Brazil. However, these also appear to be increasingly common in the Ibero-American context.

The main Latin American universities organised some of these Seminars together with Barcelona. This has allowed for an exchange of reflections and a continuous transfer of fruitful ideas, and has highlighted new ways of addressing common problems. Above all, it has provided a platform for discussion between young researchers, where they have shared endeavours, been enrichened, and consolidated a space for discussion on the main challenges our cities and territories currently face.

The idea for this publication emerged from this. Benefiting from the long journey of these Seminars and the circa two thousand papers presented over the years, this is an opportunity to reflect on the challenges our cities and territories face today. By doing this on both sides of the Atlantic, we acknowledge that we share many concerns on a common agenda, even though such an agenda has its own specificities in Europe, particularly in Southern Europe, and in Brazil, and by extension in Latin America.

The aim is to draw a "map" with the most relevant research topics in our field and illustrate it with some examples. The underlying idea is to rethink the disciplinary foundations that guide urban and city design, opening up to new challenges. Furthermore, we intend to reflect, from an academic perspective, on urban planning as an inescapable obligation.

This contribution is composed of two modestly summarized texts, evoking two distinct realities which reflect today's truly global world: one European and another South American. They are joined by papers selected in Lisbon and in São Paulo, addressing a broad range of topics, from the metropolis to public space, from urban living to mobility, from the informal to the landscape, from urban morphology to social perspectives, from territorial interventions to the publication of urbanism. These papers aim to understand and find room for innovation.

Maybe the *Global Village* metaphor has never been more accurate than it is today, where societies join forces in the fight against the pandemic, in a global coordinated effort, possibly never tested before in the known history of Humankind. Although we are sure that in the past some other shared demands have united the different peoples of the world, this has never been so strongly necessary, mainly in so far as the global scientific community is concerned. This is a fight for the survival of a society. However, we should not lose sight of what we are fighting for. We fight together for people. Not just for the abstract value of Human life, but for life in society as a whole, including its moral and ethical aspects. The topics of this book are based on this claim, on what makes it possible. We do not build our lives in a vacuum, or in distant *Invisible Cities*, but through a higher value, which represents physical life in society: the City.

We constantly make value judgements on technology and how it promotes or limits our full meaning of life. Much has already been discussed about its virtues as well as its harmful effects, and much more will be discussed. Nonetheless, it was through technology that we have, for now, been able to win some of the battles that the pandemic has forced us to fight. Over the last few months, the paradox of the importance that new technologies and social media can have in establishing a new virtual public space have also been noted. In fact, during the lockdown, almost all social activities took place in the shapeless and intangible space of the Internet. In six months, the one thousand participants of the SIIU2020 went from an in-person to on-line register, and the event took place, regardless. It was in this virtual public space, that identities, meanings, and relationships were nourished, articulated and maintained.

This book is a spin-off of the SIIU2020. Inspired by the contents of twelve research seminars, a group of researchers from the universities of Barcelona, Lisbon and São Paulo discuss the contemporary agenda of research in Urbanism. Following the conference, 35 original research papers were brought together with different perspectives about such an agenda. They were submitted to a double-blind peer review, which selected the "best papers" from both SIIU2020 locations.

Barcelona, Lisbon, São Paulo, February 2021
The Editors

Research Tracks in Urbanism: Dynamics, Planning and Design in Contemporary Urban Territories – Alegri et al (eds)

Acknowledgments

The Editors wish to acknowledge with a big debt of gratitude to all those who have given their time and experiences during the preparation of this book.

Our particular thanks go to the AEAULP - Academia de Escolas de Arquitectura e Urbanismo de Língua Portuguesa and to Professor Conceição Trigueiros.

We would also like to acknowledge the funding support by: FCT Strategic Project UIDB/04008/2020; the PhD Program in Urbanism of the Lisbon School of Architecture, University of Lisbon; the CAPES/PAEP (AUXPE 0107/2020; Process: 88881.470831/2019-01); Degree Post-Graduate Program in Architecture and Urbanism - CAPES/PROEX - (Process: 23038.001402/2021-32); UPM/CAPES/PrInt - Mackenzie Presbyterian University Internalization Program (Process: 745884P), Mackpesquisa, and Mackenzie Presbyterian University.

We greatly appreciate the help of Elisabete Rolo, Rafaela Costa, Miguel Rafael, Miguel Miranda, Alexandra Luis, Filipa Nogueira, Mafalda Casais and Caterina Anastasia.

Finally, we are profoundly grateful to Luis Miguel Ginja for his enthusiastic response and generosity that made the work so enjoyable.

Acknowledgments

[illegible]

[illegible]

[illegible]

[illegible]

[illegible]

[Re]Urban practice, towards a European research agenda for urbanism: Nature as reference, human as critical, process as fundamental

Alessia Allegri, João Pedro Costa, Luís Miguel Ginja, Melisa Pesoa, Javier Rocamonde & Joaquín Sabaté

After about 311 papers presented and published in the International Research Seminar on Urbanism that took place in Lisbon and São Paulo, and looking at the more than two thousand papers presented over the past 14 years, we wonder if there is a common research agenda between so many countries and participants. Our key object of study, the city and the urban, undergoes different fluctuations all over the world, sometimes in opposite directions: rapid growth versus restoration of the existing, searching for heritage roots versus a growing extension of a general city; concerns with poverty and poor housing versus the design of impressive leisure spaces. Thus, it appeared that it would be difficult to find common denominators in different and distant countries and research centres.

However, the seminars demonstrated how the increase of networks and personal and institutional exchanges have resulted in a fruitful and continuous transfer of ideas, showing new ways of tackling problems and challenges, and finding answers. The joint intellectual adventures between researchers on both sides of the Atlantic, and the continuous talks about shared topics of mutual interest seem to lead to a convergence, albeit little expected at first. This is what the papers presented over the past years have demonstrated. Although the ways to address problems are still site-specific, in an increasingly global world such problems affect different parts of the planet in a similar way. It is the greatest challenges that shape what we previously called the new urban question, which arises in the heat of crises. As in previous situations, this forces us to rethink the disciplinary bases in a framework with considerable uncertainties because, although these challenges refer to quite different dimensions (social, economic, political ...), they affect and are affected by urbanism.

Among the urban challenges that currently require greater attention are those resulting from climate change, the growing demand for mobility, new approaches in the design of public spaces, the explosion of cities, and the escalation of inequalities in the urban world. As we will see, these are not independent topics, in fact, they are closely related.

Like others in the past, the economic crisis at the beginning of the 21st century coincides with an urban crisis, with a new urban question. At the centre of this crisis are inequalities, greed, progressive dismantling of the welfare state, and the degradation of the quality of life of the most vulnerable social groups.

Addressing climate change is fundamental and requires learning how to design cities with environmental risks and opportunities. It is the responsibility we all have as citizens, and the deadline is fast-approaching. Yet, it is important to assess what is demanded from us as professionals and researchers in our discipline: urbanism.

There are many relevant studies, from the first warning coming from The Club of Rome, to the "The Limits to Growth", by David Meadows, and also "Our common future" (WCED, 1987), where the term sustainable development was used for the first time. We should also mention another prolific paper "Human Scale Development", by Max Neef, Elizalde and Hopenhayn (1991), not to mention others by Amartya Sen or Jürgen Habermas. If we go to the origins of the environmental concern, we should consider the pioneering studies of Humboldt, Haeckel, Odum, or Margalef. However, we should mention other more recent studies that are closer to our discipline.

It is essential to highlight the pioneering work of Ian McHarg (1971), "Design with nature", which introduces analysis by layers, even if it is just to measure the capacities of each portion of terrain to hold certain activities, or, for example, to adjust the layout of a road.

Richard Forman went further and taught us how to build a system of open spaces (Forman & Gordon, 1986; Dramstad, Olson & Forman, 1996; Forman, 2004), by describing the recovery of the Olmsted park system on a territorial scale. His studies are of great interest, even those considered of less importance. He translates landscape ecology concepts into urban terms: corridors, tiles, borders, buffers, or stepping stones. He explains how to organise them weaving connectivities and

DOI: 10.1201/9781003220855-1

how to apply clear principles to build an environmental matrix.

In another prolific research, Jeremy Rifkin (1995) gives us hope. He shows us how the great economic revolutions (and we should add the urban crises) happen by converging new energy systems and new communication technologies. The two previous revolutions (steam-printing; electricity-oil-automobile) have produced a huge carbon dioxide bill in the atmosphere, which makes it difficult to release the heat produced by the sun and threatens us with a catastrophic variation in temperature.

However, the Internet and renewable energies are a new and powerful infrastructure for a Third Industrial Revolution, which, according to Rifkin, may change the world by: transforming buildings into micro-power plants that collect and reuse such energy, the development of the hydrogen technology and other energy storage systems in buildings and infrastructures, the use of the Internet to interconnect shared energy networks, and the transition from current vehicles to other intelligent ones with clean energy.

All of this represents a true challenge for research in urbanism, working on disciplinary fringes. Numerous cities acknowledge being in a dramatic situation of climate emergency. It is necessary to look at the increasing recurrence and devastating effects of natural disasters. There are already many academic projects looking at the impacts on the environment, ecology and sustainability; however, it is still urgent to go further in the implications on the urban across the different dimensions (sustainable mobility, neighbourhoods and cities with a zero-energy record, balance with nature, endogenous development, etc.).

The search for a new symbiosis between urban and rural dimensions is generating interesting research, especially those focusing on agriculture (particularly in the peri-urban). It is vital to consider agriculture: it is part of the urban metabolism, provides energy (food), and builds more resilient territories. We should also consider the new views of Landscape Urbanism, from the south (Marot, 2020), replacing the ideas of a certain nostalgia of landscape in the urban environment, from the north, with the idea of a new urban growth, with a greater natural harmony and more resilient communities, even in situations with less economic development.

Another interesting topic is mobility, not traffic nor infrastructure, but mobility in a broader sense, as a citizenship right and at different scales. The urban scale is related to public transport, with alternative ways of moving around and the extension to the design of the public space; or with the permeability and isotropy of infrastructures, which make the city more porous and accessible to all.

For many years now, the large automobile companies have known that cars and fossil fuels will end. Thus, they have been investing millions researching other means of transportation and, consequently, other types of cities.

Reflecting on mobility in the field of urbanism is unavoidable. In fact, some ongoing studies on both sides of the Atlantic focus on walking trajectories, the decrease in traffic, or the contributions of BRTs and *ligeirinhos*.

Several studies relate personal mobility with health issues (pedestrian, cycling, among others), particularly in settings with an ageing population. This topic is likely to be further developed as a result of urban studies of the pandemic.

However, mobility at the planetary scale is equally important because it implies great and dramatic migrations, which are increasing due to climate change and social injustices. Migration has always taken place between unequal territories, and this is strongly related to another key challenge: tackling social and economic inequalities in the urbanized world.

However, before we address this issue, we first need to note the physical expressions of such an increasingly unequal city, which we call the "explosion of the city". This is a good metaphor, which describes much of recent growth both in Latin America and Europe. Although with different characteristics, we can find common aspects, and some relevant studies have analysed this phenomenon (Boeri, Lanzani & Marini, 1996; Indovina, 1990; and Portas, in Italy and Portugal).

The most vital contribution is surely from Professor Antonio Font (Font, 2007; Font, Llop & Vilanova, 1999) because it moves from description to critical interpretation, and so to the possibility of responsible interventions. He led several studies in Europe and has inspired many others in Latin America. Looking at the oil-like sprawl that was previously criticized and now longed for, his studies reveal patterns of new growth as polarisations around great infrastructures or as dispersions over the territory. He shows us that the clusters further away from the centre grow more, or how economic activities with lower added value are transferred to the periphery. He also shows how increasing prices in the centre and the relocation of employment leads to the construction of houses in more distant places, filling empty spaces, creating fragmentation and a loss of natural spaces. Finally, his work shows how tourism and temporary accommodation promote new instability in the urban centre, through processes of gentrification and *touristification*.

His research furthered the interpretation of the patterns of this new type of production of the city. New tertiary nodes emerge along the main roads, clusters and sparsely dense areas fuse with each other, multi-directional movements grow before those of the centre-periphery by dispersing population, activities, and services across the territory. His studies developed insight into the interpretation of the patterns of this new type of production in the city.

Regarding the socio-economic structure of the city, one of the most relevant urban contributions comes from another Professor who is dear to us, Bernardo Secchi. Among his several seminal texts, we highlight "The city of the rich and the city of the poor" (2015), modest in length, as it is intense and ambitious in its interpretative purpose.

Until a few decades ago, cities were places for social encounters and mobility, and urbanization was a driver for economic, social and cultural development. However, inequalities are growing dramatically in the developed world. The gap between social classes is widening and poverty and inadequate housing are growing without stop. As mobility and network efficiency increases, so too does migration between cities and countries.

In the past we placed our trust in urbanism to ensure that all citizens would live in decent conditions, but today we find that urbanism has been denied this capacity to introduce social improvements or to fight against inequality and spatial injustice. We are told that these are social and economic policy issues. There is also a higher housing turnover in a faster world with more nomadic lives, where property loses value and housing is a service. However, urbanism has a clear responsibility in this deepening inequality and should be one of the pillars of any policy aiming to reduce it, for example, in introducing basic services (schools, nurseries, transportation, parks, and housing) for everyone.

This increasing inequality is shown in different ways. One of the most obvious is related to the right to a decent house. We need to refocus on this field of study. Thus, it is urgent to recover the social mapping introduced in the 19th century by hygienists and social scientists, like Engels and Cerdà, with their statistical monography on the working class.

At a time when the social gap is widening, and the population is increasingly urban and more vulnerable, it seems inevitable that living conditions be re-mapped and, starting with academia, reflect on the responsibility of urbanism, as Secchi proposed.

This is not the first time an urban crisis emerges, and, like the previous ones, it forces us to question the scientific status of urbanism. The uncertainty and the velocity of new transformations force us to work with scenarios where a basic aspect, maybe the only permanent one, is continuous exchange. Yet, like Secchi and Rifkin point out, facing the new urban question is our obligation and our hope.

With renewed approaches, public space continues to be a core topic in urbanism research, as Jan Gehl stated: "first life, then spaces, then buildings - the other way around never works" (Gehl, 1971). At this time in history, when cities all over the world are forced to a standstill and to enormous, successive, and multiple restrictions, Jan Gehl's message has achieved the characteristic of timeless truth. Although the many topics here listed seem to be relatively "classic", the reflection and the forms of intervention proposed are different from the most canonical ones and from the known theoretical-practical interpretations.

The human being and social practices are at the core of the observed reflections, as it is also clear in the selected articles. In all of them, we also recognize a desire or a quest to find new planning paradigms. In fact, a new way to transform the city seems to have spread over more recent years: we are moving from the quest for singular clarity, obsessively searching for the perfect form, to a more open and experimental approach, where the focus is on people and their experiences.

Borrowing the tools and methods used in a laboratory, our cities are the scenario of "adaptive" urban projects. This means, transformations are gradually changing as experiences advance. Then, we observe the corresponding reactions to each application, exploring the possibilities that are awaiting us in the subsequent stages. Therefore, the challenge is to think of the city as an open system in which the unexpected, the anomalous, and the possible can happen and coexist.

Naturally, all this questions the practice of urbanism as we know it today. The problem of deconstructing an established practice is that it does not tell us how to prepare the next step. Imagining a "solid" city in constant balance with an ever-changing open system is perhaps the greatest challenge of all. Such hope is what seems to transpire from the papers in this book.

By looking at the photographs of 19th century cities and their infrastructure, we can see that streets were organised as open systems. They were shared cities: everything was specified as full or empty. Actually, the city streets have always been like this. It was in the 20th century that activities were defined and segregated, and infrastructures were designed according to function.

Before urban spaces went through a specialization dominated by the advent of the automobile, there was a kind of self-regulated internal balance within the city. The advent of the automobile completely changed the "life between buildings" (Gehl, 2011). Even the definitions themselves changed. For example, "street" nowadays, is commonly referred to as the separate lane of the pavement reserved for the vehicles, whereas before it meant a public space where several activities, aside from mobility, would take place. These new concepts were born essentially from the visions of a new city and from the pursuit of disruption with the past. It was an opportunity for change, an inflection between past and future, not only of buildings and cities, but also as a new way of looking at the act of designing a project.

> "The Plan is the generator. Without a plan, you have lack of order, and wilfulness. The Plan holds in itself the essence of sensation. The great problems of tomorrow, dictated by collective necessities, put the question of "plan" in a new form. Modern life demands, and is

waiting for, a new kind of plan, both for the house and for the city" (Le Corbusier, 1986, p. 45).

A new paradox was created in the city. We moved from the definition of an urban form, independent from its function, to its opposite: a functionalization of everything. Serving a practical purpose, space is now rationally defined by zones, indexes, and paradigms based on numbers. This is the disruption in history proposed by Le Corbusier. The problems of the future of the cities were set according to collective needs based on statistical analysis attainable by mathematics (LE CORBUSIER, 1986, p. 51). Such needs were thus detached from the urban characteristics, so clearly outlined on Giambattista Nolli's map of Rome (1701-1756). Unique to the Nolli Drawing was the representation of public space inside buildings, as part of the urban realm. There was no distinction between inside and outside; only space and mass. Here, the city is synonym of an open system, giving little importance to the separation between public or private spaces: the identification of city spaces was established by its availability, established as public regardless of the property's nature.

Much has been discussed on whether time has actually come to move away from the idea of mere planning and "controlling", conceptual in nature, and to now embrace a logic of adaptive design instead. Everything points in that direction, the aim of embracing the city project, which is the tool determining the possibility of use, attractiveness, inclusion, regardless of its macro planning. However, if we would imagine a new transformation of the street, or the city, for an open system, how could we do it in today's cities?

Maybe the answer to this question could start by examining the necessary change of paradigm, from the profession itself: architects, designers, urban planners, and decision-makers of urban spaces. Some of these changes are already in motion, but the role of the decision-makers also needs to be reconsidered. Increasingly, we must perform more of a "facilitator" and "mediator" role, rather than an imposing one. We should learn to go from the role of experts, imposing their decisions on those who try to understand a space and look to engage its users (and not just its inhabitants) in the process. Such is the idea of inclusion in the open system, not just at the scale of the streets, but also in the urban processes and its planning. Firstly, we should question traditional planning: master plans and all the tools supporting them are instruments with proven limitations. To overcome them, we need to embrace a concept, which inevitably needs to be flexible and adaptable over time to preserve its usefulness and its strength. We need to make clear that we should not disclaim 20th century urbanism. Such urbanism is fundamental for a consolidated future, one which allows us to establish links between what we were and what we want to be. Aside from this, the inclusion of a minimal quantity of it in the planning process in establishing the first notion of a public city is undoubtedly important. It is surely a fundamental step to start thinking of the city as a collective or, even, a common good.

While the challenge for the professionals might be more or less defined, its counterpart is no less important. We are talking about the need to make the city dweller be at peace with the city in itself. This is, a return to a sense of belonging that was ripped from them over the course of the 20th century.

"Looking at cities can give a special pleasure, however commonplace the sight may be", as Kevin Lynch (1918-1884) writes in the opening of his book "The Image of the City" (1960). Maybe this is all it takes for everyone to become involved once again in all aspects of the city: we just need to *look at* them and *care for* them. Cities are not just a reflection of the *here* and *now*, but are the result of a temporal process of the cultural investment of previous generations. We have the responsibility to serve as guardians of a heritage that we will pass on to future generations. Left by different people on different occasions, sometimes well-established other times not so much, the city is one of the reflections of Humankind's excellence and of everyone's effort towards a common good. Nowhere in the city can this encounter be more real than on the street, again understood in a broad sense. On the street, all our activities are interconnected: endless flows of people, in a continuous bustle, or in moments of contemplation. There are also moments of introspection and meeting with others, similar or different. It is the place where the horizontal plane meets the vertical plane and where the horizontal plane meets the horizon. In this regard, we recall Jane Jacobs (1916-2006) and her exhaustive description of the street that can be seen from her home. From her window, she observes the choreography of the city: children on their way to school, well-dressed people with briefcases on their way to the bus or the underground train. Women who, when leaving their homes, get distracted talking to each other on the street. Through the viewing of her window, it is possible to get to know those who are just passing by, as well as those arriving or leaving from near her home, from dawn and into the late hours of the night (JACOBS, 1992, pp. 51–54). This description reminds us of the choreographic adventures of *Monsieur Hulot*.

If there was a strong suspicion of the need for a change in city planning, such doubt became a certainty after the events beginning in 2020. The fast spiral embracing cities across almost the entire technological world was forced to slow down by the announcement of the pandemic by the World Health Organization on the 11th of March of 2020. The street, as the fundamental link and synonym of the city's life, became empty and remained as a spectator of a world that seemed to cease being ours. And so, we were left with admiring the empty city from our windows, seemingly a nightmare from a metaphysical painting. The landscape of the city became a distant and inaccessible scenario, and a remainder of what we had taken for granted as life in society.

* * *

When the call for the XII SIIU opened, we were far from imagining how drastically our way of life would change in such a short period of time. However, the pandemic placed spatial and planning issues on the top of the social and political agenda. Undoubtedly, the COVID-19 pandemic has placed some long-standing issues of our contemporary cities under the spotlight: territorial inequalities, over-specialization of urban areas and road infrastructure, lack of decompression zones. However, the restrictions imposed by the pandemic led most cities around the world to paralyse, and the large and successive restrictions emphasized the importance of the cities' public life and how this crisis is, above all, an urban crisis. Therefore, from a more optimistic point of view, the COVID-19 crisis led to a quick adoption of some processes that may benefit cities. The emergence of the pandemic crisis allowed the testing of new spatial configurations, noting the importance of public space, and recalibrated dwelling dimensions.

Returning to the previous discourse, the need for a change of paradigm in planning, the measures adopted by the governments and public administrations over the past months have shown, and accelerated, the process of change. This process of change is described in the papers published in this book, even though not in a totally explicit way. The dramatic changes that our daily lives had to go through over these past months are a critical challenge, and the tools for tactical urbanism have proven to be the most effective to face it.

Rapidly and unexpectedly, we had to move from pilot-projects to large-scale application. The pandemic has made us question how dwellings are developed in the large cities and metropolitan areas. Lockdown has forced us to give up the public dimension. It has become clear that the key for dwelling quality is not just indoor space (right size, functionality, residents' privacy), but is also about intermediate spaces (with balconies finally becoming valuable) and outdoor spaces (easy access to comfortable nearby outdoor spaces). All these mitigate the effects of #stayathome.

Mobility restrictions turned historical squares and city parks into remote and inaccessible places, and people were left with the streets and pavements near their residences as the only available outdoor space for months. From that point of view, the right to live in the city became a qualitative issue that considers the relationship between inside and outside spaces, where private and public space are closely connected. For a long time, city public spaces were identified as historic public squares, as well as areas with a symbolic value, becoming prime sites for events and collective meetings. In this way, public spaces were interpreted as episodes of the urban fabric, and firmly associated to those locations hosting extraordinary events.

With few exceptions, the pandemic brought to light the continuity of a functional hallmark of the modern city, where public space was thought of as a break from the constructed urban structure, a separate element (even better if fenced) from the rest of the city and its daily urban flow.

Maybe today, more than ever, it is important to rethink the traditional concept of public space in urban planning: public space, the space between the buildings, needs to be re-imagined as part of a network, interconnected and extending across all of the urban fabric, with fragments intertwining with the new fabric of the contemporary city. Fragments that hold unexpected possibilities, which is, in a way, the self-generating path of creating the city: the greater freedom given to the new, leaving a mesh for changeable uses, the greater the possibility of that space becoming a living space, that may be overwritten and adaptable.

The economy and services provided were also abruptly modified, with an impact on public space. The promoting of esplanades for a safe socialisation; increasing pedestrian areas and cycle paths; home deliveries and take away making use of active mobility; disrupting some logistics chains, which affected global activities and nurtured nearby production chains; sudden drop in tourism, which emptied parts of the city and led to the closure of activities dedicated to tourism, thus showing less resilience; making new uses of public spaces because gyms and school playgrounds moved to public lawns on the outside; the rise in remote working, replacing the need for physical mobility; albeit, it is not clear which transformations will have an impact on the city at the end of the pandemic. Are these contingencies of the situation or are they a sign of a new urban normal?

This difficult balance is perhaps the greatest challenge of the new planning: not a final vision of a defined and unchangeable future, but a process on an ever-changing reality. Designing the city as an ongoing process where the human being is the main focus and where life comes first, then spaces, then buildings.

* * *

Today, in an ever-accelerated reality, in a global information era, the cycles overlap in time. The crisis entering the 2010s had a great impact on the city and territory, adding to previous problems. This was almost immediately followed by the economic recovery, reinforcing the new paradigms of the green economy and a new way of living. Now, with circumstantial consequences, the COVID-19 global pandemic crisis is also introducing profound transformations on urban life that are yet to be completely figured out; new problems and opportunities are emerging.

Building a research agenda on urbanism to serve societal goals is an ever-complex task. Keeping focus is quite difficult, in a fast-changing reality, submerged by information, where one is permanently revaluating if former established ideas are still valid. However, some concepts are constant, and still stand. As fast as possible, research in urbanism tries to apply its disciplinary practice to present-day circumstances. It can be said that it is a [re]urban practice, with three approaches present in most contemporary studies: nature stands as reference; human is the critical factor; and the process itself is the fundamental perspective.

REFERENCES

Boeri, Stefano; Lanzani, Arturo & Marini, Edoardo (1996). *Il Territorio che cambia: ambiente, paesaggi e immagini della regione milanese*. Milano: AIM Associazione Interessi Metropolitani.

Dramstad, Wenche E.; Olson, James D. & Forman, Richard (1996). *Landscape ecology principles in landscape architecture and land-use planning*. New York: Island Press.

Font Arellano, Antonio (2007). *La Explosión de la ciudad. The explosion of the city: transformaciones territoriales en las regiones urbanas de la Europa Meridional: territorial transformations in the South Europe urban regions*. Barcelona. Ministerio de Vivienda. Col·legi Oficial d'Arquitectes de Catalunya.

Font, A.; Llop, C. & Vilanova, J. M. (1999). *La Construcció del territori metropolità: morfogènesi de la regió urbana de Barcelona*. Barcelona: Àrea Metropolitana de Barcelona, Mancomunitat de Municipis.

Forman, Richard (2004). *Mosaico territorial para la región metropolitana de Barcelona*. Barcelona: Gustavo Gili - Barcelona Regional.

Forman, Richard & Gordon, Michael (1986). *Landscape ecology*. New York: Wiley

Gehl, Jan. *Life Between Buildings: Using Public Space*. New York: Island Press, 2011. [Edição Original:1971].

Indovina, Francesco (1990). *La Città diffusa*. Venezia: Istituto universitario di architettura di Venezia.

Jacobs, Jane. *The death and life of great American cities*. Vintage Books ed. New York: Vintage Books, 1992. [Edição Original:1961].

Le Corbusier - *Towards a new architecture*. New York: Dover Publications, 1986. [Edição Original: 1923].

Marot, Sébastien (Ed.) (2020). Taking the Country's Side: Agriculture and Architecture. Lisbon: Polígrafa and Trienal de Arquitectura de Lisboa

Max-Neef, Manfred; Elizalde, Antonio & Hopenhayn, Martin (1991*). Human Scale Development: Conception, Application and Further reflections*. New York: Apex Press.

McHarg, Ian (1971). *Design with nature*. New York: Doubleday-Natural History Press.

Rifkin, Jeremy (1995). *The end of work*. New York: Putnam Publishing Group

Meadows, Dennis; Meadows, Donella; Randers, Jørgen; Behrens, Williams (1972). *The limits to growth*. New York: Universe Books.

Secchi, B. (2015). *La Ciudad de los ricos y la ciudad de los pobres*. Madrid: Libros de la Catarata.

World Commission on Environment and Development (1987). *Our common future*. Oxford: Oxford University Press.

Research Tracks in Urbanism: Dynamics, Planning and Design in Contemporary Urban Territories – Allegri et al (eds)
 ISBN 978-1-032-10788-2

Urbanism research (ongoing) in Latin America: Some keywords

Maria Cristina da S. Schicchi, Eunice Helena Abascal & Angélica Benatti Alvim

> Once the *Cosmographiae Introductio* acknowledged that the new lands, despite being somewhat isolated by the Ocean, constituted an integral part of the world, man´s sovereignty over universal reality was claimed for the first time[1] (O'GORMAN, 2006, p.179).

Urban studies on cities in Latin America (LA) have grown since the 1970s, including publication of books on the formation of cities, the relationship between intellectuals and cities, national and inter-american conferences, pan-american congresses, building of social housing and theories about the city. Studies which have employed a broader approach to more common issues of cities, often comparative in nature, range from pioneering studies such as those by Hardoy (1989), Hardoy, Morse & Schaedel (1978), Morse (1971, 1978), Segre (1977, 1989, 1999), Santos (1979, 1994), Carrión (1991, 1992), Liernur (2002) to the more recent publications of Novick (2003), Gomes & Espinoza (2009), Gorelik (2005), Gorelik & Peixoto (2019), Gutiérrez & Tartarini (2007), Almandoz (2007a, 2007b, 2013), Arango Cardinal (2012), Abramo (2012), Sambrício (2014), Velázquez & Pradilla (2014), to name but a few. Also noteworthy is the contribution of Canclini (2007), which took the debate in a new direction, not exactly analyzing issues shared by a group of cities, but introducing approaches such as the effects of interculturality and of frontiers of globalization in the Latin-American context. The flow of ideas in this case cannot be characterized as the fruits of technical or economic exchange, but rather of an *imagined globalization.* The various possible globalizations are the product of the contribution of culture, not in the shape of a domesticated land, but in their potential to address discontent, forms of resistance and affirmation of singularities.

In a bid to broaden the perspective regarding the contexts of the texts presented in this book, the meanings of some concepts underpinning debates and proposes in urban history of Latin American were explored (ALMANDOZ, 2007b).

We sought to ascertain at which point in time the search for a *Latin-Americanness* in urban thinking first emerged, charged with embracing the urban-territorial processes inherent to the region (RANGEL, 2014). There is consensus among scholars of the city, particularly in the social sciences field, that scientific research into the region took shape from the 1950/70s, mainly in the Post-War era, a time when industrialization replaced imports, promoted by the State and accelerated urbanization created a need for more in-depth studies to support urban and regional planning (VELÁZQUEZ & PRADILLA, 2014).

It was during this period that the "Latin-American City" came into being as a cultural construct, i.e. "(....) as a category of social thinking, as a figure of the intellectual and political imagination in vast regions of the continent (...)"[2] (GORELIK, 2005, p.125). According to the author, the Latin-American city not only existed in this period, but encompassed a series of figures, disciplines, institutions, as well academic, literary and political settings taking shape in the region, through parallel processes and transformations in the cities which resulted in them being characterized as a demographic, social and political problem to be tackled via the participation of numerous international professional exponents and application of the latest knowledge on urbanism techniques at each step. Although the public policies developed in Argentina, Chile, Brazil, Cuba, Venezuela, Colombia and Mexico were sometimes introduced independently and with a time gap of over a decade, the means adopted to tackle the issues and models adopted were similar, as were the institutions funding institutions and municipal entities created. Visits among the cities made by the same foreign experts commissioned to help devise of local plans and projects was also common at the time (SAMBRICIO, 2012).

Between 1950 and 1970, Latin American was regarded by the West as place where modernization could forge ahead without the onerous costs which

---

1 Puede decirse, entonces, que cuando se admitió en la *Cosmographiae Introductio* que las nuevas tierras, pese a su aislamiento por el Océano, constituían una de las partes integrantes del mundo, se reclamó por vez primera la soberanía del hombre sobre la realidad universal.

2 "(...) como categoria do pensamento social, como uma figura do imaginário intelectual e político em vastas regiões do continente (...)"

DOI: 10.1201/9781003220855-2

developed countries had to bear post-war (GORELIK, 2005). However, unlike the countries of the north, industrialization in the countries of the subcontinent did not precede but followed urbanization in most Latin-American cities, whose growth was driven by policies of appreciation of urban lands and heavy investment in the major cities (ALMANDOZ, 2008). A number of buzzwords were prevalent in discussions of the period, referring to inter-dependent phenomena: *urbanization, industrialization, modernization, centers of development, regionalization, and planning.*

There was a flourish of studies with a focus on the region and directly related to this simultaneousness and intensity of urbanization during this period. In 1950, over half of the populations of Uruguay (78%), Argentina (65.3) and Venezuela (53.2) resided in urban centers, as did Brazil and Mexico, which only failed to reached this percentage in urban area because this was offset by the sheer size of their populations, but nonetheless attained 40% urbanization (United Nations 1996, pg. 47 *apud* ALMANDOZ, 2007b, pg.116).

Not that there were no studies on Latin-American cities previously, nor that this period represented a total break with preceding decades. The historiography for the period that went before is extensive, only, according to Gorelik (2005), up until the 1940s, these studies tended to focus the national milieu, at a time when the problems faced by cities were often seen as the particular symptoms of a nation, a point also highlighted by Almandoz (2007b, pg.65-66):

> Since the last 1920s, industrial development, the demographic mobility and urban expansion were evident in the major urban centers of Latin American, with the urgent need to adopt plans implemented by local governments with the aid of foreign experts and new generations of native professionals.[3]

The author also states that many local professionals took on concomitant roles as administrators, designers and urban promoters, the same group which would oversee the first urban plans to control the growth of the cities, including Contreras in Mexico city, Maurício Cravotto in Montevideo, Carlos della Paolera in Buenos Aires, Leopoldo Martínez Olavarría in Caracas, Pedro Martínez Inclán in Havana and Anhaia Mello & Francisco Prestes Maia in São Paulo. The urban proposals for modernization were centered mainly on the capitals and acted on 3 main fronts: sanitary reform, urban renewal and projects for new areas of residential expansion (ALMANDOZ, 2007b). Representative and recurrent terms used in the period were: *Urban expansion, urban plans, renewal, urban redevelopment, migration, demographic mobility.*

From the late 1950s, studies on cities in Latin America focused predominantly on the developmentalism agenda, under the auspices of organizations such as the Organization of American States (OAS-OEA), Economic Commission for Latin America and the Caribbean (ECLAC-CEPAL) – a UN regional mission, whose head office was located in Chile. ECLAC promoted a number of studies on the economic and urban development of countries deemed as "developing countries" in the region (BIELSCHOWSKY, 2000). Besides ECLAC, Chile also hosted Ilpes (Instituto Latinoamericano de Planificación Economico y Social – Latin American Institution for Economic and Social Planning), which ran courses bringing together the leading experts in Latin America of the 1960s, such as Fernando Henrique Cardoso, Anibal Quijano, and José Medina Echavarría. With UNESCO backing, since 1957 the first headquarters of the FLACSO, Facultad Latinoamericana de Ciencias Sociales (Latin American School of Social Sciences) was also set up in Chile, which soon became a center for research and teaching, which took in a number of intellectuals exiled during the dictatorship regimes in the countries of the region. In 1967, CLACSO — Conselho Latino-americano de Ciências Sociais (Latin-American Council for Social Sciences) was set up, introducing post-graduate courses in a number of countries in Latin American, with its commissions for regional and urban development and running of a series of study missions on cities, whose goal was to collaborate in training governmental agents and fostering development of critical studies, particularly in the social sphere, of public policies.

According to Duhau López (2014), the consolidation of urban studies occurred from the 1970s, when the major Latin-American cities had already taken shape. From the perspective of metropolization, the various different interpretations of urbanization became evident (RIBEIRO, 2014), as did the consensus on their structural characteristics, some enduring to the present century, and conjure terms such as *metropolization, reurbanization, urban network, informality, residential segregation*, and *social inequality:*

> (a) the speed and intensity of migration of the population from the countryside to the city; b) the constitution of a metropolization process which creates a disarticulated urban network; c) the disassociation between urbanization and industrialization which results in a city marked by informality, irregularity and precariousness in terms of housing and urban living conditions; d) residential segregation and sociospatial inequality; e) lack of planning.[4]

---

3 Desde finales de los años 1920, el desarrollo industrial, la movilidad demográfica y la expansión urbana habían evidenciado, en las mayores urbes latinoamericanas, la urgencia de adoptar planes que fueron emprendidos por los gobiernos locales apoyados en expertos foráneos y nuevas generaciones de profesionales criollos.

4 a) la velocidad y la intensidad de la transferencia de la población del campo a la ciudad; b) la constitución de un proceso de metropolización precoz que genera una red urbana desarticulada; c) la desasociación entre urbanización e industrialización que genera una ciudad

From another perspective, the increase in urban studies only really took off from the mid-1980s[5] accompanied by the reintroduction of the study and teaching of urbanism through a more critical prism within the universities of the countries in the region, amid more stable political process of redemocratization (GONZÁLES; BAQUERO, 2013)[6]. The role of the transition governments, with the adoption of new models and ways forward for resolving urban issues called for in new "ontological and epistemological approaches for investigators"[7], as well as for urban plans and projects (CONNOLLY, 2014, pg. 803). In the field of studies on architecture conducted in the different countries of Latin America from 1985, regular Seminar on Architecture in Latin America (SAL) spearheaded by researcher Ramón Gutiérrez da Costa (2008), whose initial incentive was due to the Spanish Agency of International Cooperation (AECI), during a period of resumption in relations of cooperation between Spain and Latin America.

From the late 1980s, the crisis of accumulation of capital, the dismantling of the socialist field and the adoption in the West of the neoliberal hegemonic model, led to a weakening of the debates, in the areas of both government and urban studies. Scientific study thus developed based on four trends: homogenization of the reality and their explanations, grounded in ideas such as the "global city" and non-differentiation of regional characteristics; generalized theories, underpinned by models and concepts of cities in Latin America, produced by hegemonic countries; a focus on the local level, excluding national and regional scales, blurring historical and cultural differences; and fragmentation of knowledge in many professional fields, decharacterizing the distinct interdisciplinarity of the field of urban studies (VELÁZQUEZ; PRADILLA, 2014).

Consequently, this period marked a turning point in paradigms of scientific research, compared to the decade before, when sociological analyses of development prevailed, giving way to the theory of dependence, which addressed social-economic development and urbanization in the countries of the region. In Brazil, the main advocates of this change were researchers from the CEBRAP - Brazilian Center for Research and Planning, founded in 1969. Also in the 1970s, the idea of the efficient city for geographical and social integration and national development was revisited by authors such as Jorge Enrique Hardoy and Gino Germani, from the perspective of affirming the place "of education and of the city as modernizing agents"[8] (HIERNAUX NICOLÁS, 2014, pg. 88). During this period, one of the prominent interlocutors of these and other Latin-American researchers was Richard Morse. He carried out studies since the 1940s on the cities in the countries of LA. One of the most well-known and cited of these studies since the 1980s in Brazil is entitled "*The History of the Formation of São Paulo from the community to the metropolis*"[9] (MORSE, 1970).

It is no coincidence then, that between 1980 and 1990, a number of research institutions were created specialized in urban and regional studies: ANPUR – Associação Nacional de Pós-Graduação e Pesquisa e Pós-Graduação em Planejamento Urbano e Regional (National Association of Post-Graduate Studies and Post-Graduate Program in Urban and Regional Planning) (Brazil, 1983), Red Nacional de Investigación Urbana (National Network of Urban Research) (Mexico, 1982), which shortly after began publishing their own journals to disseminate ideas and studies in the region, the most renowned of which is the Chilean journal EURE – Revista Latinoamericana de Estudios Urbano Regionales (Latin-American Journal of Regional Urban Studies), running since 1971 (DUHAU LÓPEZ, 2014).

Duhau López (2014, pg.65-66), having studied the topics of articles published by the EURE, noted that the journal was the main vehicle for discussions of researchers which adopted the theory of global cities as their central theory, with an emphasis on large cities. He also noted that, in several other periodicals publishing articles on the LA milieu from the 1990s onwards, other perspectives and approaches regarding global cities and criticisms of the neoliberal model were addressed, such as *sustainable development* elaborated *on* by Acselrad (1999), and more general discussions on *informality*, the *informal land markets* and *residence mobility* by Dureal et al. (2002), in addition to the *representations* and *urban imaginaries* by Silva (1992).

Changing subject, the geographer Milton Santos (1994) warrants mention for his proposing of an analysis of the effects of a "selective modernity" on the

---

marcada por la informalidad, la irregularidad y la precariedad en términos de las condiciones habitacionales y urbanas de vida; d) la segregación residencial y las desigualdades socioespaciales; e) la ausencia de planeación.

5 According to Almandoz (2007), during the period studied, between 1940 and 1960, the teaching of history and urban studies were introduced into architecture course after these had overcome both artistic methods of the schools of Fine Arts and the more technical approaches adopted by the Polytechnic Schools. In the same period, foreign interest, particularly North American, in the Latin-American modern experience, led to an increase in local studies.

6 This process was not exclusive to the area of urban studies. González & Baquero (2013) noted that also in the social sciences, especially the area of Politica Science area, comparative studies on policy in the region also increased, becoming institutionalized only from the 1990s onwards.

7 "posturas ontológicas y epistemológicas para los investigadores."

8 "de la educación y de la ciudad como agentes modernizadores de un país."

9 "Formação Histórica de São Paulo: da comunidade à metrópole"

regions, with the expansion and dissolution of the city in all places (HIERNAUX NICOLÁS, 2014, pg. 108). One of these effects, as mentioned by Lencione (1998), prompted the "Mudanças na Metrópole de São Paulo (Brasil) e transformações industriais" (Changes in the City of São Paulo (Brazil) and industrial transformations) expounding on the phenomenon of *deconcentration* of the city. Another point on which some authors appear to agree is that, since the 1990s there has been growing interest in exploring the cultural dimension of the issues in LA, as highlighted regarding proposals by Nestor Canclini.

With respect to public space, Kuri (2014) observed contrasting tendencies in the construction of "public", with difference forms of identifying and belonging to the city on the one hand, while on the other, different forms of disputing and excluding citizenship. According to Tamayo "these are new urban cultures that break with the tradition of collective identity of the corporate type of major unions and propose new way of being together"[10] (2014, pg. 978). These phenomena manifest particularly in historic centers and reflect the high level of unemployment in LA (ALMANDOZ, 2008), with increased informal working and social inequality. By contrast, in the late 1990s, numerous programs and projects for redeveloping old city centers aimed at appreciating the value of these areas and removing informal traders from public spaces (CARRIÓN, 1991, 2010a, 2010b; BARROS & ARAÚJO, 2006; D'ARC & MEMOLI; BIDOU-ZACHARIASEN, 2006).

A number of authors indicate that the historiographical output of the region is sparse and highly restricted to each respective country, and there is a dearth of studies comparing cities. There are least two approaches in dealing with issues of the subcontinent. On the one hand, studies investigating major cities address shared problems such as poverty, marginalization, uncontrolled urbanization and other regional imbalances, while on the other, there is research which centers more on case studies of specific cities. However, in general, these latter studies show there is little utility in making comparisons or generalizations (GORELIK, 2005).

Currently, in is not possible to address all new contributions to urban studies and urbanism, not only because the scope of discussion has widened to include different contexts and continents, via new technologies for communicating, consolidating and sharing of ideas, but also due to the emergence of novel multi and transdisciplinary dialogic approaches which require a continual disciplinary review. However, Velázquez & Pradilla (2014, p.19) hold that:

> (…) while the number of researchers and output of studies and documents on the territory (urban-regional) have increased exponentially as a result, among other reasons, of policies of promotion, wage remuneration and production hierarchy, which have become incorporated into Latin American systems of science and technology systems and into the universities leading to dissemination of scientific output via new means of communication, this has consolidated and increased the publication and distribution of scientific output in hegemonic capitalist countries. However, the output of Latin American researchers on this topic, although broader than before, remains marginal with relatively low regional circulation.[11]

This leads us to a greater appreciation of opportunities for discussion on the academic output of researchers and students in LA that, in spite of so many local differences in objects and objectives, still spark interest through acknowledging homologous issues and increasingly involve the Ibero-american context, including the researchers of Spain and Portugal. This comes about not only for the opportunity of critical review and strengthening of shared historical ties, but also through the genuine interest in getting to know both contexts and devising joint solutions.

Since commencing the process of organizing the group of articles by researchers and post-graduate students on urbanism received for the XII International Seminar of Investigation in Urbanism in São Paulo, one issue was always present: which parameters should be established for selecting those which can most contribute to the discussion in the field of urbanism, particularly, within Latin America?

This was no easy task, proving both challenging and thought-provoking. The answer arose, partly from the large number of studies on urbanism and cities of the subcontinent. A possible observation, given the wide range of themes and approaches, was that the desire to seek answers to problems that did not always fit within traditionally established fields of urbanism, was the most consistent finding among the studies submitted.

This raised the question over the difference that this component could make with regard to more specific and specialized congresses and seminars – some traditional – for the area, particularly in view

10 "Son nuevas culturas urbanas, que rompen la tradición de la identidad colectiva de tipo corporativa de los grandes sindicatos y proponen nuevos modos de estar juntos."

11 (…) al tiempo que se multiplica exponencialmente el número de investigadores y la producción de investigaciones y documentos sobre lo territorial (urbano-regional) como resultado, entre otras causas, de las políticas de promoción, compensación salarial y jerarquización productivista, aplicadas a los sistemas latinoamericanos de ciencia y tecnología y a las universidades, y se desarrollan formas de difusión de la producción científica mediante nuevos medios de comunicación, se afianza y multiplica la circulación y difusión de la producción científica en los países capitalistas hegemónicos. Mientras tanto, la producción de los investigadores latinoamericanos sobre nuestra problemática, aunque más amplia que en el pasado, sigue manteniendo su marginalidad y relativamente baja circulación regional.

of the lack of defined lines and subtopics for submission of the articles. However, this does not appear an erroneous outcome, but rather a matter for reflecting on the path forward, i.e. where this heterogeneous group of studies was heading. In other words, whether there would be in this 12th version of the event, elements or initiatives that could provide an indication of new directions or approaches for dealing with established themes or for reviewing concepts and tackling new problems.

An initial reading of the studies yielded the answer sought. Many of the key words used revealed that, besides being linked to the themes, titles and content of the texts, terms were present that referred to much broader concepts, often not tied to a given theme, case study, spatial or temporal sample, or the themes connected with the event. These terms refer to more abstract, qualitative concepts alluding to empirical and/or subjective criteria, but which allow objects to be interpreted, review pre-established definitions, establish vertical approaches in the study, so as to reveal qualities and confer greater accuracy to the objectives; or provide emphasis on the individuals involved; or to transcend the theoretical constraints pre-established by the disciplines of origin, allowing inroads to be made into other areas of knowledge.

This marks a return to the "research map", produced by Prof. Sabaté (2013), together with the organizers of the 5th seminar in 2013, the first produced bilaterally, based in cities, Barcelona and Buenos Aires, on 2 continents. The map resulted in a constellation of topics regarded as priority for defining new avenues of discussion for subsequent events and also for discussion by the urbanism research groups involved. Four major themes were identifies at the time: City, territory, theory and history and landscape.

As a participant in the group of researchers for some years, this initial effort to synthesize the themes proved key. Given that there has been annual publication of the papers presented concomitantly on the two continents since 2013, i.e. almost a decade´s worth of discussion on the cities in Latin America and Europe, it was decided for this chapter to undertake an analysis of the period spanning from 2013 to 2020, with a sample of the articles addressing issues involving Latin-American territory. The objective of this new synthesis, however, was not to identify new major themes to prioritize, given the relatively short period elapsed between the two timepoints, but instead to extract some keywords found in the texts, which mark or span the 8 years of the event being held in Latin America.

The synthesis process entailed construction of a matrix containing the abstracts and keywords from the studies, allowing word clouds to be built for each year of the event. The statistical and visual result was surprising, and yet suggestive. The terms found for the separate years, with characteristics outlined previously, were: *Territory*, *Urbanity*, *Mobility*, *Centrality*, *Diversity*, *Landscape*, *Social Exclusion*, *Identity*, *Segregation*, and *Citizen Participation*.

In the present text, the aim was not to determine whether the key words were instruments which best reflect the issues addressed, beyond facilitating indexing tools and other study search mechanisms, since this is outside the scope of the chapter. It is also important to highlight that these words emerged from the reading and comparison against the abstracts from the articles presented. However, rather than exhaustively exploring the complexity of the theme, akin to a *Thesaurus* (TOPALOV et al., 2014), this task is more like the procedure described by "*Notions de l'urbanisme par l'usage*" (*Concepts of urbanism by use*), particular in terms of its objectives (BEAUCIRE &DESJARDINS, 2015):

> This collection of texts is also a response to the needs of our students in urbanism and planning, often faced with these concepts implicitly when analyzing or devising urban projects.[12]

1 SOME KEY WORDS

For this chapter, 5 out of the 10 key words elected were chosen to provide more in-depth definitions and the contexts in which they were employed in the papers presented between 2013 and 2020; as opposed to focusing on a sole definition for each term, different perspectives of discussion and interrelationships with the major themes cited were explored, while alerting to the importance of not naturalizing the choice of these words in our studies.

The synthesis presented below is also an attempt to establish a dialogue with the historiography on urban studies in Latin America. The process sought to highlight the variety of different themes, interests, and approaches which might contribute to a dialogue of the urban studies and of urbanistic thinking between the two continents, revealing avenues for future research.

1.1 *Territory*

Territory, in the context of the studies analyzed, is generally related to the growth and expansion of cities, to metropolization and its outcomes. This phenomenon is not time limited and is still underway in some contexts, and the expansion is addressed as a factor placing new constraints on actions for managing and structuring space. Urban issues and problems are identified based on historical studies which interweave with contemporary contexts.

12 Cette collection de textes est aussi une réponse aux besoins de nos étudiants en urbanisme et aménagement, souvent confrontés à ces notions de façon implicite lors de l'analyse ou de la conception de projets urbains.

A metropolitan territory undergoing rapid transformation is the target of many studies, subject to economic activities, which manifest in terms of logistics, growth and location of commerce and industry, increase in precarious housing and the emergence of new types of housing distributed throughout the regional territory. These movements reveal concerns over peri-urban boundaries and spaces, and shaping of a new *landscape*, calling for systemic nature and the need for design projects and planning to guide change. An intriguing debate emerges when focusing on the peri-urban development of the metropolitan outskirts as territory for envisaging possible futures. Given the diversity of cities and regions, there is an array of socio-territorial models and patterns, encountering the consolidated city, both metropolitan and rural areas, as well as the urban sprawl and different respective types of land use.

The notion of territory, during the course of the different Seminars, gives way to the concepts of deterritorialization and reterritorialization, seen in studies on the central and peri-urban areas of cities, extending the meaning of *urban* to encompass the idea of singular space and disputed territory, exposing conflicts between scales and visions. According to this perspective, the production system and insertion and conflicts of cities in a society dominated by image and consumerism should be interpreted within a globalized market, emphasizing the role of urban space and its reproduction as a locus of accumulation, bringing the notion of territory closer to that of urbanity and identity.

Put another way, recognizing a complex system involving competition between cities and territories, some studies question the current paradigm of territorial structure and management of the landscape, setting this against the growing importance of preserving the identity of places, while also emphasizing the need for understanding regional systems, in their ability to provide the population with services and facilities, beyond city limits, and new tools for analyzing recently devised hybrid human settlements.

1.2 *Urbanity*

Urbanity as an expression for the everyday dimension and processes and instruments for humanizing the public space was a focus of many of the studies. Studies sought to determine how these factors can be incorporated into urbanistic practice through a critical perspective. The theme contributed to the understanding of the growing complexity of the urban territory, which presumes interlocution of the actors in the production of urban space. The contribution of public and private actors involved in Urban Project and Plans was similarly highlighted as a form of dialogue in the construction of public space.

Some articles empirically investigated the constitution of a network of public spaces and their historic process as strategic in creating a truly inclusive environment, bringing together attributes under the notion of *urbanity*. The discussion of urbanity revolves around territory by identifying urban qualities, where public space is the core component of this new dimension. The notion applies to a group of elements and environments which make a city welcoming for all, promoting debates on how urban form can be conducive to this condition or otherwise; the forms of appropriation of public spaces, with an emphasis on everyday life, reinforce the role of public space as a place for meeting and experimentation. Urbanity portray the compact city, densification and mixed uses, concepts which complement critical analyses of the unfettered expansion of cities in LA, the precariousness of space and the exhausting of natural resources. Research methods broaden when establishing degrees of urbanity based on several indicators: urban form, diversity of uses and activities and receptiveness toward different forms of appropriation, among others.

Unregulated space allowing spontaneous alternative activities potentially leading to material changes in the overall functional and social dynamic are held as successes of urbanity and as potential factors transforming the perception of spatial qualities of urban territories. Understanding and assessing the performance of public space in terms of urbanity is a theme of permanent interest, because it fosters relationships with heritage, landscape and the preservation of identity, safeguarding the memorable places of different groups.

1.3 *Landscape*

The word landscapes depicts a multidisciplinary view, bring together physical geography, human geography, urbanism and the environment, seeking meanings and reflections on the anthropic actions in nature. This is cross-sectional construct (Santos, 1996) that integrates relations between city, society, nature and culture.

According to the phenomenological line (CORREA e ROSENDAHL, 1998; COLLOT, 1990; BERQUE, 1990), these relationships are of great importance in many studies, in that they contribute to understanding of the historic processes of cities in LA, based on the understanding that landscapes are built over time from relationships between society and nature, cultures and identities. A group of studies address landscape and cultural heritage, as a group or series, as part of the identities and territorialities of different social groups that make up cities and regions. Other studies involve theme associated with the processes of environmental perception and cognition, with cultural tourism, and seek to understand the forms of appropriation of Landscape (appreciating or depreciating it), as well as plans, projects and interventions proposed for it.

From a different perspective, the term denotes the environmental issue within a context where uncertainties of the intense rapid urbanization of cities in LA are juxtaposed with their sustainability. This is a broad theme with important evidence for debate that involves the preservation and curbing of human actions and of the environment impact on a host of natural resources. Numerous studies use the hydrographic basin as an environmental unit, an analysis which allows, by comparing landscapes, determination of the interplay of social and physical factors in its dynamic. Studies examining the problem of access to land by socially vulnerable populations integrate with environmental problems and degradation of nature and the landscape. A critical view on the lack of systemic integrated action in urban and environmental planning also feature in articles.

More recent, but no less important, studies have analyzed the physical and sociospatial phenomena which influence the dynamic of large and medium-sized cities, particularly their peri-urban areas, defined as transition areas between rural and urban regions, which have contemporary structures and contrast with the classic center-periphery dichotomy, established since the 1970s. Other analyses investigate the importance of systems of green areas and free space, the recovery of water courses and fronts, risk management (fires, earthquakes and flooding), elucidation of the effects of climate change, among other issues associated with the term Landscape. Evaluations of management of cities point to the need for planned anthropic actions that are consistent with the local specificities, favoring urban and environmental sustainability of the regions.

1.4 *Identity*

The public space is construed as being endowed with meanings. Conferring a historical value to this space confirms its multiple dimensions, representing a transformation of the notion of place and its role in creating multiple experiences in the city in the public sphere.

The phenomenological approach is found in texts emphasizing forms of settlement of urban land, construed as permeable frontiers (pores), together with the dynamics of use and identity of many different places. This also includes approaches associating experience from emotional and affective fields, translated by process of interaction in urban space and perception of this dimension and its developments on various scales and representations.

In addition, this involves focus on cultural landscape, evoking the safeguarding of memory as a construct of social identity. The concept integrates with the term cultural landscape, defined as the locus of representation of many different ways of life and of interactions between nature and culture, material and non-material heritage. In several articles, multiscale views allow extension of the approach of *memory places* (NORA, 1993) to *exurbia* territories, implying citizen participation in the planning process, in the forming of a metropolitan identity and as part of the process of governance.

Museums, as part of the formation of the image and identity of cities, on local and global levels, are also objects of analysis of leisure and tourism as fields of study of behavior and its relationship with places of sociability, appropriation and reproduction of cultural industry.

Central areas and consolidated urban fabric are objects provoking reflection on the theme of identity, particularly in face of the imminence of transformations brought about by projects for regeneration and intervention in historical areas, of possible conflicts and disputes by different economic groups, with special treatment of resulting phenomena such as gentrification, real-estate appreciation and abandonment of cultural heritage.

1.5 *Mobility*

The term "mobility" initially emerges in the SIIUs from a group of studies concerning the urban and regional space and its relationships with modes of transportation. The approach took on a new character from 2015, in studies investigating mobility as am attribute of individuals, advocating that the contemporary city should be prepared to expand the possibilities of mobility of its inhabitants with quality (SILVA, 2013) on its multiple different scales. This involves designing a more human city so that people engage in their everyday activities (or otherwise) in quality and timely manner; it a call for the right to mobility for all citizens in an equitable and sustainable fashion in the cities of LA.

A group of studies critically analyzed the "model of the Latin-American city" designed for the car as opposed to public transport (VASCONCELLOS, 2000) with low walkability; a model that has contributed to expansion, dispersion and fragmentation of cities, irrespective of size, and consequently to an increase in sociospatial inequality.

The need for integrated planning between transport and land use as part of the strategic vision for the future of more sustainable cities, was a focus of many of the studies. These investigations sought to assess Master plans and mobility plans implemented in the last 2 decades across many cities in LA, with a view to guiding development and integrated urban density along transport routes and stations. Through cases studies, qualitative assessments of the public space were carried out as fundamental support for mobility of pedestrians (including those with reduced mobility), interfacing with the urbanistic standards of the different areas of the cities (central and peripheral, including precarious settlements), revealing the absence of city policies promoting the pedestrian scale.

During the last SIIU in 2020, there was a rise in the number of studies that incorporated the issue of gender in mobility. In general, the studies critically examined the group of factors that have an unequal effect on getting around, both in the transport systems and public space, allied with the lack of public policies securing the right to mobility with gender equity.

1.6 *Final considerations*

The presence of these words transversally to the major themes makes evident some transformations in the concepts over time, in line with new technical and social demands. The fact that there were some historic homologous processes in the cities of LA, may have contributed to the consolidation of certain terms specific to this side of the Atlantic, such as *modernization, centers of development, dependence, residential mobility*, and *informality*.

Nevertheless, by the same token, no great disparity between the themes investigated by the researchers on the two continents was found, except for the fact that research into "Urban Dynamics" and its subthemes Urban Cultures, Migration, Social Justice, Insurgent Urbanism and Urban Collectives were more numerous among LA researchers. This is an indicator that the geographic and demographic conditions, as well as structural and historical socio-cultural factors of the countries and cities of LA give rise to different problems to those encountered in Europe and call for the formulation of different hypotheses and questions on the relationship between urbanism and society, while solutions require the devising of novel methods, instruments and approaches. This situation renders it strategically necessary to question the transposing of theories devised in Northern countries for application to cities of the subcontinent. This questioning also serves to promote further reflection and contribution of these problems to the debate, which appear to be of greater complexity this side of the Atlantic. As noted by Ribeiro (2014, pg. 190), it is important "not to regard our reality as the particularization of something universal" which, according to the author, is commonplace in the LA debate and has a direct bearing on the technical and methodological guidance in the field of urban studies and intervention projects in cities.

Regarding the other key words identified, *Centrality* refers to urban centers that have reached a prominent position as "main centers", "business center", "historical center", and undergo a shift to incorporate different urban dynamics at various levels. *Diversity* alludes to question of the challenge of heterodox solutions to the planning of cities; *Segregation* is the result and at the same time the explanation for the processes of fragmentation of territories and regions, whereas *Social Exclusion* and *Citizen Participation* are opposite terms representing the intractable social problems for the new conditions of governance of the cities in LA.

The overcoming of the sectoral approach of urban policies, the creation and debate of other instruments of induction, control of urban land use and settlement for promoting a city design shaped by social participation, as opposed to market forces, are set to persist in the forthcoming SIIUs, in addition to the forms of promoting greater access, in its broader sense, minimizing phenomena of segregation and gentrification of urban space. The concern over the implications for climate change, ability to support the urban and regional environment, gender issues and population aging should be themes driving scientific studies and a review of development paradigms.

The perverse effects stemming from the indiscriminate use of public highways by vehicle traffic in cities, will be increasingly the focus of critical analyses and the basis of new projects safeguarding built heritage and public spaces, among the many other problems impacting the quality of life of the population of cities in LA.

Lastly, for the upcoming editions of the event, there is likely to be increase in studies addressing the situation of the new mobility time-spaces promoted by the greater use of new technology and means of communication, a topical theme as a result of the social isolation brought on by COVID-19 pandemic. In this context, the word mobility takes on new meanings in interweaving and overlaying work activities and social relationships in the same space.

The identification of these and many other topics developed through the reading of the studies from the different SIIE, is intended to raise new hypotheses for a research agenda on both sides of the Atlantic, supported by a possible transdisciplinary dialogue.

REFERENCES

Abramo, P. 2012. La ciudad com-fusa: mercado y producción de la estructura urbana en las grandes metrópolis latinoamericanas *Revista EURE*, vol. 38, núm. 114, mayo, pp. 35–69 Pontificia Universidad Católica de Chile Santiago, Chile.

Acselrad, H. 1999. Discursos da Sustentabilidade Urbana. *Revista Brasileira de Estudos Urbanos e Regionais*. N.1, p. 79–90, São Paulo.

Almandoz, A. 2008. Despegues sin madurez. Unbanización, industrialización y desarrollo em la Latinoamerica del siglo XX. *Revista EURE*, Vo. XXXIV n. 102, pp. 61–76.

Almandoz, A. 2007a. *Entre libros de história urbana. Para una historiografía de la ciudad y el urbanismo de América Latina*. Caracas: Equinoccio/Universidad Simón Bolívar.

Almandoz, A. 2013. *Modernización Urbana en América Latina. De las grandes aldeas a las metrópolis masificadas*. Colección Estudios Urbano UC. Santiago de Chile: FADEU/PUC-Chile.

Almandoz, A. 2007b. Modernización urbanística en América Latina. Luminarias extranjeras y cambios disciplinares, 1900-1960. *Iberoamericana. América Latina – España – Portugal*. Vol. 7. Núm. 27. https://journals.iai.spk-berlin.de/index.php/iberoamericana/article/view/858

Arango Cardinal, S. 2012. *Ciudad y arquitectura. Seis generaciones que construyeron la América Latina moderna*. México: Fondo de Cultura Económica, Consejo Nacional para la Cultura y las Artes.

Barros, C. & Araujo, E. L. C. (Ed.). 2006. *Reabilitação urbana de centralidades metropolitanas: reflexões e experiências na América Latina do século 21*. Recife: Zoludesign.
Beaucire, F.; Desjardins, X. 2015. *Notions de l'urbanisme par l'usage*. Paris: Publications de la Sorbonne, 117 p.
Berque, A. 1990. *Madiance: de milieux en paysages*. (1ª ed). Paris: Reclus.
Bielschowsky, R. (Ed.) 2000. *Cinquenta anos de pensamento na CEPAL*. Rio de Janeiro. Ed Record, vol. 1.
Canclini, N. 2007. *A globalizaçao imaginada*. São Paulo: Iluminuras.
Carrión, F. 2010a. *El laberinto de las centralidades históricas en América Latina*. Quito: Ministerio de Cultura de Ecuador.
Carrión, F. & Velarde, P. et. al. 1992. *Municipio y Democracia. Gobiernos Locales en Ciudades Intermedias de América Latina*. Santiago: Ediciones SUR.
Carrión, F. (ed.). 1991. La investigación urbana en América Latina. Caminos recorridos y por recorrer, vol. 4: *Conversaciones sobre los Caminos por Recorrer*. Quito: Ciudad.
Carrión, F. 2010b. *El centro histórico como objeto del deseo. El labirinto de las centralidades históricas en América*: Ministerio de Cultura.
Collot, M. 1990. Pontos de vista sobre a percepção das paisagens. *Boletim de Geografia Teológica*, Rio Claro, v. 20, n. 39, e. 21–32.
Connolly, P. La ciudad y el habitat popular: paradigma latinoamericano. In RAMÍREZ VELÁZQUEZ, B. R.; PRADILLA COBOS, E. (Ed.) *Teorías sobre la ciudad en América Latina*. Primera Edición. México, D.F: Universidad Autónoma Metropolitana, p. 26–76.
Correa, R. L. & Rosendahl, Z. (ed.) 1998. *Paisagem, tempo e cultura*. Rio de Janeiro: Editora da UERJ, 123 p.
D'arc, H. R., Memoli, M. (Ed.). 2012. *Intervenções urbanas na América Latina. Viver no centro das cidades*. São Paulo: Editora Senac.
Duhau López, E. 2014. La investigación urbana y las metrópolis latino-americanas. In RAMÍREZ VELÁZQUEZ, B. R.; PRADILLA COBOS, E. (Ed.) *Teorías sobre la ciudad en América Latina*. Primera Edición. México, D.F: Universidad Autónoma Metropolitana, p. 26–76.
Dureau, F. et al. (Ed.) 2002. *Metrópolis en movimiento. Una comparación internacional*. Bogotá: Institut de Recherche poutr le Développement, Alfaomega.
Gomes, M. A. A. F. & ESPINOZA, J. C. H. 2009. Olhares cruzados: visões do urbanismo moderno na América do Sul, 1930-1960. In: GOMES, M. A. A. F. (Ed.) *Urbanismo na América do Sul: circulação de ideias e constituição do campo*, 1920-1960. Salvador:EDUFBA, p. 13–40.
González, R. S. & Baquero, M. 2013. A política comparada na América Latina: dilemas e desafios no Brasil. In *Revista Debates*, v.7, n.3, Porto Alegre, p. 111–126.
Gorelik, A. 2005. A produção da cidade latino-americana. *Revista Tempo Social*, São Paulo, v. 17, n.1, p. 111–133.
Gutiérrez da Costa, R. 2008. Cooperación. Un camino de ida y vuelta. *La Ciudad Viva* no1. Coordinación por la Dirección General de Vivienda y Arquitectura.
Gutiérrez, R. & Tartarini, J. 2007. *Congresos Panamericanos de Arquitectos 1920-2000. Aportes para su historia*. Buenos Aires: CEDODAL.
HARDOY, J. E. Las Ciudades de América Latina a partir de 1990. In: *La Ciudad Hispanoamericana. El sueño de un orden*. Madrid: Centro de Estudios Históricos de Obras Públicas y Urbanismo, 1989.
Hardoy, J. E. y & Morse, R.M. (Ed.) 1978. *Repensando la ciudad de América Latina*. Buenos Aires: Grupo Editor Latinoamericano/IIED América Latina.
Hiernaux Nicolás, D. 2014. La ciuades y las regiones. In Ramírez Velázquez, B. R. & Pradilla Cobos, E. (Ed.) *Teorías sobre la ciudad en América Latina*. Primera Edición. México, D.F: Universidad Autónoma Metropolitana, p. 77–121.
Kuri, P. R. 2014. La ciudad desde el espacio público y las prácticas ciudadanas. In Ramírez Velázquez, B. R. & Pradilla Cobos, E. (Ed.) *Teorías sobre la ciudad en América Latina*. Primera Edición. México, D.F: Universidad Autónoma Metropolitana, p. 916–967.
Lencione, S. 1998. Mudanças na Metrópole de São Paulo (Brasil) e transformações industriais. *Revista do Departamento de Geografia* n.12, São Paulo, p. 27–42.
Liernur, J. F. 2002. *Escritos de arquitectura del siglo 20 en América Latina*. Madrid/Sevilla: Tanais.
Morse, R. M. 1971. *La investigación urbana Latino Americana: tendencias y planteos*. Buenos Aires: SIAP.
Morse, R. M. 1978. Los intelectuales latinoamericanos y la ciudad (1860-1940). In: Hardoy, J. E. & R. Morse (Ed.). *Repensando la ciudad de América Latina*. Buenos Aires: Grupo Editor Latinoamericano/IIED América Latina, p. 91–112.
Morse, R.M. 1970. *Formação Histórica de São Paulo: De Comunidade a Metrópole*. Trad. complementares Antonio Candido. São Paulo, Difel.
Nora, P. 1993. Entre memória e história: a problemática dos lugares. *Projeto História*, São Paulo, n.10, p. 7–28.
Novick, A. 2003. El urbanismo en las historias de la ciudad. *Revista Registros*, Mar del Plata, p. 5–26.
O'Gorman, E. 2006. *La invención de América: Investigación acerca de la estructura histórica del Nuevo Mundo y del sentido de su devenir*. 4. Ed., México: Fondo de Cultura Económica.
Peixoto, F. A. & Gorelik, A. (Ed.) 2019. *Cidades sul-americanas como arenas culturais*. São Paulo: SESC.
Ramírez Velázquez, B. R. & Pradilla Cobos, E. (Ed.) 2014. *Teorías sobre la ciudad en América Latina*. Primera Edición. México, D.F: Universidad Autónoma Metropolitana, 2014.https://www.researchgate.net/publication/330203099_Blanca_R_Ramirez_Velazquez_y_Emilio_Pradilla_Cobos_CompsTeorias_sobre_la_ciudad_en_America_Latina
Ramírez Velázquez, B. R. & Pradilla Cobos, E. 2014. Presentación. In Ramírez Velázquez, B. R. & Pradilla Cobos, E. (Ed.) *Teorías sobre la ciudad en América Latina*. Primera Edición. México, D.F: Universidad Autónoma Metropolitana, p. 15–25.
Rangel, R. L. 2014. Las teorias urbanas, um tema transdiciplinario no neutral. In In Ramírez Velázquez, B. R. & Pradilla Cobos, E. (Ed.) *Teorías sobre la ciudad en América Latina*. Primera Edición. México, D.F: Universidad Autónoma Metropolitana.
Ribeiro, L. C. Q. 2014. Las metrópoles del pensamento latino-americano. Refelxiones para uma teoria del processo de metroplización. Traducción Carlos Márquez. In Ramírez Velázquez, B. R. & Pradilla Cobos, E. (Ed.) *Teorías sobre la ciudad en América Latina*. Primera Edición. México, D.F: Universidad Autónoma Metropolitana, p. 176–252.
Sabaté Bel, J. 2013. Un punto de inflexión en los seminarios de investigación en urbanismo. A: Seminario Internacional de Investigación en Urbanismo. In *V Seminario*

Internacional de Investigación en Urbanismo, Barcelona-Buenos Aires, junio 2013". Barcelona: DUOT, p. 11–14.

Sambricio, C. (Ed.) 2012. *Ciudad y vivenda en América Latina 1930-1960*. Madrid: LAIMPRENTA CG.

Santos, M. 1979. *O Espaço Dividido. Os Dois Circuitos da Economia Urbana dos Países Subdesenvolvidos*. Río de Janeiro: Francisco Alves (Coleção Ciências Sociais).

Santos, M. 1994. *Técnica Espaço Tempo. Globalização e meio técnico-científico-informacional*. São Paulo: Ed. Hucitec.

Segre, R. 1999. *América Latina fin de milenio, raíces y perspectivas de su arquitectura*. La Habana: Editorial Arte y Literatura.

Segre, R. 1989. *Arquitectura y Urbanismo de la Revolución Cubana*. La Habana: Editorial Pueblo y Educación.

Segre. R. 1977. *Las estructuras ambientales de América Latina*. México/España/Argentina/Colombia: Siglo Veinteuno Editores.

Silva, A. 1992. *Imaginarios urbanos. Bogotá y São Paulo, cultura y comunicación urbana en América Latina*. Bogotá: Tercer Mundo Editores.

SILVA, F. N. 2013. Mobilidade urbana: os desafios do futuro. *Cadernos Metrópole* [online], vol.15, n.30, pp. 377–388. http://www.scielo.br/scielo.php?script=sci_arttext&pid=S2236-99962013000200377&lng=en&nrm=iso

Tamayo, S. 2014. Crítica de la ciudadanía y movimentos sociales urbanos. In Ramírez Velázquez, B. R. & Pradilla Cobos, E. (Ed.) *Teorías sobre la ciudad en América Latina*. Primera Edición. México, D.F: Universidad Autónoma Metropolitana p. 968–1041.

Topalov, C.; Bresciani, S.; Coudroy De Lille, L.; D'arc, H. R. (Ed.) 2014. *A aventura das palavras da cidade através dos tempos, das línguas e das sociedades*. 1ª edição, São Paulo: Romano Guerra.

Vasconcellos, E. A. 2000. *Transporte urbano nos países em desenvolvimento. Reflexões e propostas*. 3a. ed. São Paulo: Annablume.

Portugal

L'Architecture d'Aujourd'hui special issues on Latin America, 1945-1975. A look at its contents and the context of its production

A. Avila-Gómez
HiCSA, Panthéon-Sorbonne University; Université Paris 1 Panthéon-Sorbonne, France

ABSTRACT: During the first half of the 20th century, the dissemination of images of urban and architectonic projects – built or unbuilt – in Latin America, had a special reception in some European journals, whose global reach promoted the recognition of the spatial modernization achieved by some Latin American cities. Along with these images, new urban narratives – whose origins were in European and North American schools – were also transmitted. This is made evident through the professional connections and networks built by Latin American architects and urban planners who studied in Europe and in the United States, and later applied their newly acquired knowledge in their countries of origin. In this regard, this article analyzes the content from special issues of *L'Architecture d'Aujourd'hui* that were devoted to Latin American cities and/or their architecture, emphasizing the context – both professional and editorial – in which those thematic issues were created.

1 INTRODUCTION

During the second half of the 20th century, the main French architectural magazines became a medium for reflection on technical, artistic, intellectual, and theoretical production. Based on the production of visual and critical material whose contents permanently enriched the material culture of other disciplines such as urbanism, construction, or design, the importance of these periodicals even allowed them to leave aside language barriers and settle into the professional imagination and popular culture.

Among these publications, *L'Architecture d'Aujourd'hui* (AA), founded in 1930 by André Bloc (1896-1966) and Marcel Eugène Cahen (1883-1930), quickly occupied a privileged space, bringing a series of material and conceptual innovations that quickly positioned it at a national level (alongside other traditional publications such as *L'Architecture* and *La Construction Moderne*), and at an international level. From its first issues, *L'Architecture d'Aujourd'hui* offered its readers a broad panorama of the architecture and urbanism produced on all continents, and it was thus that from the inter-war period, projects built in Latin American cities increasingly populated the pages of the magazine.

Three specific questions have served as a starting point for this research:

In what editorial and professional context were the special issues on Latin America prepared by *L'Architecture d'Aujourd'hui* produced?

What kinds of contents and discourses on urbanism and the city were included in these issues?

What kind of connections fed the exchanges between Latin American and French professionals related to the preparation of these issues?

Although there are studies that have partially analysed the presence of Brazilian (Braschi, 2016) and Mexican (Noelle, 2009) architecture in some of the issues that make up this corpus, the set of publications on all the countries of the region prepared by the successive editorial teams of *L'Architecture d'Aujourd'hui* requires a more detailed look. This detailed look contributes to draw a global panorama to which some historians of architecture such as Ana Esteban Maluenda (2017; 2019) and Maria B. Camargo Capello (2011) have already provided valuable interpretations and elements of analysis. In this sense, the work carried out by Esteban Maluenda offers us the study of interesting quantitative aspects such as, for example, the calculation of the number of pages devoted to Latin American projects between 1943 and 1969 in a universe of nine

DOI: 10.1201/9781003220855-3

European magazines that include two Spanish[1], two British, two French, and three Italian. It also shows the comparison of the distribution of the number of pages devoted to the architecture of each Latin American country in each of the nine magazines of the corpus defined by the author.

2 THE INTERNATIONAL DIMENSION OF A FRENCH MAGAZINE

Since its creation, *L'Architecture d'Aujourd'hui* has distinguished itself from other prestigious French publications such as *L'Architecte* or *La Construction Moderne*, by its initiative to offer the reader a current panorama of recent architecture produced abroad, including modern or more moderate trends, both in purely architectural projects and in urban projects and large-scale interventions. The magazine favored exchanges and contacts with professionals from countries on all continents, thus weaving an international network that permanently fed the contents and debates presented in its pages.

To this end, trips abroad were organized – almost always on an annual basis –, which included not only meetings with colleagues from other latitudes and visits to emblematic sites, but also the opportunity to discuss with local authorities: this happened initially in 1932 in Poland and the USSR; in 1933 in Italy; and in 1935 in Czechoslovakia, Hungary, and Austria. In addition to this, there were the activities promoted by the *Réunions Internationales d'Architectes* (RIU) also created by the magazine in 1932, whose first general secretary was Pierre Vago, its *rédacteur en chef*.

Based on these initiatives, the magazine became a platform for architects from other countries to present projects and current issues in architecture and urban planning, considered to be of general interest in an increasingly interconnected world. As architecture historian Hélène Jannière (2002) pointed out: "Thanks to study trips, special issues, and a very efficient network of correspondents, *L'Architecture d'Aujourd'hui* was able to offer the reader an unprecedented world panorama of contemporary architecture."

Since its reappearance in 1945, after the interregnum imposed by the Second World War, *L'Architecture d'Aujourd'hui* resumed the publication of special issues devoted to modern architecture and urbanism in a particular country, or in a region of the world made up of several countries. Thus, during the period between 1947 and 1979, the French magazine published a total of 28 issues devoted to one country: the first of which to Brazil and the last to China [Table 1].

In total, 18 countries were selected to be a focus during this period: eight in Europe, three in Asia, two in North Africa, two in the Middle East, two in Latin America, and one in North America. Among them, only three countries were selected on three occasions: Brazil (in 1947, 1952, 1960), the United States (in 1953, 1965, 1971), and Japan (in 1956, 1961, 1966).

A special case is that of the issues prepared according to a geographical or even geopolitical criterion: this is the case of the three issues dedicated to the Nordic countries; the two dedicated to Latin America (in 1951 and 1974); and those dedicated to North Africa, Black Africa [sic], and the Third World (in 1968) [Table 2].

Table 1. Special issues of AA dedicated to a single country, 1946 – 1979.

Countries	Special issues dedicated to each country	Total issues
Brazil, USA, Japan	**3**	**9**
Mexico, Spain, Italy, Israel	**2**	**8**
USSR, Switzerland, Austria, Iran Great Britain, Denmark, Poland, Portugal, China, Tunisia, Morocco	**1**	**11**

Source: Own elaboration from data of *L'Architecture d'Aujourd'hui*.

Table 2. Special issues of AA dedicated to regions, 1946 – 1979.

Region	Year	Title of the issue	(No. issue)
	1954	Pays nordiques	(54)
Nordic countries	**1961**	Pays nordiques	(93)
	1967	Architectures nordiques	(134)
Latinamerica	**1951**	Urbanisme en Am. Latine	(33)
	1974	Amérique Latine	(173)
Africa	**1955**	Afrique du Nord	(60)
	1957	Afrique Noire	(70)
Third Word	**1968**	Tiers Monde	(140)

Source: Own elaboration from data of *L'Architecture d'Aujourd'hui*.

1 On Latin American students at ESA and the IUUP in the first half of the 20th century, see especially the articles by Andrés Avila-Gómez about Latin American students at the Institut d'Urbanisme de l'Université de Paris (2019); and at the École Spéciale d'Architecture (2020).

3 LATIN AMERICA AND ITS COUNTRIES AS A CENTRAL THEME IN SPECIAL ISSUES

L'Architecture d'Aujourd'hui's interest in current architectural and urban issues in Latin American countries was already evident in the second half of the 1930s, when Eduardo Vasconcelos Pederneiras in 1936, and Mario Pani Darqui in 1937, became the magazine's correspondents in Brazil and Mexico respectively. During the magazine's first decade, the correspondents mainly fed the *Architecture à l'étranger* section, but Jannière stressed that, before the war: "This worldwide network of correspondents had not had a systematic impact on the coverage of the global scene that the publication was intended to show: thus, for example, despite having correspondents in Brazil, South Africa, Mexico, and New Zealand – all of them permanently featured in the magazine's credits – these were countries to which no specific, really important articles were devoted."

At the end of the war, the magazine had about twenty foreign correspondents, and by 1950, 24 countries were represented, four of them Latin American: Argentina (R. Moller), Brazil (Maria Laura Osser), Mexico (Vladimir Kaspé), and Venezuela (Carlos Raúl Villanueva). Two decades later, in 1970, correspondents were quoted in 34 countries, six of them Latin American: three in Colombia (Vieco, Reyes and Botero); two in Brazil (Maria Laura Osser and R. M. Peres); and one in Ecuador (Barragán Dumet), Mexico (Vladimir Kaspé), Uruguay (Luis García Pardo), and Venezuela (Carlos Raúl Villanueva).

During the period we are dealing with here, which corresponds to the so-called *Trente Glorieuses* of sustained economic growth in the most developed countries, a total of seven special issues of *L'Architecture d'Aujourd'hui* were dedicated to Latin America: two issues on the region; two on Mexico; and three on Brazil [Table 3].

To this was added the preparation of other thematic issues which, in their contents, gave special importance to certain countries in the region. As is the case of the issue on construction in warm countries published in 1956 which presents Venezuela; the issue on urbanism in large capitals published in 1960 which presents Cuba and Venezuela; and the issue on the work of the Brazilian architect Oscar Niemeyer published in 1974 which presents his projects in Brazil.

Tabla 3. AA numbers on Latin America and its countries, 1946 – 1974.

Year	Title of the issue	(No. issue)	Issue director
1947	Brésil	(13-14)	Persitz, Osser
1951	Urbanisme en Am. Latine	(33)	Persitz
1952	Brésil	(42-43)	Bloc, Piro, Hanning
1956	Californie. Venezuela. Construction en pays chauds	(67-68)	Bloc, Fischer, Diamant-Berger, Villanueva,Kaspé
1955	Architecture mexicaine	(59)	Persitz, Kaspé, Valeix
1960	Brésil	(90)	Bloc, Osser, Diamant-Berger
1963	Mexique	(109)	Bloc, Kaspé, Diamant-Berger Nerubay
1974	Amérique Latine	(173)	Cousin, Lepetit

Source: Author, from data of *L'Architecture d'Aujourd'hui*.

4 THE TWO ISSUES DEDICATED TO URBAN PLANNING IN LATIN AMERICA: 1951 AND 1974

In the text by Paul L. Wiener and Josep L. Sert entitled 'Conditions générales de l'urbanisme en Amérique Latine,' which opens the issue on Urbanism in Latin America published in 1951 [Figure 1], the two authors stated bluntly that "*en Europe et de par le monde, la connaissance de l'Amérique Latine est limitée et confuse.*"[2] A map serves to indicate the nine cities for

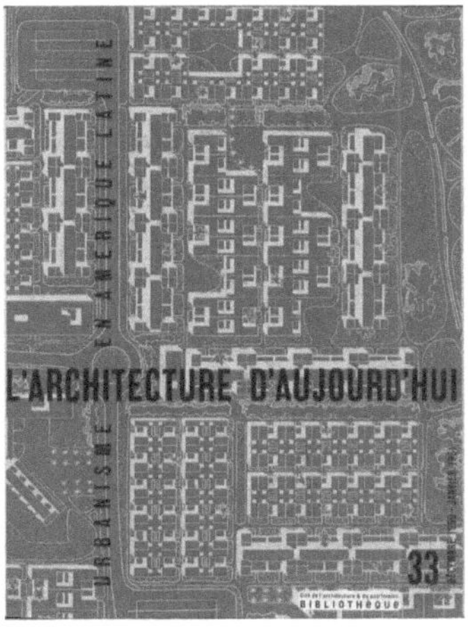

Figure 1. Cover of No. 33 entitled *Urbanisme en Amérique Latine*. Source: AA, January 1951.

2 On the dissemination of the Latin American architectural panorama in Spain during part of the same period analyzed here, see the article by Ana Esteban Maluenda (2012)

which the projects shown in the issue were conceived: two in Colombia (Medellín and Tumaco), two in Peru (Lima and Chimbote), two in Brazil (Rio de Janeiro and São Paulo), two in Argentina (Buenos Aires and Bahía Blanca), and one in Venezuela (Caracas).

In their text, Wiener and Sert affirmed that, throughout those last decades, the École des Beaux-Arts de Paris had exercised a dominant influence on the practice of Latin American urbanism because "Many cities have had their own Baron Haussmann, whose achievements have generally proved more dire than fortunate. Large avenues, squares, monuments, and expensive 'Versailles' gardens have not succeeded in solving the underlying problems, nor have they served to hide the sordid working-class housing or the slums that proliferate around them."

They blamed the new evils which are overwhelming Latin American cities on the growing American influence to the detriment of French models, stating that "the benefit obtained from this change sometimes appears to be doubtful. The American film industry, in particular, is responsible for the construction of numerous skyscrapers in at least some strange locations; sports clubs outside the city; Hollywood-style garden cities." And yet, when compared with North American developments, some of the characteristics of Latin American cities are praised by the two authors: "Latin American cities are generally better designed than North American cities from this point of view, since they generally have a clearly defined public center. The square, with its church and public buildings, is still the meeting place par excellence, and the inhabitants continue to wander the 'paseo' or promenade. The main roads of modern American cities do not correspond to the 'civic' needs of their inhabitants."

The projects presented in the issue are extensively illustrated and commented on: color plans [Figure 2], aerial photographs and black and white photographs, and descriptive and analytical texts. Particularly noteworthy are the sharp aerial photographs that show the configuration of the cities in question. These examples of *nouvel urbanisme* were designed by a generation of architects and urban planners, among which we highlight Maurice Rotival (1892-1980), Paul L. Wiener (1895-1967), Josep L. Sert (1902-1983), Affonso E. Reidy (1909-1964), and Enrique Gebhard (1909-1978), followed the general guidelines set out by the *Congrès Internationaux d'Architecture Moderne* – CIAM at the famous Charte d'Athènes in 1933.

A quarter of a century later, under the title *Amérique Latine*, the last issue of the corpus cited here was in turn the first issue published during the interregnum (1974-1977) in which the architect Bernard Huet occupied the role of *rédacteur en chef* of the

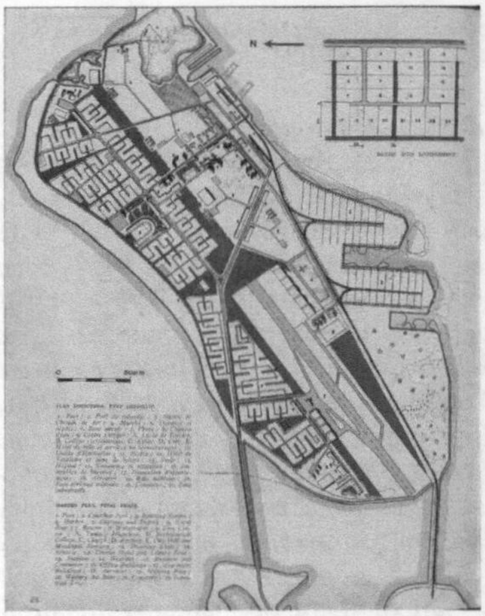

Figure 2. Pilot Plan for Tumaco (Colombia) by Wiener and Sert. Source: AA, no. 173, June 1974, p. 24.

magazine, at a time when the publication was experiencing profound internal conflicts and radical changes with respect to its editorial line and contents. A leading figure in the renewal of architectural teaching and practice in France, Huet had trained at the École des Beaux-Arts and at the atelier of Louis Kahn at Yale and was also a fervent reader of authors such as Benjamin, Lukács, Bloch, Gramsci and other Marxist *déviants*.

In his memoirs, Pierre Vago, former editor-in-chief of the magazine, describes the Huet period as follows: "[…] in my opinion, this was not the best choice. *L'Architecture d'Aujourd'hui* then followed all the fashions, adopting the pseudo-progressive syringe typical of left-wing pseudo-intellectuals, with the pretension of being up to date just by using the ideological verbiage described by Baudouï"[3] (Vago, 2000, 426).

A panorama is presented that is radically different from the optimistic and almost propagandistic tone latent in the texts of 1951: the reader of *L'Architecture d'Aujourd'hui* finds on this occasion a visceral criticism of the urban reality present in the national capitals of the region.

The cover of the issue [Figure 3], produced by Pierre Lepetit, *maquettiste* of the magazine, is not exempt from the usual clichés that are fed in Europe about the region: the color photograph shows a composition in which a bottle of "Mariachi" tequila stands out, the label of which bears a photo of a group of Mexican charros; a machete; a couple of green jalapeño peppers; and a box of Cuban cigars on which one is lit and half consumed.

3 In Europe and throughout the world, knowledge of Latin America is limited and confused.

Figure 3. Cover of no. 173 entitled *Amérique Latine*. Source: AA, June 1974.

At the bottom of the image, the multiform and colorful wooden blocks taken from a children's building set, evoke the bars and blocks of houses so characteristic of modern urbanism: this staging was no doubt inspired by the last scene of Jean-Luc Godard's film *Deux ou trois choses que je sais d'elle* (released in 1967), in which about thirty boxes and packages of detergents, cigarettes, and everyday food products are arranged on a lawn, evoking the organization of the volumes of a *grand ensemble* [Figure 4].

A few pages later, we find a black and white reproduction of the same photograph on the cover, but this time accompanied by the following text: "If these pages have not been clear enough to illuminate the spirit of the reader who is reluctant to decipher the riddle on the cover, here is the answer: 'Everyone sells his salad as he can.' The bourgeoisie and its state apparatus distribute the social product, justifying the class reality of such distribution, the market. Both in its own sense and figuratively, the people are at the mercy of the scavengers."

In this controversial issue which announced the new editorial direction of the magazine, the authors of the central texts on Mexico, Lima, and Caracas,

Figure 4. Final plan of *Deux ou trois choses que je sais d'elle* by Jean-Luc Godard, 1967.

the three big capitals analyzed on this occasion, are respectively Jean-Pierre Cousin, Juvenal Baracco, and Marco Negrón who express in their writings a line of thought similar to that of the British architect John F.C. Turner.

In his text *Dépendance, architecture et université*, Baracco goes beyond the debate on the city and the role of urbanism, and questions the role and nature of the education of Latin American architects and urban planners: "The fundamental problem is undoubtedly the modification of the structure of knowledge, which implies the revision of the medieval concept of university, as well as the foundations of new criteria to define the services that should be returned by those who possess knowledge, which will necessarily lead to a re-examination of what is understood by 'professionalism' and its supposed 'specialties.' [...] Only practice can disperse the pile of dead leaves left by the academy, to give back to the student his role as the true protagonist of the creation of the Third World university." Likewise, the Peruvian architect tackles the dichotomy of Urbanism and/or Self-management; and through the case of Villa El Salvador, he makes a new reading of the concepts of "residential group," "neighborhood," "sector," "district," always resorting to French references in different fields: "Paraphrasing Jacques Lacan: WHAT IS THE WISH OF THE URBANIST, in this whole policy-political operation in which he is the topographer?".

It was not until 1987 that the magazine devoted a special issue to Latin America, and once again, it was about Brazil (No. 251).

5 SOME ELEMENTS ABOUT THE MONOGRAPHIC ISSUES ON BRAZIL AND MEXICO

Brazil is the Latin American country that, after the exhibition entitled *Brazil Builds*, held at the MoMA in New York in 1943 and accompanied by the bilingual (English-Portuguese) catalogue [Figure 5] produced by Philip Goodwin with photographs by George Kidder Smith, became positioned on the international cultural scene (Del Real, 2012).

In the subsequent years, an unusual interest was aroused among American and European publications specializing in architecture and urbanism: the magazines *The Studio* and *The Architectural Review* devoted special issues to it, in October 1943 and March 1944 respectively; and in September 1947 *L'Architecture d'Aujourd'hui* consecrated a large double issue [Figure 6] based on the selection of works made by Goodwin for the 1943 exhibition (Braschi, 2016; Camargo Capello, 2011).

The black and white photograph of the northwest façade of the Ministry of Education and Health (designed by Costa, Niemeyer, Reidy, Vasconcelos, Leão, and Machado) in Rio de Janeiro, which serves as the cover for the issue of *L'Architecture*

Figure 5. Cover of the Brazil Builds catalogue. Source: Goodwin, 1943.

Figure 6. Photo of the Ministry of Education and Health on the cover of no. 13-14. Source: L'AA, September 1947.

d'Aujourd'hui published in 1947, is similar to that with which the second part of the 1943 MoMA catalogue begins [Figure 6]. This choice confirms a fact that is verified when reviewing the contents of both publications: they give priority to architectural projects, while urban and larger-scale projects are conspicuous by their absence.

In 1953, just one year after the appearance of the second issue on Brazil, a special issue entitled *Habitations* presented two photographs that illustrated the news about the exhibition *Architecture brésilienne contemporaine*, held at the Musée d'Art Moderne in the French capital, underlining the link with the São Paulo architecture biennial that awarded the Grand Prize to Le Corbusier in its first edition. On this occasion, *L'Architecture d'Aujourd'hui* also reproduces an extract from the conference given by the renowned art critic Mario Pedrosa (1900-1981) in which he dealt with the current state of modern architecture in Brazil.

Around the same time, the 8th Pan-American Congress of Architects, held in Mexico in 1952, presented the visit of three eminent French architects: Pierre Vago, Eugène Beaudouin, and Robert Lebretc as French delegates invited to the meeting. From this experience and on the initiative of correspondent Vladimir Kaspé[4] with the close collaboration of the Society of Mexican Architects and the Mexican College of Architects, the first of two special issues dedicated to this country [Figures 7 and 8], published by the French magazine in the period analyzed here, were produced.

Meanwhile, in the United States the most influential exhibition of the time was taking place: the MoMA in New York presented on this occasion, in November 1955, the exhibition *Latin American Architecture Since 1945*, accompanied by a catalogue of the same name by the architecture historian Henry-Russell Hitchcock (Del Real, 2007).

In 1956, although the title of the special issue of *L'Architecture d'Aujourd'hui* on Architecture in Warm Countries mentions only Venezuela, and the state of California, several pages are also devoted to recently built projects in Mexico, Brazil, and India. Once again, architects and correspondents Carlos

Figure 7. Cover of No. 59 on Mexico Source: AA, April 1955.

4 In the original text: "[…] ce ne fut pas, à mon avis, un choix heureux. L'Architecture d'aujourd'hui suivait toutes les modes, adoptait le charabia pseudo-progressiste des pseudo-intellectuels-gauchisants et voulait être à la page en utilisant ce que Baudouï appelait sa logorrhée idéologique."

Figure 8. Cover of No. 109 on Mexico Source: AA, September 1963.

Raúl Villanueva and Vladimir Kaspé advised the issue, in which self-promotion continued to be a common practice: this is how three projects led by Villanueva open up the Venezuelan case, notably the deployment given to the 2 de Diciembre and El Paraíso residential units. The text that introduces in double page the project of Villanueva's own house, highlights the familiarity of the architect with the French magazine's editorial staff: "The house of our friend and correspondent, the excellent architect Carlos Villanueva, conceived from the premises of the traditional Spanish house [...] Collectors and hosts like Mr. Villanueva play a key role in the dissemination of modern ideas; and the rapid evolution that these have had in Venezuela during the last years, is due in good part to his efforts."

The September 1957 issue devoted to young architects from around the world includes projects by professionals from 22 countries, four of them Latin American: Argentina (Gerardo Clusellas); Brazil (Adolpho Rubio Morales, Rubens Carneiro Vianna, Ricardo Sievers, David Liebeskind); Cuba (Frank Martínez); and Mexico (Enrique Castañeda Tamborrel, Manuel Teja Oliveros, Juan Becerra Vila, Manuel Rosen, Héctor Velázquez Moreno, Ramón Torres Martínez, Carlos Zetina, Salvador Ortega Flores, Ignacio Medina Roiz, and Felipe Salido Torres). In the introductory text, Mario Pani highlights the French influence: "Since approximately 1930, Mexican architecture has developed characteristics that are entirely its own; whereas before, it used to follow foreign examples and especially those produced in France [...]." Further on, the text that accompanies the Brazilian examples: "Le Corbusier is undoubtedly the one who has most influenced Brazilian architecture; and that is why young people are currently following the path laid out by their elders, always with the concern of producing a national architecture, less committed than in Mexico, but well defined."

The second issue on Mexico appeared at the end of 1963, when the VII Congress of the *Union Internationale des Architectes* (UIA) was held in Havana on the theme of *Architecture in Underdeveloped Countries*. This was the first of the UIA meetings held in Latin America: in 1969 it took place in Buenos Aires with the theme *Architecture as a social factor*, and in 1978 in Mexico City with *Architecture and national development*.

An almost idyllic vision of Latin American reality characterizes the texts of the issue of *L'Architecture d'Aujourd'hui*, as is also the case in the subsequent testimonies of the European protagonists of those meetings that brought together architects and urban planners from all over the world. Recalling that UIA Congress in Havana, Pierre Vago wrote in his memoirs: "There would be so much to tell about the stimulating Cuban days! The charm of the famous open-air cabaret. The visit to Ricardo Porro's interesting School of Plastic Arts. The great poverty that did not decimate the confidence and enthusiasm in the exultant future that was promised. The omnipresence, clearly superfluous, of armed militiamen and militia women – even our friendly minister [Osmany Cienfuegos, Minister of Construction] who never abandoned the tremendous revolver he carried in his belt, nor the machine gun he carried in his car."

The initial part of this second issue maintains the tendency to make these specials dedicated to "distant cultures," brief history lessons: the reader finds widely illustrated texts entitled *Architecture hispanique*, *L'architecture de la Nouvelle Espagne*, *Le Baroque et le Néo-classique*, *L'architecture mexicaine de 1955 à 1963*, *Évolution de l'enseignement de l'architecture au Mexique*, and *Évolution de l'urbanisme au Mexique*. In the latter, a double-page spread signed by architect Domingo García Ramos, the author stressed that "for twenty years, in all the existing architecture schools in Mexico, courses in urbanism have been opened to prepare young students to face the problems of the future, always considering the study of economic, social, and administrative disciplines, etc."

Although most of the issue is devoted to projects located in Mexico City (a full page aerial photograph shows the main recent urban and architectural projects), it is interesting to note the presentation of some of the projects conceived within the framework of the National Border Program [*Programa Nacional Fronterizo* (PND)] for cities such as Matamoros, Piedras Negras, Tijuana, Nogales, Ensenada, Mexicali, Ciudad Juárez, Nuevo Laredo, and Reynosa.

As the young Mexican architect Irving Nerubay points out in a brief article, a group of projects by the PNF directed by the architect Mario Pani had been exhibited in the Mexican section of the exhibition organised by the Musée d'Art Moderne in Paris on the occasion of the *XXVI Congrès Mundial d'Habitation, d'Urbanisme et d'Aménagement du Territoire* held at UNESCO in September.

Among the projects for the Mexican capital is the restructuring of the Nonoalco-Tlatelolco sector carried out by Mario Pani and associates, in an operation that involved the relocation of the inhabitants of the Guerrero neighborhood to the three new units of Nonoalco-Tlatelolco, constituting a model for "extirpating" the *bidonvilles* located between the north-eastern zone and the city center: "This complex is in fact a ville nouvelle in the interior of the city: 28% of the land is reserved for construction; 50% for parks and gardens; 22% for roads." The style of the drawing that shows what should be the Plaza de las Tres Culturas [Figure 9] is striking: the pre-Hispanic culture captured in the pyramid of Tlatelolco; the Hispanic heritage captured in the Church and the Colegio de Santa Cruz; and the modern architecture that characterizes the new buildings, among them the imposing Torre de Tlatelolco [Figure 10].

Figure 9. Drawing of Plaza de las Tres Culturas, Nonoalco Tlatelolco project. Fountain: L'AA, September 1963.

Figure 10. The Tower of Tlatelolco under construction. Source: AA, September 1963: L'AA, September 1963.

6 CONCLUSION

The French view of Latin America at the beginning of the second half of the 20th century was initially conditioned by the Anglo-Saxon approach proposed both through exhibitions and through publications and analyses in specialized magazines.

The presence of Latin American students in French schools of architecture and urban planning during the first half of the 20th century is closely related to the importance that France had as a reference in architectural and urban issues, both in practice and theory and in teaching. A significant number of professionals to whose projects *L'Architecture d'aujourd'hui* is dedicated have studied in France at institutions such as the École Nationale Supérieure des Beaux-Arts (ENSBA), the École Spéciale d'Architecture (ESA), or the Institut d'Urbanisme de l'Université de Paris (IUUP).[5]

In addition to standing out in the practice of the profession as urban planners and architects, Carlos Raúl Villanueva (ENSBA registration no. 7824, student at the atelier of Gabriel Héraud), Mario Pani (ENSBA registration no. 9113, student at the atelier of Georges Gromort), and Vladimir Kaspé (ENSBA registration no. 8964, student at the ateliers of Jules Godefroy, Eugène Freynet, and Albert Ferran) were undoubtedly the main *passeurs* in Paris who, through the media coverage of their own projects and those of their closest colleagues, fueled the interest of French and European readers, and those from all continents, in the most recent developments in Latin American cities. Graduates of ENSBA in 1928, 1934 and 1935 respectively, their contacts with former colleagues in ENSBA's ateliers and with figures from the French professional and publishing circles ensured a high profile that was also extended to other periodicals such as *Architectural Design* and *Bauen + Wohnen*.

To conclude, we present the translation of a fragment of the curious poem entitled *Il était une fois* [Once upon a time], published in the special issue of *L'Architecture d'Aujourd'hui* devoted in 1974 to *Amérique Latine*. The text undoubtedly evokes the fascination of French culture and institutions in general, and of the arts and architecture, among the successive Latin American generations who tried to train in France, and ideally in Paris.

Once upon a time (excerpt)

Don Pedro's son was seduced
by what the architect did
And he said: Dad, I want to do
beautiful things
And the father replied

5 On the growing role of young Mexican architects returning from studying in France, and of Kaspe himself – a French nationalized Mexican – in the Mexican professional scene of this period, see Louise Noelle's article about Architects and French architecture in Mexico during the 20th century (2012). On Kaspe's early collaboration with Mexican magazines, and other aspects of transfer during the second quarter of the 20th century, see also Georg Leidenberger's article about Three Mexican architecture magazines between 1923 and 1950 (2012).

I am glad you want to become
someone in life
I will send you to Europe to study
Don Pedro's son learnt French and went away
Was a bohemian then an apprentice
Learnt orders and watercolors
Knew Rome, Greece, and Paris
And returned
Returned to discuss cornices
with clients
And boudoirs with clients
He joined the Ministry
And had to admit that we did not understand
the artists
And he asked: why cannot we
teach here what is taught in Paris?

REFERENCES

Avila-Gómez, A. 2019. Éstudiantes latinoamericanos en el Institut d'Urbanisme de l'Université de Paris, 1923-1941. *Revista de Arquitectura* 21(2): 44–56.

Avila-Gómez, A. 2020. Éstudiantes latinoamericanos en la Ecole Spéciale d'Architecture, 1900-1939. *Revista Legado de Arquitectura y Diseño* 15 (27).

Braschi, C. 2016. Écritures locales, lectures globales. L'art et l'architecture brésiliens dans les revues spécialisées internationales, 1943-1964. *Marges* 23(2): 35–46.

Camargo-Capello, M. 2011. Arquitectura moderna en Brasil y su recepción en los números especiales de las revistas europeas de arquitectura, 1940-1960. *Revista de Arquitectura* 17(23),41–51.

Del Real, P. 2007. Building a Continent: MoMA's Latin American Architecture Since 1945 Exhibition. *Journal of Latin American Cultural Studies* 16 (1): 95–110.

Del Real, P. 2012. Building a Continent: The Idea of Latin American Architecture in the Early Postwar Graduate School of Arts and Sciences, Colombia University.

Dictionnaire des élèves architectes de l'École des beaux-arts de Paris (1800-1968). In *AGORHA*. 2011-. [Base de données en ligne]. Paris: Institut national d'histoire de l'art. Disponible sur http://www.purl.org/inha/agorha/001/7.

Esteban-Maluenda, A. 2012. Transmisión bipolar: la difusión de la arquitectura moderna latinoamericana en España, 1949-1968, *Revista 180* 29: 24–29.

Esteban-Maluenda, A. 2017. Brasil exporta. La arquitectura moderna brasileña y la latinoamericana en las publicaciones periódicas europeas después de la II Guerra Mundial. In Esteban-Maluenda, A., Medina, V., Sánchez-Carrasco, L., Martín Soria, A. (eds.). *Rutas iberoamericanas. Contactos e intercambios en la arquitectura del siglo XX.* Madrid: Mairea.

Esteban-Maluenda, A. 2019. Intereses divididos: la arquitectura moderna latinoamericana en las revistas europeas después de la Segunda Guerra Mundial. *Bitácora Arquitectura* 43: 34–47.

Goodwin, P. 1943. *Brazil Builds: Architecture New and Old, 1652-1942*. New York: The Museum of Modern Art.

Jannière, H. 2002. *Politiques éditoriales et architecture « moderne ». L'émergence de nouvelles revues en France et en Italie, 1923-1939*. Paris: Editions Arguments.

Leidenberger, G. 2012. Tres revistas mexicanas de arquitectura. Portavoces de la modernidad, 1923-1950. Anales del Instituto de Investigaciones Estéticas xxxiv (101): 109–138.

Noelle, L. 2009. La arquitectura mexicana en las publicaciones periódicas del siglo XX. *Bitácora Arquitectura* 19: 12–17.

Noelle, L. 2012. Arquitectos y arquitectura francesa en México, siglo XX. *Villes en parallèle* 45-46: 240–260.

Vago, P. 2000. *Pierre Vago. Une vie intense*. Brussels: Éditions Archives d'Architecture Moderne.

Research Tracks in Urbanism: Dynamics, Planning and Design in Contemporary Urban Territories – Allegri et al (eds)

Types of land occupation in the urban periphery. Ibagué, Colombia, 1935

A. Francel, P. Hormechea & C. Uribe
E-ARC, Department of Architecture and Design, Universidad del Tolima, Colombia

ABSTRACT: The characteristics of land occupation in the urban periphery of Ibagué in 1935 are the historical material evidence of the conjunction of several factors such as the continuing growth of the city according to the preceding urban principles, in the absence of urban policies. Such policies changed from a strong control of the state over the city, to an indeterminacy or liberation of the land, in connection with the ongoing transformation towards economic liberalism. As a consequence, a regulatory gap was generated whose free interpretation resulted in the distortion of the colonial urban grid and generated an anarchic heteromorphism of the city in the mid-twentieth century that even today characterizes the urban planning exercise.

1 INTRODUCTION

Towards the end of the 19th century, Colombian cities experienced accelerated growth as a consequence of migrations from the countryside to the city derived from social conflicts and from the search for opportunities in the midst of a weak economy (Clavijo Ocampo, 1990; 1998; 2004). Many of these migrants settled on the periphery and generated types of occupation that characterized the transformation of rural to urban land ownership and defined the current city. This transformation process has not been studied from architecture and urban planning perspectives. This work addresses such a study for the city of Ibagué. For that purpose, the oldest parcel plan of Ibagué's urban periphery of the beginning of the 20th century, drawn up in 1935, was selected. From this plan, the identification and analysis of land division characteristics in the urban periphery were studied, as well as the land's models of occupation and its impact on the dynamics of the modern city. The study was limited to the sector between streets 23 to 26, and between the first and third avenues (the name avenue refers to "carrera" in Colombian Spanish, meaning an urban road perpendicular to streets).

The analysis performed led to the generation of graphs that helped the identification of urban and architectural settlement patterns in the sector. Likewise, the overlapping of the city map and hydrographic charts, taking into account the conditions of marginality and sustainability derived from the urban process, was analyzed. Based on this analysis, the following questions were addressed: What political, social and economic elements determined the urban configuration of the periphery of Ibagué in 1935? How did the types of land occupation of the urban periphery of 1935 affect the configuration of the inner periphery of the modern city? How are inertial morphologies and neocolonial architecture of the early twentieth century articulated in the modern city?

2 METHODOLOGY

The first phase of the research consisted of a documentary review on the planimetry and notarial acts from three sources: the Municipal Historical Archive, the Darío Echandía's Cultural Center of Banco de la República and the General Archive of the Nation.

The second phase consisted of a digital redrawing of plan number 23 from the Urban Plan Ibagué Futuro of 1935, aiming to generate graphical analyzes, preserve historical material and promote its social dissemination.

The third phase involved the search of secondary sources for the reconstruction of the context of the generation of plan 23 included in the report of the Society for Engineering and Construction (SIDEICO) of 1935. In that report, the international congresses, their incidence in the generation of national regulations, and their impact at the local level are analyzed.

The fourth phase was a comparison between urban characteristics in the colony and in the republic in order to determine the elements that could generate changes in the conception of the city. The fifth phase presented the geometric proportion analysis of the parceling, orientation, dimensions and arrangement of the buildings. The sixth phase is a conceptual synthesis on the incidence of these characteristics in the modern city.

DOI: 10.1201/9781003220855-4

3 RESULTS

Seven types of parceling were identified in the sector, beginning from a large *hacienda* with a central construction that was transformed according to the subsequent subdivision of the properties and the colonization of the frontal part of each parcel, to establish a relationship of control and protection before the roads. Morphologies derived from a model of economic liberation of the land were observed, showing clear differences with the colonial urban model, but keeping some of its characteristics. This scenario presented types of neocolonial architecture and urbanism, on which remodeling intentions derived from the developmental paradigms of the early twentieth century in Colombia and Latin America were incorporated (Almandoz, 2007). The detection of these characteristics led to the development of the Urban Plan Ibagué Futuro of 1935, analyzed in this document, as part of a national urban planning system that responded to urban conflicts derived from accelerated urban growth (Delgado, 2006).

However, observing this Plan, it can be seen that its proposals in urban regulations were not executed and continue being an unresolved matter, evidencing the purpose of economic liberalism to center on self-regulation of social actors through market supply and demand conditions. Consequently, the government as an urban planner becomes a secondary actor and the city grows guided by economic interests, relegating community concerns.

4 DISCUSSION

Each one of the thematic sections corresponds to the analysis of the political, economic and social components evident in the urban morphology and the architectural typologies analyzed from plan 23 of the Urban Plan Ibagué Futuro of 1935. This is one of the few plans preserved from the 1935 report and the only one that presents the land distribution characteristics of the urban periphery at that time. It represents a window to the phenomenon of urban growth in a marginal area without urban planning and driven by the new economic dynamics of the country (Figure 1).

Figure 1. Plan 23 of the Urban Plan Ibague Futuro (1935). Redrawn: Zuled Clavijo (2014). Source: Municipal Historical Archive of Ibagué. Research seedbed in the History of Architecture and Urbanism of Ibagué (HARQUI). Research group Architecture and City Studies E.ArC.

4.1 *Measuring the lots*

The first evident characteristic in plan 23 of 1935, is the irregularity of the shapes, dimensions and sizes of the lots. That is a relevant feature, given that in 1921, in the Belén neighborhood, located in Ibagué's northwestern periphery, urban planning had been implemented using blocks and lots with similar measures. However, there was a great difference between the configuration of the Belén neighborhood, which was of private origin, and the eastern urban periphery, whose urbanization process was directly in charge of the municipal government. In this sense, it is necessary to observe two tendencies of territorial intervention.

The first tendency, as in the Belén neighborhood, corresponds to the implementation of international policies adapted to national and local contexts. It is inspired by the garden city movement, proposed by Ebenezer Howard in 1898, whose essential characteristic is the search for integration of the virtues of both the countryside and the city, keeping distance from the metropolis or main city. This last fact permits the conservation of the characteristic quiet life in small towns (Kostof, 2004). This urban concept was transformed into the park neighborhoods in Latin America (Gómez, 2015). It consists of urban areas moderately far from the dynamics of metropolitan areas, with lines of trees parallel to the roads, gardens and backyards, providing a space of tranquility and delight for former local elites (Francel, 2015).

The urban trends in Belén are directly linked to the professional practice of Arturo Jaramillo Concha and to Alberto Manrique Martin, who worked with the firm Casanovas and Manheim, in charge of importing to Colombia the trends of the historicist architecture of the German school (Arango, 1990; Ortiz Gaitán, 2005; Reyna, 2010). In addition, Manrique promoted the implementation of the deco avant-garde, using simple geometry patterns and gradually taking distance from ornament (Midant, 2004). In this sense, Belén synthesizes an approach that searches for technical criteria to determine measures, materials and spatial characteristics of a construction. Those concerns and practices led to the configuration of modern architecture (Gropius, 1965; Alvarez & Francel, 2018).

Belén was created as a new neighborhood, whose developers implemented an urban model with the freedom afforded by the distance and independence from the *hacienda*, in a place where there were no roads or previous constructions. To make the most of the terrain conditions, they set up lots with similar measurements and some size compensation patterns according to their location in the neighborhood (central or

Figure 2. Urbanization project for the Belén neighborhood (1921). Designed by Manrique Martin & Co. Copy Made by Guillermo Silva (1984). Digitization by Andrés Francel (2015).

Figure 3. Centenario Park - Belén neighborhood. Unknown author (1986). Source: number 148 of the index of the visual memory file of the cultural center Darío Echandía of Banco de la República.

peripheral). Central lots were smaller, while those located to the north and south were larger, to compensate for the unusable land near the Chipalo and Los Piojos streams. Finally, the areas surrounding the Quebrada de los Piojos were used to create the Centenario park, the largest public area in the city center, which followed the model of the Central Park in New York and the National Park in Bogotá (Ariza Zuluaga, 2010; Francel, 2015) (Figure 2).

In contrast, the eastern urban periphery was a space for non-regulated growth of the city. The municipal government approved the sale of lots without a planning process and thus without an equitable subdivision. Lot sizes depended on the economic capacity of the buyer and regulations did not define the characteristics of development beyond the continuity of the main road and the presence of a sidewalk. In addition to this lack of urban regulations, the roads were spontaneously extended, following the colonial layout of the city, without the government paving or conditioning road construction efficiently. These characteristics are observed in plan 23, in which the contours of the roads are irregular, lacking sidewalks and finishing abruptly in the water source shores, without criteria of harmony or compensation, as was the case of Belén.

This situation led to the channeling of streams in the periphery and the city blocks, overlapping their layout and generating a conflictive and mutually exclusive relationship between city and nature. On the contrary, Belén's neighborhood's natural conditions were exploited to strengthen the urban concept of park neighborhood, derived from the garden city movement, generating a space of harmony between the existing natural environment and the built one (Figure 3).

The first pattern of land occupation in the urban periphery is, therefore, the superposition of the urban grid – of colonial origin – over the natural environment in which two streams crossed, the Sillón to the north, and the Hedionda to the south. Both margins became spaces of socio-spatial segregation that impacted the quality of life then and even in present times, causing diseases derived from the accumulation of wastewater near homes.

4.2 *The emplacement of buildings*

The rural origin and agricultural mode of production of the eastern periphery of 1935 is evident in the location of a large construction in the center of the block between streets 25th and 26th. This is the first pattern of land occupation, looking for a general control of the property, of big dimensions.

In a second period, the *hacienda* is subdivided. The new lots preserve the occupation pattern towards the center or bottom of the parcel and the buildings decrease in size in relation to the dimensions of the property.

The third pattern of occupation consists in the conservation of the main construction inside the lot, to which another one, of smaller dimensions, is added and located towards the road.

The fourth pattern involves the construction of small facilities towards the road, without fully occupying the front of the lot, in which the preservation of certain rural characteristics of isolation of the house, both from the street and from the neighbors, is observed.

The fifth pattern consists of an increase in the occupation of the lot by small buildings grouped on the street border, which evinces the incremental process of housing.

The sixth pattern is the occupation of the entire front of the lot by a single construction, with a clearly urban purpose that demarcates the limit between the public space of the street and the private space and dynamics of the house and the backyard.

The seventh observed pattern is the urban-consolidation occupation, which colonizes both the limit of the lot and the street, with one wing towards the site, in an "L" shaped model (Figure 4). The most consolidated lots in the studied area show

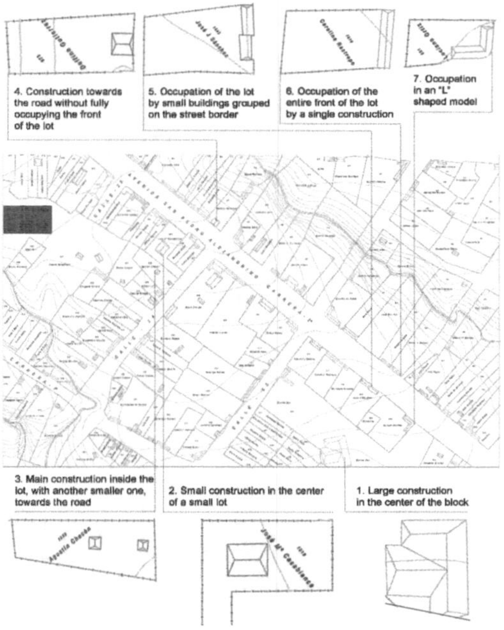

Figure 4. Models of land occupation, according to plan 23 of the report of the *Sociedad Industrial de Ingeniería y Construcciones* (SIDEICO) S.A., 1935.

multiform building groupings that include the seven observed patterns, with transformations from L-shape to T-shape, U-shape and other derived shapes depending on the shape of the blocks and the lots (Figure 5).

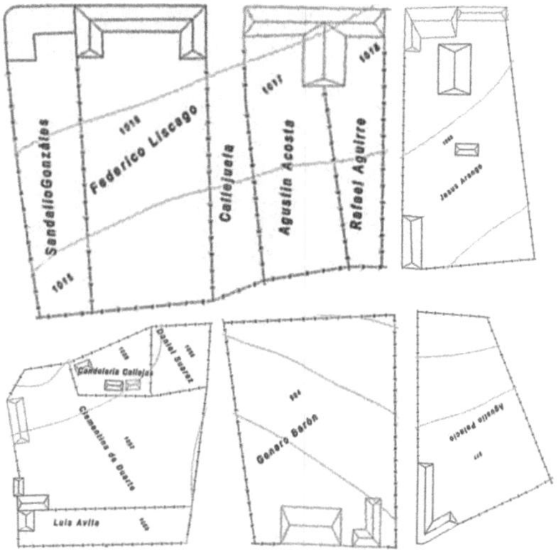

Figure 5. Mixed occupation models.

4.3 *Morphology*

In the configuration of the blocks, changes that took place in the urban layout can be observed. Those changes were a result of the deficiency of urban regulations and the state's weakness in generating proposals for urban development. When a liberation of land use occurred, according to the payment capacity of each buyer, without the existence of a normative substrate or a plan that projected the future city, the city grew indifferent to its natural environment and with inconsistencies regarding traffic, connectivity and the ideal location of the inhabitants with respect to topography and water sources (Figure 6).

This phenomenon of heterogeneous reconfiguration of the urban layout reflects free market tendencies based on the "let do, let pass" (*laissez faire, laissez passer*), a doctrine that opposes government interference in economic affairs and delegates to the market the organic functions of social regulation (Brebner, 1948). Consequently, the periphery of 1935 reflects the model of land appropriation and urban configuration of economic liberalism. Even more, in the elaboration of the urban plan Ibagué Futuro of 1935, the friction between the search for urban control by the project firm and the interests of liberation of the land by local elites is evident (Francel, The urban plan Ibagué Future 1935, 2017).

The verification of this absence of state power to control urban growth is observed in the comparison between urban drawings generated during the colonial period (Durán Rocca, 2006) and the republican period (Espinosa Abad, Calle Medina, & Jaramillo Medina, 2002). The exercise of central power during the colonial period was reflected in a homogeneous layout, with the urban power mechanisms clearly established. The main square on the 10th Street and Carrera 3 was the axis of urban development (Francel, 2015), whose

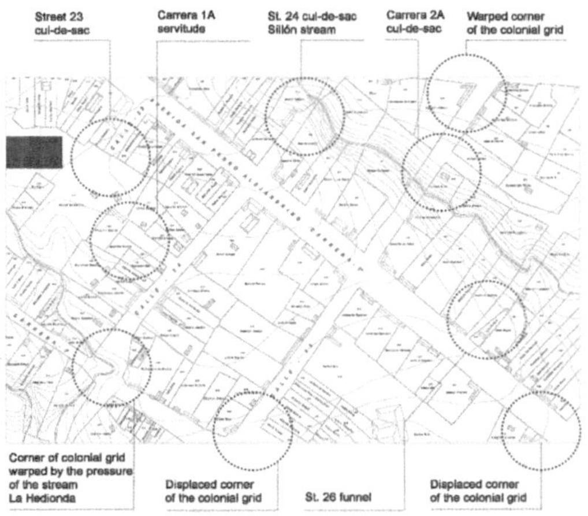

Figure 6. Morphological inconsistencies in the studied area.

end was the cemetery, in *calle* 21 (Francel, 2017). The power struggles during the republican reconfiguration (Saldarriaga Roa, 1997; Ariza Zuluaga, 2010;) were reflected in urban dimorphism, which lasted until the emergence of proposals to plan cities in the early 20th century (Botero Herrera, 1996; Alba Castro, 2013), as a consequence of the expansion of the government's presence in social dynamics, as part of the welfare state (Noguera, 1998). This power vacuum period in urban administration led to the generation of an inner periphery (López Kapstein, 2010), a degraded space inserted in the center of the city and originally marginal (Francel, 2017).

To a large extent, this *laissez-faire* tendency (letting go, letting pass), was fostered by the implementation of the railway (Patiño, 1990), which operated as a superior policy (Díaz Ángel, Muñoz Arbeláez, & Nieto Olarte, 2010), coming from the national government and impacting on various populations that did not generate plans to articulate this great infrastructure transformation with the organic growth of its urban layout (Francel, 2017). The layout of the railways and the construction of the railway station, led to a liberalization of the lots in the contiguous areas (Honorable Municipal Council, 1916; 1920; 1926). Afterwards, successive and late prohibitions to urbanize due to the danger posed by the road were issued (Francel, 2013), and finally the observation on barriers and socio-spatial degradation in the sector was finally agreed upon (González Calle, 2006) (Figure 7).

In this way, it can be seen that the occupation of the land in the urban periphery of 1935 was influenced by the following facts: (i) the state weakness supported by *laissez-faire* tendencies; (ii) the continuation of colonial settlement patterns without urban regulation; (iii) the superimposition of the urban layout ignoring water sources and (iv) the gradual coupling of rural practices to urban practices that shaped the positions and forms of the constructions with respect to the dimensions of the lots, the subdivisions and the generation of a sequence of facades

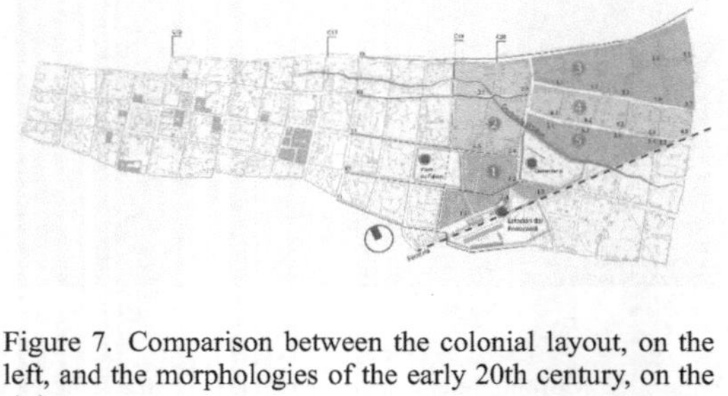

Figure 7. Comparison between the colonial layout, on the left, and the morphologies of the early 20th century, on the right.

5 CONCLUSIONS

The study of the characteristics of land occupation in the urban periphery of 1935, allows identifying the power relations that are reflected in the urban configuration of the contemporary city. The liberalization of the real estate market without a regulation that strengthens the common benefit, the sense of coherence with the environment and the protection of basic rights, leads to the generation of unhealthy urban environments that are constantly degrading in a downward spiral in which the solution is usually focused on urban renewal, that is, on the demolition of what already exists and the planning of new city sectors.

However, these patterns of land occupation lead to multitude of owners in an area, reducing the possibility of real estate operations to manage major transformations. Therein lies the social and economic fabric that has been in place for decades, which requires in-depth analysis because it explicitly implies relationship ties between the inhabitants of a specific environment and a configuration of environments.

In opposition, the generation of urban planning criteria of a particular origin, covered also for *laissez-faire* tendencies, configured a coherent urban environment in the Belén neighborhood, respectful of the environment and with outstanding environmental and pedestrian conditions. It is clear, in this process, that the peripheral condition is mediated by the economic capacity of the social group of inhabitants. While in Belén they sought comfort by moving away from the city and designing a landscaped space, in the eastern periphery the inhabitants were poor people, indifferent to the conflicts generated by the invasion of the water sources and without environmental compensation mechanisms to avoid the precariousness of the urban space.

REFERENCES

Alba Castro, J. M. (2013). El plano Bogotá Futuro. Primer intento de modernización urbana. Anuario Colombiano de Historia Social y de la Cultura, 179–208.

Almandoz, A. (agosto de 2007). Modernización urbanística en América Latina. Luminarias extranjeras y cambios disciplinares,1900-1960. Iberoamericana, VII (27), 59–78. Obtenido de http://www.iai.spk-berlin.de/filead min/dokumentenbibliothek/Iberoamericana/2007/27-Almandoz.pdf

Alvarez, M., & Francel, A. (2018). Historical patrimonial study and three-dimensional restitution of the Tolima soft drink factory (1945-2016). História (São Paulo), 37 (1),1–26. doi: DOI: http://dx.doi.org/10.1590/1980-436920180000000017

Arango, S. (1990). "La persistencia de los estilos". Historia de la arquitectura en Colombia. Bogotá: Universidad Nacional de Colombia.

Ariza Zuluaga, C. (2010). De la arquitectura doméstica republicana a la moderna. Primeras décadas del siglo

XX. 1 (1).Obtenido de http://www.revistahistorik.com/descargas/arquitectura_domestica.pdf

Ariza, C. (julio-octubre de 2010). La exposición del centenario de la independencia en Bogotá. Revista Virtual de Investigación en Historia, Arte y Humanidades, 1, 1–3.

Botero Herrera, F. (1996). Medellín 1890-1950. Historia urbana y juego de intereses. Medellín: Colección Clío Editorial, Universidad de Antioquia.

Brebner, J. B. (1948). Laissez Faire and State Intervention in Nineteenth-Century Britain. Journal of Economic History (8), 59–73. doi: DOI: https://doi.org/10.1017/S0022050700090252

Clavijo Ocampo, H. (1990). Historia económica y social del Tolima, 1600-1930. Ibagué: Universidad del Tolima.

Clavijo Ocampo, H. (1993). Monopolio fiscal y guerras civiles en el Tolima, 1865-1899. Bogotá: Biblioteca digital Luis Ángel Arango. Obtenido de Biblioteca Digital Luis Ángel Arango, en:.

Clavijo Ocampo, H. (1998). Economía, bonanzas de tiempo en tiempo. En Colombia, país de regiones. Tomo 3. Bogotá: Centro de Investigación y Educación Popular. Obtenido de http://www.banrepcultural.org/blaavirtual/geografia/region3/a5.htm

Clavijo Ocampo, H. (2004). Educacion politica y modernizacion del Tolima en el centenario salesiano en la ciudad musical de Colombia (1904-2004). Ibagué: Pía Sociedad Salesiana y Universidad del Tolima.

Delgado, M. (2006). Morfología urbana y conflicto social. Las medidas antigueto como políticas de dispersión de pobres. En R. Bergalli, & I Rivera Beiras, Emergencias urbanas (págs. 133-169). Barcelona: Anthropos.

Díaz Ángel, S., Muñoz Arbeláez, S., & Nieto Olarte, M. (2010). Ensamblando la nación: Cartografía y política en la historia de Colombia. Bogotá D.C.: Universidad de Los Andes, Facultad de Arte y Humanidades, Facultad de Ciencias Sociales, Departamento de Historia, Centro de Estudios Socioculturales e Internacionales (CESO), Ediciones Uniandes. Obtenido de https://books.google.com.co/books?id=M6A7DwAAQBAJ&pg=PA23&lpg=PA23&dq=1601+Mapa+del+r%C3%ADo+Magdalena+desde+su+desembocadura+hasta+m%C3%A1s+arriba+de+Mariquita&source=bl&ots=DFl3u_0v0V&sig=M1WmlpNduMoOyjwwdie3pmF_US0&hl=es&sa=X&ved=2ahUKEwjV2-1jITdAhXKu

Durán Rocca, L. (2006). La malla urbana en la ciudad colonial iberoamericana. Apuntes. Revista de estudios sobre patrimonio cultural, 19(1),30–7. Obtenido de http://revistas.javeriana.edu.co/index.php/revApuntesArq/article/view/9048

Espinosa Abad, P., Calle Medina, M. I., & Jaramillo Medina, C. (2002). La cité cuencana: el afrancesamiento de Cuenca en la época republicana (1860-1940). Cuenca, Ecuador: Monsalve Montero. Obtenido de 9978-14-077-8.

Francel, A. (2013). Cuatro décadas de arquitectura ibaguereña (1904-1940). Ibagué: Universidad de Ibagué.

Francel, A. (2013). Cuatro décadas de arquitectura ibaguereña (1904-1940). Ibagué: Universidad de Ibagué.

Francel, A. (2015). Belén: una mujer, un barrio, una historia. Ibagué, 1920-1925. Ibagué: Caza de Libros Editores.

Francel, A. (Diciembre de 2015). La calle del Comercio de Ibagué (Colombia), 1893-1950. Un estudio sobre sus transformaciones arquitectónicas y conceptuales derivadas del modelo industrial en el tránsito de la Colonia a la República y las primeras manifestaciones del Art Déco. Dearq (17), 56–73. doi: http://dx.doi.org/10.18389/dearq17.2015.04

Francel, A. (2015). Metodología y reflexiones para la reconstrucción de la historia urbana del barrio belén de Ibagué (1921–1925). V Simposio Colombiano de Historia Regional y Local HistoReLo (págs. 1–13). Ibagué: Universidad Nacional de Colombia.

Francel, A. (2017). El plan urbanístico Ibagué Futuro 1935. Ibagué: Universidad del Tolima.

Francel, A. (2017). Historia y patrimonio de la periferia interior de Ibagué. Ibagué: Premio de investigación en patrimonio. Alcaldía de Ibagué. Caza de libros editories. Plan municipal de estímulos. Obtenido de https://www.academia.edu/35357626/Historia_y_patrimonio_de_la_periferia_interior_de_Ibagu%C3%A9

Francel, A. (2017). Kintsugi urbano en la avenida del ferrocarril. Ibagué, Colombia (1921-1970). IX Seminario Internacional de Investigación en Urbanismo, 1–18.

Francel, A. (2017). La superposición de cartografía histórica como método de análisis morfológico y toma de decisiones urbanísticas. Ibagué, Colombia, 1935-2016. urbe. Revista Brasileira de Gestão Urbana (Brazilian Journal of Urban Management), 9 (2), 21. doi: DOI: 10.1590/2175-3369.009.002.AO10

Gómez, A. (mayo de 2015). Los barrios parque y de fin de semana en el mapa de la expansión metropolitana. Buenos Aires, 1910-1950. EURE, 159–185. Obtenido de http://www.scielo.cl/pdf/eure/v41n123/art07.pdf

González Calle, J. L. (2006). De la ciudad al territorio. La configuración del espacio urbano en Ibagué 1886-1986. Ibagué: Aquelarre.

Gropius, W. (1965). The new architecture and the Bauhaus. Cambridge: The Massachusetts Institute of Technology.

Honorable Concejo Municipal. (1916). Por el cual se autorizan los terrenos para la construcción de la estación del ferrocarril. Caja 401, legajo 4, documento 6, folios 152-154r. Ibagué: Archivo Histórico de Planeación Municipal. LIbro X. Copiador de comunicaciones oficiales.

Honorable Concejo Municipal. (29 de marzo de 1920). Por la cual se ceden unos terrenos para el trazado de vías del ferrocarril. Escritura 214 de marzo 29 de 1926. Ibagué: Archivo Urbanístico de Planeación Municipal.

Honorable Concejo Municipal. (1926). Por el cual se autizaron los lotes para la construcción de la estación del ferrocarril. Caja 401, legajo 4, folios 152-154r. Ibagué: Archivo Histórico Municipal de Ibagué.

Jaramillo Medina, C. (2002). La cité cuencana. El afrancesamiento de Cuenca en la época republicana (1860-1940). Cuenca, Ecuador: Monsalve Moreno.

Kostof, S. (2004). The city shapped: Urban patterns and meaning througth history. Londres: Bulfinch Press.

López Kapstein, P. (julio-agosto de 2010). Vulnerabilidad y periferia interior. Cuaderno de Investigación Urbanística (71), 7–119. Obtenido de https://www.researchgate.net/profile/Paula_Kapstein/publication/256405575_Vulnerabilidad_y_Periferia_Interior_Cuaderno_de_investigacion_urbanistica_num_71/links/0046352274a1f2c84b000000/Vulnerabilidad-y-Periferia-Interior-Cuaderno-de-investigacion-urbanis.

Midant, J.-P. (. (2004). Diccionario Akal de arquitectura del siglo XX. Madrid: Ediciones Akal.

Noguera R, C. E. (1998). La higiene como política. Barrios obreros y dispositivo higiénico: Bogotá y Medellín a comienzos del siglo XX. Anuario colombiano de historia social y de la cultura (25), 188–215. Obtenido de http://www.bdigital.unal.edu.co/20542/1/16693-52270-1-PB.pdf.

Ortiz Gaitán, A. E. (2005). Arquitectura, historia e historias. Bogotá: Universidad la Gran Colombia. T/M editores.

Patiño, V. M. (1990). Ferrocarril del Tolima. En V. M. Patiño, Historia de la cultura material en la América equinoccial (págs. Tomo III. Vías, transportes, comunicaciones. Capítulo VIII, parte II). Bogotá: Instituto Caro y Cuervo. Obtenido de http://www.banrepcultural.org/blaavirtual/historia/equinoccial_3_transportes/cap8c1.htm.

Reyna, M. S. (2010). La historia de frente. Arquitectura de Bogotá. Bogotá: Letrarte editores.

Saldarriaga Roa, A. (1997). La imagen de la iglesia y del Estado en la arquitectura republicana. Bogotá: Revista Credencial Historia, Edición 86.

Sociedad Industrial de Ingeniería y Construcciones (SIDE-ICO). (1935). Informe sobre el plano de Ibagué, levantado en 1935. Catálogo de arrendatarios de ejidos, áreas de los lotes, etc. Ibagué: Archivo Histórico Municipal.

Sociedad Industrial de Ingeniería y Construcciones (SIDE-ICO) S.A. (1935). Informe sobre el plano de Ibagué, levantado en catálogo de arrendatarios de ejidos, áreas de lotes, etc,. Ibagué: Honorable Concejo Municipal de Ibagué.

Research Tracks in Urbanism: Dynamics, Planning and Design in Contemporary Urban Territories – Allegri et al (eds)

Fast urbanization and the commodities global market – The example of Parauapebas - Brazil

A.P. Godinho
FAUP, Faculty of Architecture of the University of Porto

ABSTRACT: Drawing from the notion that our contemporary world is organized into a large "urbanized global archipelago" made up of "islands" that directly communicate and engage with one another in various commercial and financial exchanges, this article seeks to identify this phenomenon in the Brazilian Amazon, specifically in the city of Parauapebas, in the state of Pará. It examines the impacts of iron ore production in Carajás from an analysis of the physical and morphological characteristics of the Parauapebas urban landscape in three separate phases, which in turn produced three distinct "urban typologies": the construction of the Vila de Carajás Company Town at the beginning of mining operations; the spontaneous urbanization that emerged along the PA-275 national highway; and the "financed" urbanization phase that accompanied the commodities boom in the first decade of the 21st century.

1 INTRODUCTION – GLOBAL URBANIZATION

The seminal contributions of neo-Marxist urbanologists such as Henri Lefebvre, David Harvey and Manuel Castells have produced excellent new analysis categories and methods of the aspects of capitalism in contemporary urbanization processes. Drawing from this perspective, modern cities are regarded as the spatial materialization of social processes associated with capitalist modes of production, including the accumulation of capital and class struggle (Brenner & Keil, 2014).

Historically, the division of global society by the binary system of urban versus rural was at the core of theoretical discussion in all areas of knowledge. The last few centuries have borne witness to a domination of urban areas over rural spaces, emphasized by the division between intellectual work and manual labor, and between the areas of capitalist market control and goods production. Urban areas have assumed a predominant position in relation to the countryside, consolidating characteristics and creating stigmas for these two concepts: the "urban" and the "rural" (Monte-Mór, 2014).

However, urbanization can no longer be understood through the binary system of city and countryside, or the city and the rural hinterland. There are capital appreciation processes that mobilize land, that involve the construction of socio-technical networks for the transportation of energy or goods, large equipment for organizing converter or extractive industries, and urban allotment, and urbanization operations that transform and produce territories where conventional descriptors lose their taxonomic rigor and explanatory power.

Since the beginning of industrialization in the second half of the 19th century, the development of cities has largely organized itself around the dominant role of concentric urbanization, with suburbs orbiting a nuclear urban core. In Cédric Price's popular "the city as an egg" metaphor, this mode of urbanization corresponded to the "fried egg" (CCA). In this metaphor, large metropolitan areas are composed of a downtown marked by various polarizing functions, and extensive peripheries that are largely residential in nature and are duly infused with transportation infrastructure systems used by people to commute to and from work daily.

Present-day approaches signal the transition of this model to the "scrambled egg", a pattern of unrestricted urbanization that is continuous, fragmented yet organized in mono-functional zones, and that mixes all types of activities. With the monopoly of centrality gone, several downtowns emerge, amassing different functions and linking with arterial infrastructure systems of increasing territorial size. As Henri Lefebvre notes, we have entered into the era of large urban areas, of "extended urbanization":

"The *urban fabric* grows, extends its borders, corrodes the residue of agrarian life. This expression, 'urban fabric,' does not narrowly define the built world of cities but all manifestations of the dominance of the city over the country. In this sense,

DOI: 10.1201/9781003220855-5

a vacation home, a highway, a supermarket in the countryside are all part of the urban fabric. Of varying density, thickness and activity, the only regions untouched by it are those that are stagnant or dying, those that are given over to 'nature'" (Lefebvre, 2014, pp. 38, 39).

The arrival of the 21st century is characterized by the emergence of a series of metropolises of global importance in tandem with the observable phenomenon of urban network sprawl. Author J. Gottman's book "Megalopolis: the urbanized northeastern seaboard of the USA" (1967) offers a prescient perspective of this phenomenon. Large global metropolitan areas, such as New York and London, retain the fundamental role of capital control, but continuously compete for importance in the network of relationships that deepen with globalization while widening and diversifying the elite club of global cities, as noted by Saskia Sassen:

"(...) the concept of the global city brings a strong emphasis on the networked economy because of the nature of the industries that tend to be located there: finance and specialized services, the new multimedia sectors, and telecommunications services. These industries are characterized by cross-border networks and specialized divisions of functions among cities rather than inter-national competition *per se*. In the case of global finance and the leading specialized services catering to global firms and markets — law, accounting, credit rating, telecommunications — it is clear that we are dealing with a cross-border system, one that is embedded in a series of cities, each possibly part of a different country. It is a *de-facto* global system" (Sassen, 2005, p. 40).

Areas that were once isolated become connected to international business centers, and the world can now be understood as an "archipelago" composed of "islands", or general regions that act according to the speculation of international futures markets (Bassens & Van Meeteren, 2014). Presently, interactions between these "islands" can take place in various configurations to the detriment of the classic downtown-periphery logic.

In this way, ore-producing cities in the Amazon or jean production areas in Bangladesh are connected in real time — by way of the global financial system — to the New York Stock Exchange or to public tax policies of the European Union. These territories begin to influence simultaneously and depend directly on the global movement of capital. The speed at which these processes take place is the speed of financial capital, much faster than that at which we are accustomed to building cities, territories, and social and community identities.

As author Marc Augé notes, the concept of identity of a given community and the relationship between all urban spaces are classified by anthropology (Augé, 1994). Biancarelli adds that "in this way, the construction of a subject's identity inevitably passed through the construction of a place's identity" (Biancarelli, 2019).

"Local" and "global" become categories that can no longer be distinguished and regarded as mutually exclusive. In fact, when we assert that something is local, we can only track the location of where a given phenomenon takes place. Upon closer inspection, this phenomenon appears to be the direct result of a process of globalization. Let us imagine, for example, that in a future commodities market the trend of increases in the price of iron is regularly demonstrated. If, in light of this, the price of iron ore increases in the global market, the mine could respond by intensifying operations and hiring more workers. Occasionally, the opposite can also occur. The expectation surrounding this sequence of events can entail the immediate decision to begin an urbanization operation in Parauapebas. A drop in prices or cancellation of investment can have the same effect, but in the opposing direction.

2 PARAUAPEBAS

The city of Parauapebas is an example of the type of region directly connected to the globalized financial system. Located in the southeast of the state of Pará in Brazil's North region, the city lies within the *Amazônia Legal* (Brazil's Legal Amazon - Term coined by the Brazilian government in 1948 to officially designate all seven states in the north of the country that include part of the Amazon rainforest) and sits upon the Carajás Mountains, the largest mining province in the world (Figure 1). Due to its extremely high concentration of iron, the region's mining potential was discovered in 1967 by then-state company Vale do Rio Doce (VRDC), which, following privatization in 1998, was changed to Vale. Since its discovery, the entire region has shown itself to be of paramount economic and strategic importance to the Brazilian government.

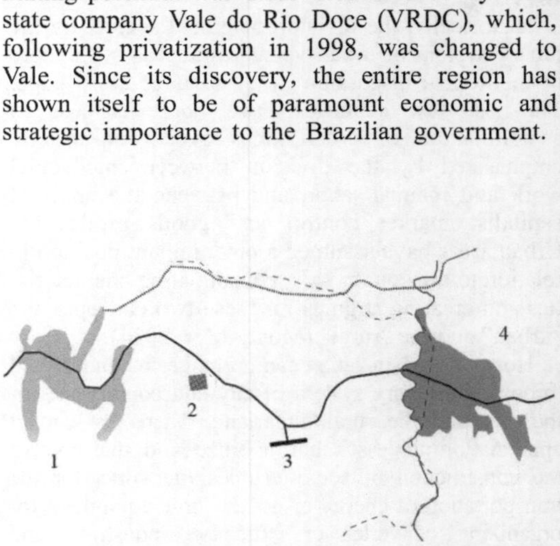

Figure 1. Carajás Mountains Map. Source: Author. 1- Iron ore mine; 2 – Company Town; 3 – Company airport; 4 – Parauapebas urbanization area.

Historically, the Amazon has always been the scene of a series of migrations and the promise of "*El Dorado*". Extractivist exploitation of gold, wood and fruit such as Brazil nuts and cacao began following the first colonial expeditions of the 17th and 18th centuries. The end of the 19th century saw the beginning of the cycles of rubber (then nicknamed "black gold") (BBC, 2009), which lasted until the mid-20th century. These were followed by a set of policies imposed by the Brazilian federal government during its dictatorship in which the slogan *integrar para não entregar* ("integrate in order not to abdicate") predicted the occupation of the Amazon.

This period saw the intensification of incentives to migrate to all parts of the Amazon region, which was regarded as an "empty", "unoccupied" corner of the country. The majority of migrants were from the country's Northeast region, and as such the government sought to solve two problems at once: unite the "landless people" – Northeasterners (*nordestinos*) who struggled with regional drought – to the "peopleless land" of the Amazon region, in total disregard of the local populations (Becker, 1982).

Portrayed in Cacá Diegues' iconic 1982 film *Bye Bye Brasil*, the characters of the Rolidei Caravan depart the Northeast towards the Amazon region in search of better opportunities. "Pineapples there are the size of a jackfruit and the trees are the size of skyscrapers", says one of the characters, encouraging the taming of the mythical rainforest and of its modern *El Dorado*.

Large infrastructure works were built specifically for the purpose of demarcating this occupation and promoting "national integration". Major incentivizing factors of large-scale migration included projects such as the BR-163 National Highway, which connects the states of Mato Grosso and Pará from North to South; the Trans-Amazonian Railway, designed to link the Northeastern coast to the westernmost portion of the country in the state of Amazonas; the *Projeto Grande Carajás* (Greater Carajás Project, PGC), among others.

The *Projeto Grande Carajás* was conceived in the 1970s as a production complex for ore mining in the state of Pará. Included in program were the mining facility for iron ore extraction in Carajás (Parauapebas); the construction of the Carajás Railroad for transportation of ore production to the Port of São Luís in Maranhão; the construction of the Tucuruí Dam for supplying electrical energy to the entire new complex; a bauxite improvement plant in Barcarena, next to Pará's state capital; and highways connecting urban areas and production locations.

The Amazon region is uniquely complex in its socioeconomic structure. It represents the union of various local and global interests: indigenous populations and migrants, civil movements such as the MST (Landless Workers' Movement - *Movimento dos Trabalhadores Rurais sem Terra*), illegal logging and mining, transnational corporations linked to steel and agribusiness, and international environmental organizations, among other actors.

"From peasant tenants, the annihilation of indigenous peoples, the Araguaia Guerrilla War, and the phenomenon of Serra Pelada to the implementation of the Grande Projeto Carajás, never before has there been a comparable search for a benchmark of organized resistance to the order of capital" (Coelho, 2015: 17).

The growth of Amazonian cities has accelerated, specifically in regions with access to commodity production, allowing us to better understand urbanization as a consequence, rather than a driving force, of the changes to and use of local landscapes (Richards & Vanwey, 2015). In other words, "non-urban" production (or that which exists outside of what we commonly understand as the city) is what attracts new inhabitants to neighboring cities.

Given this context, this article identifies three "periods" in the history of the city of Parauapebas that resulted in three distinct typologies of urbanization, each with specific characteristics.

3 VILA DE CARAJÁS – COMPANY TOWN (1984/1986 – CURRENT)

In the early 1980s, the Companhia Vale do Rio Doce (CVRD) commissioned modernist architect Joaquim Guedes, of São Paulo, to design a town for the company's residents within the perimeter of the environmental reserve where the mining operation was located, around 25 km from the city of Parauapebas. Access into the reserve was only permitted with prior authorization (Figure 2).

The town was a veritable urban oasis in the middle of the Amazon rainforest and a region lacking in services. Based on a model of an octagonal garden city in a 306-hectare complex, there are 1,346 houses of identical design without fences separating them from the streets where approximately five thousand residents live (Repórter Brasil). There

Figure 2. Company Town (Google Earth 2020).

are also 70 commercial buildings and public facilities such as parks, a hospital, a theater, a movie theater, schools and daycares, restaurants and a multisport club. Additionally, an airport was constructed with direct flights between the mining complex and the Southeast of Brazil, where the financial center controlling the company's operations is located.

Within the mold of a company town, the longest-serving and most senior employees that went to work at the company's new operation were received with first-world accommodations. In order for the architectural and urbanistic pattern to be maintained, no construction was permitted in the town, and all administration and security was conducted by the company in a sort of corporate condominium where the distinction between public and private was muddled. Quality public spaces prevailed in this nearly idyllic slice of city, featuring wide sidewalks and squares with recreational equipment for children to play and for adults to perform physical exercise.

4 THE SPONTANEOUS CITY/HIGHWAY-AVENUE PA-275 (1985 – 2005)

Under Projeto Grande Carajás, National Highway PA-275 was built to connect the Carajás Mine to Marabá, the largest city in Southeastern Pará. The town of Parauapebas initially grew as an informal settlement next to the entrance to the Carajás National Forest along the PA-275 highway, which eventually became a central thoroughfare and main artery of the city (VALE). In 1985, mineral extraction activity in Carajás was already underway. The Carajás Railroad, responsible for transporting mineral production directly to the Port of São Luís in the state of Maranhão, also hauled ore to the port for export and brought hundreds of new migrants to the city each week.

Parauapebas became a municipality in 1988, and was no longer a district of Marabá. It grew from a foundation of scattered population centers with poor access to basic public services and infrastructure. In addition to connecting Marabá to the core of ore extraction, highway PA-275 played an important role in providing supplies and local transportation, effectively becoming a "highway-avenue". New neighborhoods and urban settlements sprang up over time, always connected in some way with the city's central structural backbone.

The resident population was composed mainly of the third-party workforce contracted by Vale, as well as the organic growth of other economic sectors, such as commerce and services that emerged from the city's expansion. Throughout the 1990s, occasional instances of intervention by public authorities and the company sought to consolidate and provide structure to the central area of urbanization. The focus was primarily along the highway-avenue, which included public spaces for use by residents – practically nonexistent in the rest of the city – such as sports courts, skating rinks and gardens.

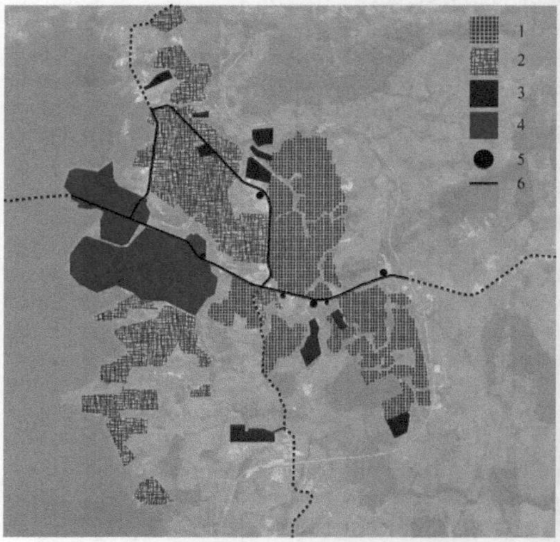

Figure 3. Parauapebas City Map. Source: Author. 1 – New settlements; 2 – Informal Settlements; 3 – Public housing; 4 – Central area of urbanization; 5 – Shopping Centers and department stores; 6 – Highway-avenues.

The mining produced a historically contradictory effect on the city. On one hand, it created a large collection of resources from royalties that the Vale do Rio Doce had to pay to the municipality, resulting in opportunities for greater public investment. On the other, this same financial contribution was particularly attractive to increasingly younger residents, heightening the need for infrastructure (Mesquita, 2014).

Since the beginning of the city's formation, we can observe a clear separation between formal spaces planned by the company and the public sector, and informal spaces organized in the urban core by collective investment of the poor migrant population (Melo & Cardoso, 2014).

During the 1980s and 1990s, the Parauapebas urban area was consolidated in a notably precarious and irregular manner.

5 THE COMMODITIES BOOM AND THE CONSTRUCTION OF THE CITY IN THE AGE OF FINANCIALIZATION

In the middle of the first decade of the 21st century, Brazil became an "emerging market", thanks to a set of political and economic factors. Fueled by rapid economic growth in China, which became the country's largest trading partner, the prices of commodities rose at breakneck speed. Since 2001, Vale shares have been traded on the New York commodity futures exchange NYMEX.

Between 2002 and 2006, Vale decisively entered the Chinese market, increasing exports from 17.5 million metric tons to 77.9 million and making the company the largest supplier of iron ore to China (VALE – Figure 4).

"The substantial increase in iron ore and pellet production at the Carajás Complex signaled the intensification of the Brazilian economy's neo-extractivist character (...). The price of USD $52.92 in October of 1998 reached its apex in February of 2011 at USD $256.24 and then decreased to USD $171.96 by February 2014 (...). In markets where transactions have little or no regulation, such as official fees or minimum exchanges, commodities futures contracts rose from USD $418 million in 2001 to USD $2.6 trillion in 2011" (Coelho, 2015. 85, 97).

Figure 4. Carajás iron ore mine: Source: Tuca Vieira.

Boosted by this cycle of international economic prosperity (in addition to being one of the largest global producers of ore, soy and petroleum, among other commodities) and policies of social inclusion and budget allocation to disadvantaged groups, millions of people were brought into the consumer market, further dynamizing the Brazilian economy. However, as defined by André Singer, this was "weak reformism", which, despite bringing about fundamental transformations – such as a significant increase of minimum wage and government programs for bolstering income –, did not allow for structural changes in society and the economic system. According to Tádzio Coelho, it was more similar to predatory neo-extractivism than a new period of effective development in the country (Coelho, 2015).

The speed of the financial market and of the fluctuations of international stock exchanges on which the iron ore produced at Carajás depended was much greater than the speed at which "real life" took place in the city. The creation of territory both physically and in terms of identity takes place slowly, and requires a planning and assimilation process that is often pushed aside by the speed of capital. Concurrently, and beyond meeting the need for settlement of the growing migrant population, the construction of the city becomes a product that is speculated at the same speed as financial capital. As Guilherme Wisnik notes:

"Now, in the vortex of global hyper-urbanization, the land itself is seeing its obsolescence rate accelerate. And architecture, an art historically linked to *firmitas*, to solidity, finds itself facing an identity crisis" (Wisnik, 2019).

Against this backdrop, the population growth rate of Parauapebas exploded during the first decade of the 21st century. The city experienced a demographic leap from 70,000 inhabitants in 2000 to nearly 150,000 in 2010 and 208,000 in 2019 (estimate – IBGE), representing growth of 297%. The city's revenue has also kept pace with this growth:

"The municipal budget of Parauapebas was BRL $1.34 billion in 2014, one of the largest in the state of Pará. In addition to being one of the leading export municipalities in Brazil, Parauapebas boasts the highest GDP of Pará (BRL $19,897,435), more than the GDP of the city of Belém (BRL $19,666,725). Surprisingly, the municipality of Parauapebas also had the highest GDP per capita in Brazil in 2011 (BRL $124,181.23), higher than Vitória (BRL $85,794.33), Brasília (BRL $63,020.20), São Paulo (BRL $42,152.76) and Rio de Janeiro (BRL $32,940.23). Nevertheless, this serves to demonstrate not only the fragility of GDP per capita as an indicator, but also the concentration of revenue resulting from mining, given that the population of the city suffers from a lack of basic public services. Only 8.1% of families in the city have access to a sewage system. It is estimated that the population living in informal settlements in Parauapebas is around 50,000 individuals" (Coelho, 2015, 166).

This new wave of investments and the promise of employment led thousands of people to not only settle in Parauapebas but also to invest in its real estate market, increasing the speed of construction of new settlements and, consequently, appreciation of land and real estate. It also sped up the dynamics of urban precarity observed in the two prior decades. The logic of the accelerated financial market and of ore speculation was reflected in the growth of the city itself. The response to this demand was mixed: condominiums, luxury real estate projects and shopping centers for the wealthier classes, in contrast to the possibility of access to social housing subsidized and financed by the government through the program *Minha Casa, Minha Vida* (My House, My Life) and informal settlements (Melo, 2015).

In a seemingly contradictory manner, although the demand for different types of real estate projects and varying economic access was very high, speculation tended to separate property values from the local reality, bringing them closer to that of mining markets in international stock exchanges than the purchasing power of local families. In the city, the result was an increasing amount of new, precarious housing stemming from informal settlements (Figure 5).

Figure 5. Informal settlement. Source~: Author; Figure 6 – PMCMV. Source: Parauapebas Municipality; Figure 7 – Alvorá Parauapebas condominium. Source: Published by condominium.

As a consequence, the construction of the city's identity was compromised. Currently, several new real estate projects of varying types and social objectives are emerging, whose goal of sale and resale overrides the need for building dignified communities.

Another important factor driving the city's real estate market – and Brazil in general in the 21st century – is the State, by way of the federal home financing program *Minha Casa, Minha Vida* (PMCMV – Fig. 6).

The implementation of PMCMV favored penetration of the real estate sector in peripheral areas once considered "second class" spaces. Following implementation of PMCMV and allotments of high and medium-income planned districts near informal settlements, a general appreciation of the value of informal lands began to take place, along with a dispute between the real estate sector and the marginalized population, which reclaimed floodplain land in order to gain access to urbanization (Melo & Cardoso, 2014).

In the social and geographic apartheid observed in Parauapebas and many other Brazilian cities, high-income real estate development projects with controlled access (shopping centers, closed condominiums – Fig. 7) are in direct conflict with precarious housing from informal settlements, allotments without infrastructure and government housing neighborhoods in which issues of dignified civility and urbanity are left by the wayside.

6 CONCLUSION

In its 30 years of existence, Parauapebas has grown by 336%, has expanded its GDP by 144% from 2008 to 2011, has surpassed the GDP of Pará's state capital and is Brazil's export champion (Mesquita, 2014).

Simultaneously a victim and a beacon of globalized capitalism, the city is full of contradictions. On one hand, mining has generated growth in GDP per capita and exports, and government policies have fostered income distribution and have increased the rate of employment, creating a new driving force of migration and hope. On the other, the same uncontrolled mining accelerated deforestation of the Amazon rainforest, advanced the removal of indigenous communities, and widened social inequality, leading to a total lack of physical sanitation and housing structure as well as precarity and inexistent basic public services.

Parauapebas is in itself an "archipelago" shaped by the very distinct "islands" that compose it. None of the typologies offered seem to account for the precarious issue that is urbanization: the company town, that is, the perfect city, is unachievable for the general population. Relegated to a pattern of uncontrolled growth, the city raises serious questions about the quality of life. The notion of the globalized city appears to further accelerate this process as other actors become involved, suggesting that we must continue to ask ourselves what the possibilities of urban planning are in this context.

REFERENCES

Augé, M. 1994. *Não-Lugares: introdução a uma antropologia da supermodernidade*. Campinas: Papirus.

Bassens, D. & Van Meeteren, M. 2014. World cities under conditions of financialized globalization: towards an augmented World City hypothesis. In Progress. In *Human Geography*.

Becker, B. 1982. *Geopolítica da Amazônia: A Nova Fronteira de Recursos*. Rio de Janeiro: Jorge Zahar Editores.

Biancarelli, L. 2019. Haut lieu de traitement du corps dans la cité. Doctoral thesis. Paris: Sorbonne.

Brenner, N; Keil, R. 2014. From Global Cities to Globalized Urbanization. *Glocalism: Journal of Culture, Politics and Innovation* (Vol. 3, p.4).

Coelho, T. P. 2015. *Projeto Grande Carajás: Trinta anos de desenvolvimento frustrado* (A Questão Mineral no Brasil vol. 1). Marabá: Editorial iGuana.

IBGE - Instituto Brasileiro de Geografia e Estatística.

Lefrebvre, H. 2014. From the city to urban society. *Implosions, Explosions* (Ch. 2, p. 37–38). Berlin: Jovis.

Melo, A. C. 2015. Modernização e transformações recentes nos processos intraurbanos no Sudeste do Pará. Master's thesis, UFPA. Belém: 2015.

Melo, A. C. & Cardoso, A. C. 2014. Cidade para quem? O descompasso entre políticas ambientais e urbanas na periferia do capitalismo. *Cadernos de Arquitetura e Urbanismo* (v.21, n.29).

Mesquita, J. 2014. Riqueza com prazo de validade. *Exame* (v. 4, n. 43, p. 12–13).

Monte-Mór, R. L. 2014. What is the urban in the contemporary world? *Implosions, Explosions*. Berlin: Jovis.

Richards, Peter & Vanwey, Leah. 2015. Where Deforestation Leads to Urbanization: How Resource Extraction Is Leading to Urban Growth in the Brazilian Amazon. *Annals of the Association of American Geographers*.

Sassen, Saskia. 2005. The Global City: introducing a Concept. *Brown Journal of World Affairs* (Vol XI, Issue 2).

Wisnik, G. 2019. *Dentro do Nevoeiro*. São Paulo: Ubu.

ELECTRONIC SOURCES

BBC Brasil, accessed in January, 2020 - https://www.bbc.com/portuguese/noticias/2009/07/090722_amazonia_timeline_fbdt#integrar

CCA – Canadian Centre for Architecture, The City as an Egg - https://www.cca.qc.ca/en/search/details/collection/object/420807. Accessed in January, 2020.

REPÓRTER BRASIL. Parauapebas: entre o céu e o inferno. https://reporterbrasil.org.br/2007/01/parauapebas-entre-o-ceu-e-o-inferno/. Accessed in January, 2020.

VALE. *Nossa História*. http://www.vale.com/hotsite/PT/Paginas/conheca-mais-sobre-historia-carajas-maior-mina-minerio-ferro-mundo.aspx. Accessed in January, 2020.

Research Tracks in Urbanism: Dynamics, Planning and Design in Contemporary Urban Territories – Allegri et al (eds)

Urban context's role in the emergence and development of social innovation

A. Rivera Vinueza
4CITIES, Erasmus Mundus Master in Urban Studies, Vrije Universiteit Brussel, Belgium

ABSTRACT: Social Innovation (SI) is understood as disruptive socio-political processes that attend basic needs and improve quality of life. SI is the manifestation of resistance to exclusion mechanisms. These inclusion efforts have the challenge fo transcending benevolent actions into long-term social change. SI in the labour market is studied in Malmö and Medellín using the C.S.I analytical framework examining their socio-political Context, Strategies for social inclusion, and labour inclusion Initiatives. This research produced two theses. (1) The transformative potential of initiatives is more complex than the vehement advocacy for scaling up/out suggests. It resides on their capacity to recognise and attend causes of exclusion, their SI features, and their development on a three-stage evolution path constructed here. (2) A re-examination of urban governance relations suggests the consideration of a fourth element, which devises the link between SI and peacebuilding. The study shows how the urban context is fundamental in determining these outcomes.

1 INTRODUCTION

"The Social Innovation (SI) phenomenon" is expanding across geographies, sectors, and fields. It is also inspiring many to act as social innovators or entrepreneurs. In Europe alone, there have been 474 EU-funded projects related to SI since 1990 until the date of writing (EC CORDIS 2020). SI-DRIVE, one of them, mapped more than 1,000 social innovation initiatives around the world (Figure 1) (Howaldt et al. 2016). Undeniably, the proliferation of these initiatives helps fulfil immediate needs and improve the quality of life of many, but can such socially innovative thinking and the multitude of initiatives around the world not just alleviate, but *resolve* social problems? It seems the velocity at which the SI concept is adopted makes its definition nebulous and its impact hard to measure.

This study investigates initiatives that promote the inclusion of specific disadvantaged groups in the labour market. It explores (1) their socially innovative features and (2) their potential to transform exclusion mechanisms. Studying the two contrasting cities of Medellín and Malmö sheds light on how the urban context plays a role in determining those outcomes (1 and 2). The challenge is the magnitude of the causes and the conditions of possibility and action that vary considerably depending on each contextual sphere. This work helps explain the initiatives' current challenges and opportunities, identifies their potential for a more transcendental effect, and formulates theory on how it could be achieved. The following chapter disentangles the meaning of SI and starts the journey.

2 THEORETICAL FRAMEWORK

2.1 *Evolution and theory of social innovation*

There is a misconception in thinking that the term "social innovation" is new. While its meaning has a component of novelty, the term itself can be associated, for example, with the American revolutions, dating back to 1770s-1826, which were acts of social change by definition. Then, trans-disciplinary theories of social change considered social power dynamics as inherent to human evolution (Marx, Tarde, Durkheim, Weber, Fielding Ogburn, Schumpeter). However, after WWII, there was a bifurcation in the conceptualisation. Other authors, started to conceive SI as changes in management, organisations or legal processes (Coleman, Conger, Drucker, Gershuny). Therefore, a debate emerged between the understanding of SI as community development with a social purpose and SI as organisational development for an economic one.

This study approaches SI from a human and community development perspective (Moulaert et al., 2010; Jessop, 2013; Novy, 2018). Here, SI refers to new or disruptive socio-political processes that attend basic needs and improve the quality of life of

DOI: 10.1201/9781003220855-6

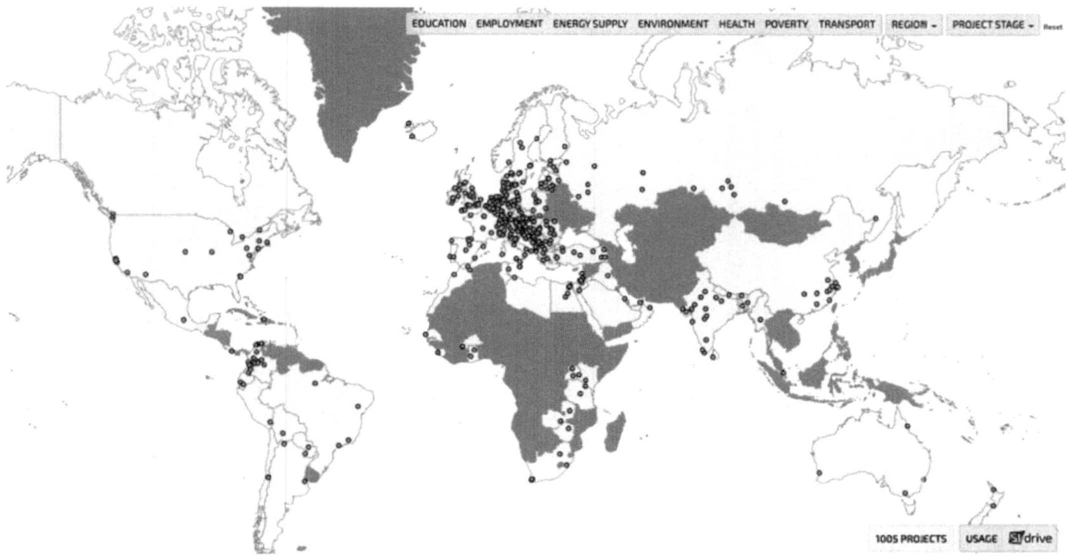

Figure 1. 1005 social innovation initiatives around the world surveyed by the SI-DRIVE project. Howaldt et al. 2016.

disadvantaged groups. It is the manifestation of resistance to social exclusion mechanisms and a proposal for a better scenario. A good starting point of this approach to SI theory are social exclusion-inclusion dynamics (Moulaert, 2010). On one hand, the complex mechanisms that create and sustain exclusion are exacerbated by the lack of attention to social needs, and on the other, the reactions from the affected community emerge as inclusion initiatives that try to mitigate the *effects* of structural, political, and cultural forces that produce social exclusion (Jessop et al., 2013).

In the case of the labour market, social exclusion refers to long-term unemployment due to deeply rooted personal and/or societal issues and invisible barriers that systematically deny the possibility to enter the market. Social inclusion, as a reaction, is the socio-political process fighting to reconfigure power relations in favour of equal opportunities in a more just society. Figure 2 visualises this constant creation process of SI and its three main elements: (1) satisfaction of basic needs, (2) collective empowerment which can achieve political mobilisation, and (3) changes in social (power) relations (Moulaert et al., 2013). Here, "empowerment" is a "process of construction and transformation of the subject" (Imagina Madrid, 2019), related to *consciousness of power* and *autonomy.* It emerges from conditions in the urban environment that enable citizens to gather and mobilise.

Then, socially innovative initiatives (SIIs) can be defined as locally based initiatives, such as grassroots movements, community organisations or other forms of empowered collectives, that have gained momentum in the fight against an unsatisfactory condition manifested, in this case, as social exclusion in the labour market. SIIs aim at attaining social inclusion through the collective empowerment of disadvantaged groups in the labour market.

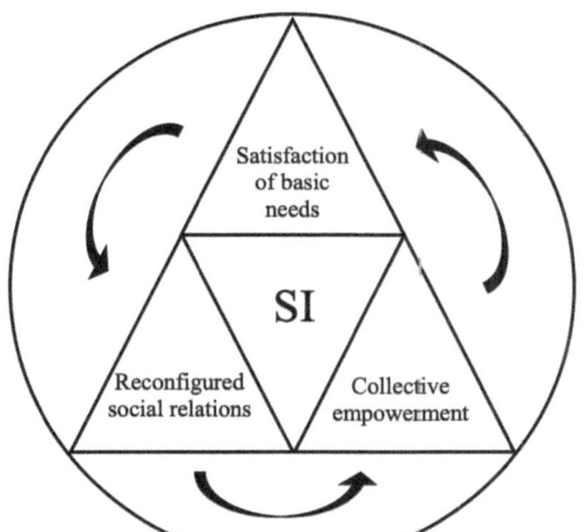

Figure 2. The triad of social innovation in an environment of creation. Author's elaboration based on Moulaert et al., 2013.

2.2 *Problem statement*

Within the capitalist framework, there is rampant competition in the labour market. The idea is the maximisation of all resources to avoid inefficiencies in the economic system. The accompanying neoliberal discourse promotes the logic that (1) producing more, better and faster with more entrepreneurial actors will bring "progress", and (2) that such economic development equates social development. The result is a public policy that creates legal and political apparatuses of institutions and financing channels that propel a frenzied economic machinery with questionable social gaps (Theodore et al., 2012).

This logic and its derived infrastructure, creates mechanisms of social exclusion for the "less competitive" individuals in the market. Depending on the context, they could be the low-skilled, homeless, immigrants, disabled, elderly, LGBT, women, victims of violence and natural disasters, indigenous groups, etc. They are often victims of discrimination and encounter resistance to enter the labour market. At the same time, several social inclusion initiatives are raised, but seem ineffective to transform the *status quo.*

The traditional approach is to attend individuals' struggles through training and capacity building programs. They aim to "level-up" the skills of the "excluded population" to become "employable" and therefore, "included" in the labour market and society. However, this approach reacts only to the *effects* [unemployment], not to the *causes* [discrimination, lack of education and opportunities, and the overall "transformation of labour" (Esping-Andersen & Castells 1999)]. Therefore, the interest of this study is on SIIs that promote empowerment and try to attend causes of exclusion. The following methodology was developed to examine their challenges and potential to transcend benevolent actions towards long-term social change.

3 METHODOLOGY

The critical realist concepts of the *empirical,* the *actual,* and the *real* are borrowed to explore and analyse social reality (Bhaskar, 2008). They conform a system of three layers linked to each other through causal relationships. The *empirical* are the observable experiences, the *actual* how the observable is actualised or the factors influencing it, and the *real* are the underlying mechanisms or existing assumptions of what is possible. Despite their apparent clear distinction, these are "overlapping domains" (ibid) more in concentric circles in which the *real* is the largest circle encompassing the *actual,* and this one encompassing the *empirical.* This three-dimensional philosophical framework is the backbone of the three research questions (Figure 3). Each one with an aim: describe, explain, and explore, which in this study cannot be separated from each other but rather built upon.

This investigation adheres to the "radical humanist research paradigm" (Burrell & Morgan 1979). The view is that social reality is constructed by individual perception but with "inter-subjectively shared meanings". The perception of social reality is influenced, if not dictated, by power relations of the *status quo*. As evidenced by the subject of study, the "sociology of radical change" is concerned with "overthrowing or transcending the limitations of existing [unjust] social arrangements" (ibid). Therefore, to explore this notion of social reality

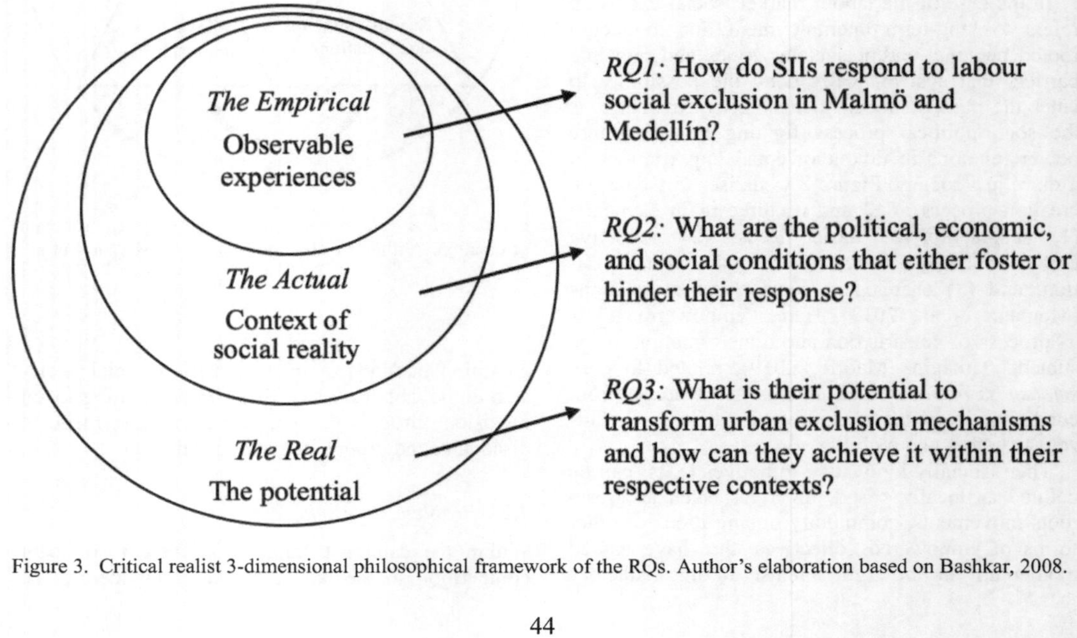

Figure 3. Critical realist 3-dimensional philosophical framework of the RQs. Author's elaboration based on Bashkar, 2008.

a qualitative research methodology is used and multiple-case study as strategy of inquiry.

The case is labour exclusion-inclusion dynamics and the units of analysis are the SIIs in each city. The methods used were participatory observation and in-depth interviews that targeted specific local actors. A non-probability sampling method was used, specifically quota sampling to ensure the engagement of all sectors, and snowball sampling to follow lines of collaboration and partnerships between actors while, at the same time, evaluating their close links or lack thereof. Interviews were held with administrators of the SIIs, government officials, educational organisations, training centres, NGOs, business owners, "innovation hubs", foundations, grassroots organisations, activists, residents, and when possible, with individuals in a situation of exclusion. A total of 21 interviews were conducted in Malmö and 20 in Medellín, between 2018–2019.

The cities for the study: Malmö, Sweden and Medellín, Colombia, were chosen for various reasons. First, the geographic stretch permitted studying social innovation in two very distinct contexts. The historical differences in socio-political, economic, and cultural terms are abysmal. This presents a great opportunity to take a deductive approach: testing the universality of the "human and community centred" approach to SI theory (Moulaert et al., 2010, 2013), and comparing results from the two different places. Within their respective national containers, the cities have been stigmatised for violence and crime but have also been reframed as "innovative" cities. They are ideal cases because they face significant challenges of social exclusion, specifically in the labour market. Simultaneously, there are many efforts to remediate that condition through projects, programs, and initiatives that promote social inclusion. However, while the cities present similar characteristics ideal for comparative urban research, their abysmal differences in contextual spheres demand higher levels of analysis.

The C.S.I. Analytical Framework was created to structure a multi-level analysis. First, at the macro-level, the *Context* reviews national systems of governance, urban regimes and local politics. It helps recognise how national containers influence local activity and shape the way actor networks develop. This is useful to avoid "the localist trap" (Moulaert, 2010), and see the empirical only as evidence of bigger underlying causalities. Then, the mezzo-level analyses five *Strategies* of social inclusion in each city: (1) instruments to promote inclusion, (2) actors' partnerships and collaborations, (3) urban planning strategies, (4) empowerment through culture and the role of civil society, and (5) promotion of economic development. Lastly, the micro-level zooms-in on the concrete response of socially innovative *Initiatives* at the local level. The analysis uses the "ALternative MOdel for Local Innovation" (ALMOLIN) that was developed in the EU project SINGOCOM for the comparative analysis of cases of social innovation (2001-2004).

4 RESULTS

4.1 *Case study 1: Malmö, Sweden*

Malmö's population's (347,825) average age is 36 years old (Malmö Stad, 2018). In 2016, it became Sweden's fastest-growing city (Statistics Sweden, 2018) in part, due to international migration. 31% of Malmö's residents are foreign-born with the largest groups from Iraq and ex-Yugoslavia states (Malmö Stad, 2018). In the context of the welfare state with a long tradition of the Socialist Democratic Party in power, there are path dependencies such as the *Million Homes Program,* an ambitious national project to build one million new dwellings between 1964–1974 (Hall & Vidén, 2005). Those housing estates served to settle various migration waves (Anderson, 2014). Today they are "vulnerable areas" like the Rosengård neighbourhood, experiencing socio-spatial segregation, and social and labour exclusion, which can also be pre-conditions for the emergence of SI.

The efforts for social inclusion in Malmö are primarily led by the public sector via short-term, EU-funded, top-down projects. The City of Malmö often partners with companies and universities in multi-level institutional alliances, but collaboration with civil society collectives is rare. Urban planning strategies seem to have a marked economic-spatial differentiation of neoliberal developments by the waterfront and projects to transform "problem areas" into "innovation areas" while trying to avoid gentrification. Social entrepreneurs attempt to close the gap between labour demand and supply through traditional methods of job training and matchmaking. However, the SII *Yalla Trappan* (YT) in Rosengård is "meant for [immigrant] women who are far from the labour market and short of education and work experience"[1]. They create their own inclusion utilising skills they already have such as cooking, teaching, cleaning, and sewing. Having similar backgrounds (the Balkans or Middle-Eastern countries), socio-economic situation, and shared goals, these women have formed a support network to attain financial and social inclusion.

The analysis with the ALMOLIN model showed SI emerged from a situation of long-term exclusion from income-generating activities and Swedish society in general. The concentration of cases in Rosengård indicates a collective need, which transformed into a labour inclusion project lead by "the excluded" themselves and supported by institutional alliances. YT offered an oasis of familiarity in an unknown land. A Palestinian woman reported: "I felt

1 Interview by the author with *Yalla Trappan* Directive on November 14, 2018.

that I haven't got a country of my own. As if I didn't belong to any country at all... [but now] I feel at home in Sweden... and like a family at Yalla Trappan"[2]. YT started a cycle of material and social construction. The material creation of a physical place for encounter and the collective creation of food, textile garments and the services provided are material proof of their capacities. These collective economic activities then yielded a space for social construction through individual empowerment and collective identity building.

4.2 *Case study 2: Medellín, Colombia*

Medellín has a population of 2,549,537 inhabitants (Alcaldía de Medellín, 2019). It is a receptor of internal migrants from the Pacific, Caribbean, and Central regions of Colombia displaced by violence and looking for economic opportunities in the city (DANE, 2018). Medellín has a marked history of violence and narcotraffic during the civil war in the 1970-1990s. In the 2000s, Colombia received USA aid to eradicate drug cartels through "Plan Colombia". It was considered a military success, yet bloody military operations wiping armed groups had inevitable(?) collateral damage due to the embeddedness of civil society in the conflict (Mejía, 2016).

Nonetheless, the violent past has slowly been left behind through an impressive city transformation accomplished through political alignment and multilevel governance alliances. The resilience of civil society has been fundamental in the reconstruction of imaginaries, both promoted and facilitated by the progressive vision of various consecutive city administrations. There is a philosophy of shared leadership between a strong local government in alliance with public and private companies. An important urban strategy for social inclusion is social urbanism: heavy public investment in social and cultural infrastructure specifically in the most disadvantaged areas of the city. Other strategies assist citizens in cultural and economic development.

The main initiative studied in Medellín, *Escuela Para la Inclusion* (EPI) is a trade school dedicated to long-time unemployed, hyper-excluded population groups e.g. elderly, disabled, LGBT, victims of violence, demobilised ex-insurgents, and ethnic groups. The EPI supports first the social and emotional development and then provides trade skills to gain employability. The EPI director summed it up as "*primero hay que formar el ser que el hacer*"[3]. The analysis through the ALMOLIN model showed SI emerged from persons being caught in a downward spiral of life conditions of exclusion. Harsh past experiences, discrimination and fierce labour market competition stack up in various layers of disadvantages. The trade school aims to generate individual empowerment and identity (re)construction as pre-conditions for marketable and employable skills.

An additional initiative was examined. *Casa Kolacho* (CK) is a youth collective conformed in the 13th commune, the most violence-stricken area in the city. It emerged from the need to find alternatives for youth besides drug dealing or joining the lines of armed groups patrolling their streets. CK evolved as a small organisation that supported the development of artistic talents and, through art, offered a route of hope for local youngsters. Based on the characteristic sense of belonging and resilience of citizens, CK fights for a space and a place for the realisation of their new imaginaries. It is a pioneer organisation of "social rescue" in the commune by (re)constructing the social and urban tissue of the neighbourhood.

4.3 *Comparative results*

4.3.1 *National and urban contexts*

Sweden, a pure social democracy, and Colombia, conservative by Spanish heritage with some liberal tendencies, have structural differences that create variations on labour market regimes and safety nets of social services in case of economic hardship. The excluded population in Malmö are long-term unemployed immigrants with cultural and linguistic barriers to integration, but most are recipients of social welfare to cover their basic needs. The excluded in Medellín are disadvantaged in market competition due to age or disabilities, discrimination or limited social capital and struggle economically.

Both cities have received economic migrants and war refugees. Malmö accommodated international migration in vacant housing estates, while in Medellín, internally displaced persons arrived to informal settlements. Hence, the reasons for migration waves and the situation upon arrival are also different. The resulting socio-spatial segregation of transition areas creates more accentuated discrimination by place of residence in the Colombian than the Swedish case. This evidences how the urban malleable form also shapes social constructs and confines possibilities for inhabitants. The SIIs examined here emerged to fight social inequalities and exclusion. They seek to return power to those dispossessed of it.

Both *Yalla Trappan* (Malmö) and *Escuela para la Inclusion* (Medellín) are forms of institutionalised social inclusion efforts. Contrasting them, it becomes apparent that they utilise different approaches: YT is a space that links people in an oasis of familiarity while EPI provides tools to change a situation. YT is a permanent place while EPI is transitory. YT is very

2 Interview with a Palestinian woman employed at *Yalla Trappan*, quoted in Björnståhl and Hartman 2012.

3 Spanish for: "First you have to educate the being, then the doing". Interview by the author with *Director of Escuela Para la Inclusion* on August 13, 2018.

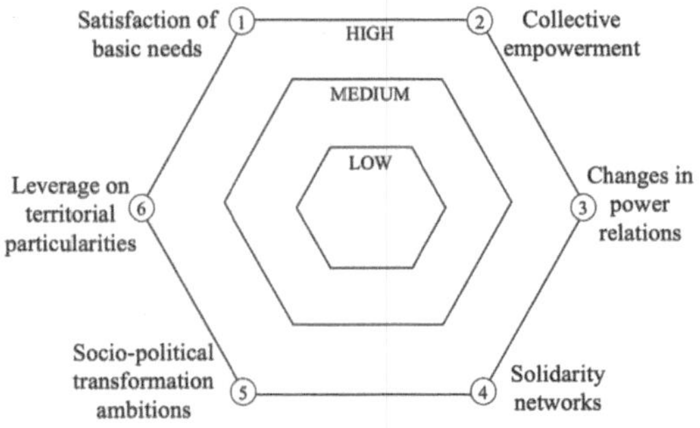

Figure 4. Generic model for the analysis of social innovation features on the initiatives. Author's elaboration.

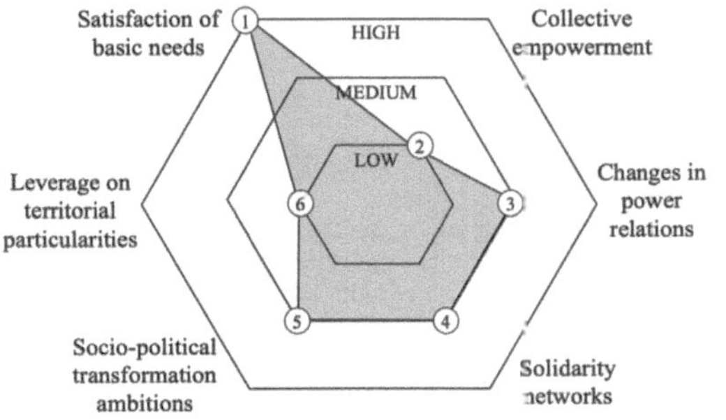

Figure 6. Social innovation features of '*Escuela para la Inclusión*' (Medellín). Author's elaboration.

localised to Rosengård while EPI seeks city-wide benefits. In addition, while both initiatives have obtained labour inclusion (employment), YT has achieved social inclusion only within its collective, not in Swedish society. Conversely, EPI attains social inclusion in society, but it lacks collectivisation.

4.3.2 *Hexagon analysis of SI features*

The hexagon (Figure 4) illustrates six key characteristics of SI and their levels of presence in the SIIs (low, medium or high). A scenario of low SI would be visualised by the smallest hexagon, and of high SI by the largest hexagon. The six key characteristics of SI were identified from the literature reviewed and employed to construct radar charts to visualise how socially innovative these initiatives are. The key features corresponding to the numbers in the hexagons are: (1) the satisfaction of basic needs, (2) collective empowerment, (3) changes in social (power) relations between the state, market and people, (4) building solidarity networks, (5) having socio-political transformation ambitions and (6) leverage of territorial particularities. The variations of these SI characteristics in each initiative yield geometric figures with varying areas according to their degree of SI. Then, the figures are juxtaposed for a general comparison.

As seen in the resulting figures of YT (Figure 5) and EPI (Figure 6), both initiatives satisfy the basic needs of their vulnerable populations. However, YT, by offering an alternative place of work for immigrant women, focuses more on (2) collective empowerment, (4) solidarity networks and (6) leverage on the territory, features that are rich in social capital. EPI is quite the opposite. The trade school of soft and hard skills spends less energy on building social capital and more on individual empowerment by providing tools for building character, and ultimately (3) changing power relations and (5) constructing ambitions for socio-political transformation.

When the additional initiative in Medellín, CK, is examined through the same lenses of the hexagon analysis, it complies with almost all of the key characteristics of SI (Figure 7). The main *motive* for CK was changing governance relations in the commune, shifting power away from terrorist groups and empowering the youth. CK uses graffiti and street art as a *social and urban transformation tool*. This initiative of civilian resilience, along with two decades of government soft and hard power strategies to recover the area, has positively reconfigured power relations between the state, people and terrorist groups in the neighbourhood.

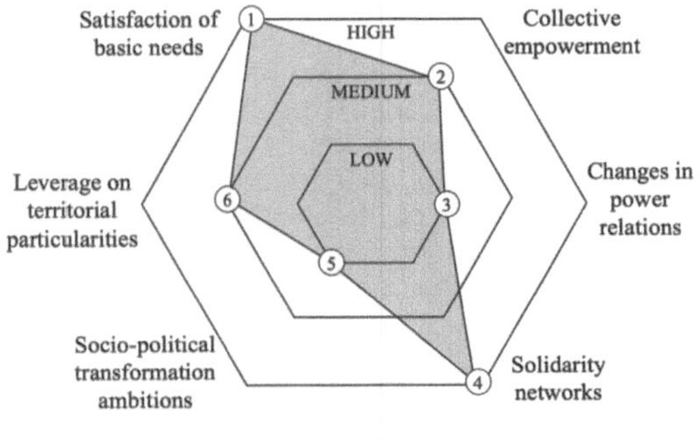

Figure 5. Social innovation features of *'Yalla Trappan'* (Malmö). Author's elaboration.

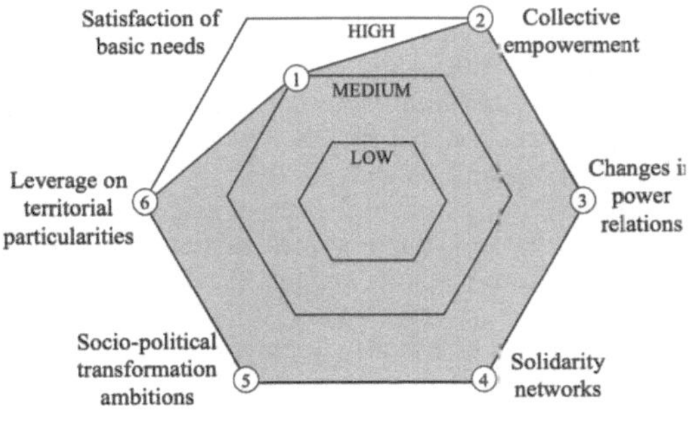

Figure 7. Social innovation features of '*Casa Kolacho*' (Medellín). Author's elaboration.

Table 1. Three-stage evolution path of the socially innovative initiatives and their accomplished stages. Author's elaboration.

STAGES*	*Yalla Trappan* (Malmö)	*Escuela para la Inclusión* (Medellín)	*Casa Kolacho* (Medellín)
1. Positive conditions of possibility from the urban and larger context	Lack of individual empowerment (language, social, and cultural barriers). The women do not appropriate urban/social space	Social and urban transformation (decrease of violence in the city) + Individual empowerment (recover trust)	Social and urban transformation (decrease of violence in the city) + Individual empowerment (recover trust)
2. Emergence of an empowered collective: Inclusion efforts initiated by 'the excluded' themselves	Solidarity network but immigrant women *are not* empowered themselves but under the roof of YT	Missing the empowered collective due to transitory nature of EPI	CK shares vision and goals. The collective formed has a sense of belonging, accomplishment and brotherhood.
3. Support from institutions in the form of projects or programs	Support from city government, Malmö University and IKEA. Role of 'providers' of inclusion and empowerment. Paternalistic	Support from the Secretary of Social Inclusion, CUSO International (NGO), and several private companies	Antagonistic relationship with the state due to scars from violent past. Instead alliance with the private sector e.g. Tucan Cafe

* White cells mean the stage was fully accomplished and present in the SIIs trajectory. Shadowed cells mean conditions are not present or the stage was not (fully) accomplished.

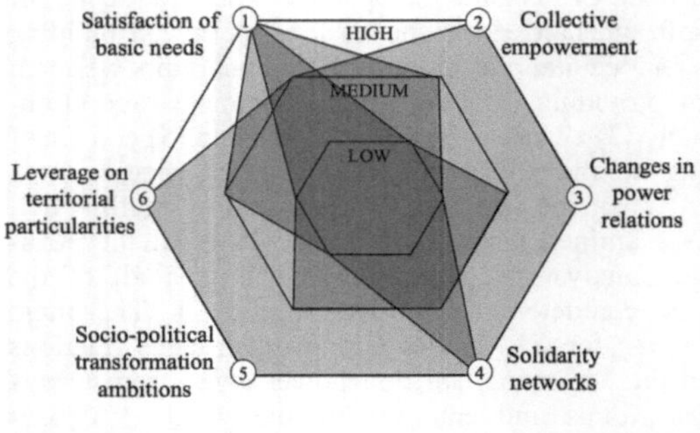

Figure 8. Social innovation features of '*Casa Kolacho*' (Medellín). Author's elaboration.

A juxtaposition of the three initiatives' SI features (Figures 5-7) leads to Figure 8, which facilitates a visual comparison. In the quest for more specific and decisive factors for an initiative to be more or less socially innovative, various conditions of the context were evaluated and compared. The results suggest that the initiatives' inception points and evolution path are essential factors. An in-depth analysis of these factors yielded the construction of Table 1: a three-stage evolution path for SIIs.

While none of the initiatives considered in this study fulfilled all three stages of development*, these comparative results suggest that the fulfilment of these three stages is very important. On one hand, YT and EPI both display certain socially innovative features but not all, and not to the same level (Figures 5-6). When the initiatives are analysed through the three-stage evolution path, it becomes noticeable how they have skipped some of the stages of development. On the other hand, the hexagon analysis revealed CK as the most socially innovative of these three initiatives, and it has also developed so far following the identified path. These comparative findings suggest there might be a correlation between the initiative's three-stage evolution path (Table 1) and the six key socially innovative features (Figure 8).

5 DISCUSSION AND CONCLUSION

Retaking the three-dimensional philosophical framework (Figure 3) of the research questions (RQs), now it is possible to answer:

5.1 *RQ1 refers to the empirical: How do SIIs respond to labour social exclusion in Medellín and Malmö?*

The SIIs in Malmö and Medellín first identify the need of "the excluded" and try to provide what is missing: a space for interaction and increasing social capital, education and training or personal support. To grow or to get started, SIIs pursue institutional collaborations with other sectors, especially the private one for financial resources, exposure, and marketing. The stage of development of each SII influences the scope of its agency and the role other actors take when interacting with it.

5.2 *RQ2 refers to the actual: What are the political, economic and social conditions that either foster or hinder their response?*

First, political constraints are a discontinuity of priorities due to changing urban regimes. Neoliberal urbanism and explicit stratification systems accentuate socio-spatial inequalities. Conversely, social urbanism purposefully addresses social needs. Second, hindering macroeconomic conditions are labour market restructuration leading to unemployment and forced migration, the effects of inflation and inadequate monetary policy. At the microeconomic level, limitations come from being subject to supply and demand dynamics, inconsiderate taxation schemes, and high dependency on public funds. Positive economic conditions are the availability of EU funds and inter-sectoral alliances that facilitate in-kind donations and cost-cutting techniques. Lastly, in the social realm, hindrance comes from discrimination, stigmatisation, lack of personal drive and empowerment, and loss of social capital. Cultural differences can be roadblocks or opportunities for interaction depending on the adopted attitude. Positive social conditions are circuits of solidarity, collective empowerment, nourishing social capital, and an active and resilient civil society to draw new imaginaries.

5.3 *RQ3 refers to the real: What is the potential of SIIs to transform urban exclusion mechanisms and how can they achieve it within their respective contexts?*

When creating social inclusion solutions there is a tendency to take the current state for granted, opting for a "problem-oriented approach". But "critical realism urges us to shift our main attention from the actual events to the mechanisms that cause these events" opting for a "potential-oriented approach" instead (Novy & Stigendal, 2018).

To examine the transformative potential of initiatives, this study used two tools: (1) radar (hexagon) charts to analyse the SI features of three initiatives and (2) constructed a three-stage evolution path to help explain the emergence and development of SIIs. The results suggest that the consecutive order of these stages is relevant for urban and social transformation: (first) leverage on the existing conditions of possibility offered by the urban context, (second) collectivisation of empowered "excluded individuals" through common identity and shared goals, and (third) institutional support to develop, without becoming dependent on paternalising top-down assistance for subsistence.

Following that specific evolution order would not only imply a positive change from individual to collective empowerment, but would also mean that their surroundings are changing to allow the thriving and sustenance of the initiative. Therefore, when the surrounding environment is tapped to change, the potential to transform that contextual sphere is higher. SIIs alone cannot be transformative; they are utterly subject to the political, economic, and social environment, as an encompassing sphere. This does not mean that they are subject to the current sphere, just that it is needed to shake it.

5.4 *Conclusion*

The first thesis of this study is that it is necessary to divorce from the idea that to potentialise local SIIs it is vital to scale up or out. While scaling or replicating an initiative may spread good, if it is not attentive to the causes of exclusion, it would not be transformative. Many more and bigger YTs and EPIs would not halt the continuous generation of labour exclusion, nor attend why excluded vulnerable groups exist in the first place. Therefore, it is questionable if scaling should always be desired as it is assumed by mainstream social innovation programs and incubators. Instead, the transformative power of well-intentioned initiatives seems to be determined by their *nature*: the social innovation features that they display (Figure 4), by their *development*: attending the missing conditions of possibility to be able to follow the 3-stage evolution path in consecutive order (Table 1), and by their *purpose*: the deployment of tailor-made actions to dismantle the physical, economic, social, and mental barriers that exclude.

A second thesis is proposed based on critical reflections from the Latin American case. Medellín's history of urban governance beyond the state requests the consideration of terrorist groups as a fourth (deplorable but real) element in the city's struggle for power. In contexts where armed groups have a heavy territorial presence, the traditional "state-market-civil society" triangle is only partially useful to describe and analyse urban power relations. Furthermore, and consequently, a positive reconfiguration of power through multilateral efforts, especially from the state and civil society, can transform social exclusion mechanisms and the entire city by recovering stability, autonomy, quality of life and, ultimately, peace. This suggests the theoretical relation between social innovation and peace, and thus the practical recognition of socially innovative initiatives as tools for peacebuilding. This link has the potential to be explored and applied in many contexts of conflict around the world.

As a critical reflection, it can be said that current SI theory is not perfectly fit to assess urban struggles in search of social justice. For SI theory to be applicable at a global level, it would need to consider realities in non-traditional, politically unstable, contexts. Furthermore, its theoretical apparatus should be expanded to include a greater variety of actors, reasons for exclusion, and the basic rights people could be excluded from. There are also entirely new aspects to consider such as (1) personal circumstances and level of individual empowerment of the excluded population and (2) the large role of the urban context (physical, institutional and cultural dimensions) in creating positive or negative conditions of possibility for the emergence, or suppression, of social innovation.

To conclude, this study wants to evince that "the Social Innovation phenomenon" does have the potential to transform mechanisms of social exclusion and resolve pressing social problems if its contextual sphere moves in that direction. Neither initiatives alone, nor their scaling up or out would suffice. Mechanisms of social exclusion have multiple and complex layers with origins in fabricated physical, economic, social and mental barriers. Thus, the transformative potential lays on SIIs that mobilise actor networks and resources, i.e. part of the urban context, to attend the causes of their own exclusion: to break down fabricated and normalised barriers and their oppressive structures.

REFERENCES

Alcaldía de Medellín 2019. http://www.medellin.gov.co

Anderson, T. 2014. Malmo: A city in transition. *Cities* 39: 10–20. https://doi.org/10.1016/j.cities.2014.01.005

Bhaskar, R. 2008. *A Realist Theory of Science*. London and New York: Routledge.

Björnståhl, M. & Hartman, E. 2012. *The Road to Yalla: A book about the women at Yalla Trappan*. Malmo: Kira Förlag

Burrell, G. & Morgan, G. 1979. *Sociological Paradigms and Organisational Analysis: Elements of the Sociology of Corporate Life*, 1st ed. Heinemann Educational Books.

DANE 2018. Dirección Nacional de Estadística de Colombia. https://www.dane.gov.co

EC CORDIS 2020. List of EU projects on social innovation since 1990. https://cordis.europa.eu

Esping-Andersen, G. & Castells M. 1999. *La Transformación del Trabajo*. Colomers: La Factoría Cultural.

Hall, T. & Vidén, S. 2005. The Million Homes Programme: a review of the great Swedish planning project. *Planning Perspectives* 20: 301–328. https://doi.org/10.1080/02665430500130233

Howaldt, J. et al. 2016. Comparative Analysis (Mapping 1)-Mapping the world of social innovation: a global comparative analysis across sectors and world regions (D1.4). Dortmund: SI Drive Project.

Imagina Madrid 2019. Entre la celebración y el conflicto (Final presentation and evaluation of the "Imagina Madrid" Project). Madrid: 23 April 2019. Acción Cultural Española.

Jessop, B. et al. 2013. Social innovation research: a new stage in innovation analysis? In Moulaert, F. et al. (Eds.) The International Handbook on Social Innovatio: 110–130 Edward Elgar Publishing Limited

Malmö Stad 2018. Facts and statistics about City of Malmö. https://malmo.se/Fakta-och-statistik/

Mejía, D. 2016. Plan Colombia: an analysis of effectiveness and costs. *Foreign Policy at Brookings*: 1–17

Moulaert, F. et al. 2010. *Can Neighbourhoods Save the City? Community Development and Social Innovation*. New York and Abingdon: Routledge.

Moulaert, F. et al. 2013. *The International Handbook on Social Innovation*. Cheltenham and Northampton. Edward Elgar Publishing Limited

Novy, A. & Stigendal, M. 2018. Founding transdisciplinary knowledge production in critical realism: implications and benefits. *Journal of Critical Realism*. Volume 17, Issue 3: 203–220. https://doi.org/10.1080/14767430.2018.1514561

Statistics Sweden 2018. Malmö population composition statistics. https://www.scb.se

Theodore, N. et al. 2012. Neoliberal Urbanism: Cities and the Rule of Markets. In: Bridge, G. & Watson, S. (Eds.), The New Blackwell Companion to the City: 15–25. Wiley-Blackwell.

Research Tracks in Urbanism: Dynamics, Planning and Design in Contemporary Urban Territories – Allegri et al (eds)

Urban space and mobility: Adjustment towards more sustainable urbanism

C. Araujo-Lima & H.E. Nichele
Postgraduate Program of Urban Planning, Federal University of Paraná, Brazil

ABSTRACT: The paper proposes a discussion on the access to mass transit, within the urban plan scope, whose implementation in the context of the social and economic-political process has created inequalities, including spatial ones. There is a need to renew urban design and transport system guidelines, together with social policies, to increase physical accessibility to transport for the majority of users, located in peripheral areas. The analysis started from four assumptions that support urban configuration and mobility as factors of sustainability. The goal is to understand urban conditions around terminals, comparing them with other areas of the city to recognize options for improvement. Urban space was read (Del Rio, 1990; Lynch, 1997) and cartograms were made using QuantumGIS, Microsoft Office, Word, Excel, and Sketch-Up. The results suggest the implantation of a new transport mode and densification of the occupation, seeking a better consonance with the users' needs and environmental imperatives.

1 INTRODUCTION

The environmental crisis makes it more urgent to rethink cities, including territory consumption (Coalition for Urban Transitions, 2019). It is known that mobility conditions are one of the factors that make urban expansion feasible (Ojima & Hogan, 2009). However, as is typical of the capitalist system, there are inequalities in the built space of cities and in the access to the mobility system. (Harvey 2005, 2012; Ojima & Hogan, 2009). This paper discusses the condition of inequality in access to public transports in contrast to the urban configuration (Santos et al., 2017). The objective is to understand Urban Design conditions, in order to verify more sustainable occupation and mobility possibilities. Mobility is essential in urban daily life. Nevertheless, it is important to make greater use of public transport to improve social and environmental conditions, optimizing land use, and avoiding low densities (Renner, 2016).

The paper covers the territory of Curitiba, the eighth largest Brazilian metropolis in population, located in the Southern Region of Brazil. The city polarizes the Central Urban Nucleus (CUN), formed by 14 municipalities (out of a metropolitan total of 29). In this core live 93% of 3,654,960 metropolitan inhabitants (COMEC 2006, IBGE 2019). Figure 1 (section 4) shows the integration between land use, public transport systems, and road networks form the conceptual framework of the Master Plan of Curitiba since 1966 (IPPUC, 2004). The Urban Design proposal is linear, through five structural axes consisting of a system of three lanes each with-different speeds, the central lane being exclusive for buses. The blocks close to an axis constitute a structural sector where high density and varied uses were planned according to urban legislation (Curitiba, 2000). The structural axes also include transport terminals, which concentrated amounts of bus lines. Numerous international awards have highlighted this urban planning project and the bus system known as "Bus Rapid Transit" (BRT), especially since the 1990s. The city has become a worldwide reference in planning (Ruano, 1999; IPPUC, 2004; Pinderhughes, 2004; Renner, 2016). Meanwhile, the urban settlement proves to be under a socioeconomic segregation process with the low-income population located in a less urbanized periphery, while there are urban voids in structured areas (Albuquerque, 2007; Lima & Kuzma, 2017; Lima et al., 2018). The set of blocks around four peripheral transportation terminals is analyzed in this work, focusing in their spatial and socioeconomic characteristics, and location in relation to the central areas. The research questions how urban areas close to transport terminals located in peripheral areas are configured, since it is where most users of the system live. This research is based on four assumptions:

a) Public transport is a solution for everyday mobility needs and, like the spatial segregation factor in Curitiba, it demands greater integration between housing and mobility policies (Cheng & Chen, 2015; Lima et al., 2018).

DOI: 10.1201/9781003220855-7

b) Increasing the number of users of public transport promotes urban sustainability by reducing vehicle circulation and the emission of contaminants (Banister, 2011).
c) Housing contiguity to basic services and public facilities, such as transport, promotes an increase in public system users (Banister, 2007).
d) Urban space is a product of the socioeconomic, political, and cultural dynamics of society (Corrêa, 1995).

Based on the problem, objective, and assumptions, a theoretical synthesis of the main concepts is presented as follows.

2 URBAN SPACE, MOBILITY, SUSTAINABILITY

Urban space is considered from an urban planning perspective for its built and social-environmental qualities as a result of the social process (Del Rio, 1990; Corrêa, 1995). Urban spaces combine different uses whose diversity or intensity varies according to the historical, socio-economic, technological or cultural moment. These conditions result in different features of neighborhoods, which, as a whole, present greater or lesser homogeneity (Carlos, 2007). The variety of uses, intensity, and frequency in the built space can mean beneficial options for living in the city, such as cultural or landscape values. Nonetheless, it can also denounce the need for government action to meet basic needs, such as mobility (Vasconcellos, 2010). Accessibility to the transport system is an item within the "Right to the City", social justice, and sustainability concepts, in their broad sense. Ojima and Hogan (2009) point out urban and social structure impacts due to the challenging accessibility conditions for the poorest, which demands public policies, such as Urban Design. A situation that gets worse in a climate crisis context (Wallace-Wells, 2019).

In Brazil, urban mobility follows specific federal and local laws (Brasil, 2012). However, reality shows a mismatch between mobility conditions and populational demands. The vehicle fleet in the country grows (DENATRAN, 2020), congesting public roads. At the same time, public transport loses efficiency while the economic cost of maintaining streets and parking lots in cities is significant (Vasconcellos, 2010). Different elements can improve urban occupation and achieve satisfactory density, diversification of uses, and socio-environmental quality. The concept of Transit Oriented Development (TOD) (Calthorpe, 1993) proposes densification policies around large transport stations to create a mutual benefit between transport users and local buildings users. This concept needs to be adapted, since it was idealized for cities in the American West in the 1990s (Cervero et al., 2009; Cervero & Dai, 2014). One possibility would be to combine Calthorpe's thinking with urban planning instruments of the Brazilian legislation, such as Preemption Law, Consortium Urban Operation, and Special Areas of Social Interest (Brasil, 2001). There may be more low-income conurbations around transport terminals, relying on higher density and avoiding the expansion of urbanized space.

3 METHODOLOGY

The research aims at a practical application. The paper develops a part of a more comprehensive research work, using a previously defined methodology, from where the theoretical and empirical criteria for the use of the 250-meter buffer comes from, and the application of the usual practices of Urban Design is defined in four stages (Lima, 2012, 2015; Lima & Kuzma, 2017):

a) adoption of a conceptual and theoretical framework;
b) spatial and non-spatial data collection (Del Rio, 1990; Kohlsdorf, 1996; Lynch, 1997), with Google Street View image capture;
c) data manipulation (information classes, categories, criteria, calculation), using AutoCad SketchUp, GIS, QuantumGIS, and Microsoft Office (Word, Excel) to elaborate cartograms and tables;
d) conclusions about spatial, socioeconomic, locational, and mobility conditions.

The study comprises four fragments around peripheral terminals of the Curitiba public transport system. The use, built area, and the number of floors were combined to determine the density: each 60 m^2 of residential and mixed-use represented a house with 3.3 people. For other uses (commercial, industrial, services, institutional, and leisure), each 60m^2 marks the use of one person. The green areas were surveyed, both as masses of vegetation and crowns of isolated trees, to calculate the green area per inhabitant in the clippings (Lima, 2012, 2015). Figure 1, in the next section, shows the study clipping: in circles the analyzed terminals (periphery) and in squares the central terminals as a comparison reference.

4 RESULTS ANALYSIS

The urbanization process of Curitiba, including as a metropolis node, accelerated from the 1970s (Moura & Magalhães, 1996) when the Town Hall implemented the transport system. In this process, the periphery concentrated less valuable land and the low-income population, a place where the primary transport users live outside the main transport axes area (Lima et al., 2018). From this point, we initiated the analysis of a broader context of the city's urban evolution (Figure 1), followed by the focus on the fragments.

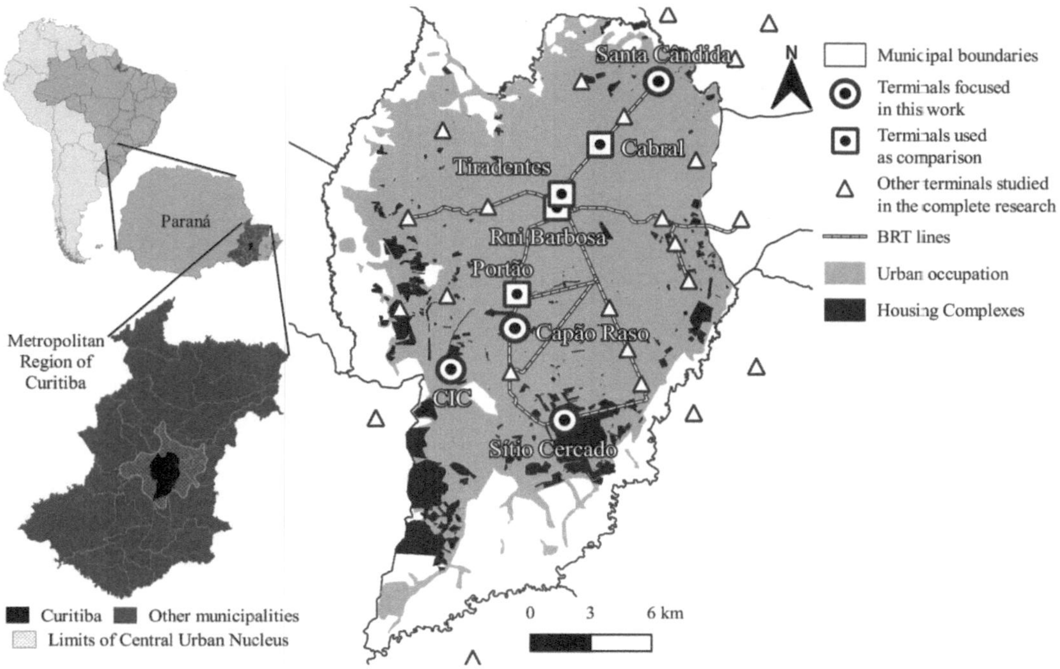

Figure 1. Evolution of urban occupation, and location of terminals and low-income housing complexes – Curitiba and Central Urban Nucleus. Source: Authors (based on Lima et al., 2018).

4.1 *General spatial, socioeconomic, and locational characteristics*

The implementation of the bus system followed an original logic of the urban occupation of Curitiba, in the Northeast-South direction. The first BRT lines connected Portão (South) to Santa Cândida (Northeast) in an itinerary of over 20 km. Between 1970 and 1980, occupation expanded to the east and southeast of the metropolitan region, and Curitiba implemented the Boqueirão structural axis (Souhteast) in 1977. In 1979, a 44 km circular bus line was created, connecting 28 neighborhoods, changing the linear proposal made in 1966. In the flooded valley of the Iguaçu River, the southeast areas were parceled and occupied with a medium-high density, despite the socio-environmental vulnerability. In the 1990s an east-west BRT axis was also implemented in areas not yet settled (Garcez, 2006; URBS, 2020).

Therefore, several factors characterized the urban evolution of the city, such as the physical and environmental basis, road connections, and metropolitan infrastructure (airport, industries, etc.). That made the BRT south axis stronger than other directions (IPPUC 2004). In the Center, the urban fabric is older and embraces two points of concentration of bus lines: Tiradentes Square (the city's geographical center) and Rui Barbosa Square, where income is higher and the age range is more advanced than in other districts. In a circle immediately after the center, there are several consolidated districts. Cabral and Portão terminals, examples of these areas, have large population density and are located in a ring of three to five kilometers from the geographical center. On a second level, far more than seven kilometers from Tiradentes, Santa Cândida and Capão Raso terminals occupy areas older than those around CIC and Sítio Cercado (third level), which are in the proximity to low-income housing complexes. This spatialization makes clear that the occupation is immensely affected by land costs and commuting, while time, a third factor, is tolerated due to socio-economic limits (Lima & Kuzma, 2017).

The main transport axes occur in five of the six land cost ranges. Tiradentes and Rui Barbosa, the main bus lines' concentrations in the central area, with the higher land cost, do not offer physical integration, which harms the users. The Cabral and Portão terminals are in districts of second and third economic value categories and offer reasonable functional conditions. Furthermore, Capão Raso and Santa Cândida, both located in structural sectors, have a land cost higher than the places around the Industrial District (CIC) and Sítio Cercado, low-income areas. Therefore, the locals over the structural axes are the ones with higher land value than other regions, thus creating a contradiction in the simple logic of distance from the geographical center. To summarize, residential use occupies 51%

(average) of the areas closer to terminals, considering the 250-meter buffer, and points out this significant type of use. There is mixed-use in the second place, which usually conjugates residential and commercial activities, followed by commercial and institutional types. Capão Raso, in the Structural Sector, stands out with 30% of the area dedicated to mixed-use, due to economic activities. Santa Cândida, at the end of the Structural Sector North, does not have a substantial mixed-use percentage. Conversely, in a lower-income location, CIC and Sítio Cercado present a less improved urban design quality surrounded by low-income housing complexes. Table 1 contains synthesized information of the eight fragments: on the left, the four peripheral fragments (Santa Cândida, Capão Raso, CIC, and Sítio Cercado) and, on the right, the more central ones, used as parameters of comparison (Tiradentes, Rui Barbosa, Cabral, and Portão).

As data presented in Table 1 shows, the most central fragments (Tiradentes and Rui Barbosa) have the highest land cost, larger proportions of mixed and commercial uses, larger density, and a bigger amount of bus lines. Cabral (North) is in a Structural Sector surrounded by upper-middle-class districts and predominantly residential use. In the South, Portão is also in a Structural Sector but characterized by mixed-use. In the four focused fragments, densities are three times smaller than the two most central ones. In CIC and Sítio Cercado, the density is higher and the locations are outside the structural sectors. The presence of green areas is low in all the eight locations. The average in the four peripheral fragments is 2.45 m² per resident, which means 4% of the 60 m² officially existing on average by the city (Curitiba 2018). In the terminals, the number of transport users is high in all of them, except Santa Cândida, a situation that could be explained by its location at the end of one structural axis and by its low number of metropolitan bus lines. The existent connections will be analyzed ahead in the text with the study of a new mode capable of improving mobility in the metropolis.

4.2 *Characteristics of the peripheral fragments*

This section explains the specific characteristics in the four peripheral fragments, in alphabetical order: Capão Raso, CIC, Santa Cândida, and Sítio Cercado. For this, we applied the analysis categories (section 3), focusing on land use and correlation of factors.

4.2.1 *Capão Raso*

Located in the South Structural Sector, this fragment presents district center's characteristics, with expressive mixed-use, represented by commercial activities like a popular shopping center. There are small stores and residential buildings in the surrounding area as well. The terminal has the highest numbers of users and lines in the studied areas, which evidences the potential for large densities incorporating the morphology's adaptation to more residents and activities both in occupied and empty areas. Figure 2 shows the digital model we made for the fragment, as an example to understand.

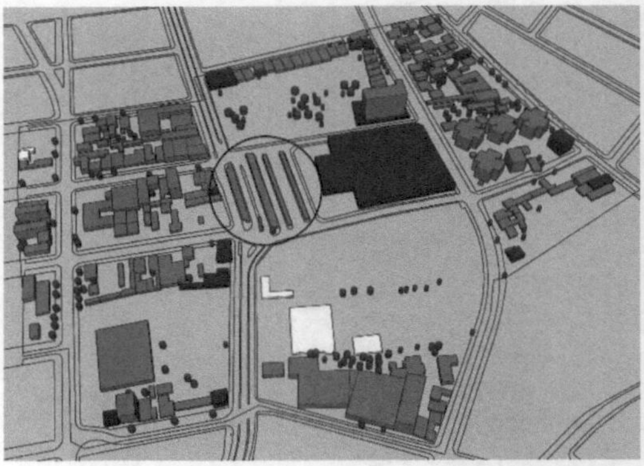

Figure 2. Model of the Capão Raso fragment. Source: Authors.

4.2.2 *CIC*

This fragment is characterized by housing complexes of small and dense land plots, with smaller land costs, located in the Industrial District (Southwest). The lines' frequencies and itineraries make the terminal an intermediary point between Curitiba and the nearby city (Araucária) and a spot of distribution of minor bus lines. The land use stands out by the residential typology, whose density would be improved by a larger diversification and vertical standard, at least in the main streets and lots closer to the terminal. The commercial, mixed, and institutional uses are present in the terminal's immediate surroundings, which counts 18 bus lines and 59,912 daily users. Green areas are almost nonexistent and the dimension of the lots is compatible with on-foot mobility.

4.2.3 *Santa Cândida*

It is the terminal with the lowest quantity of users and lines among the four studied, since it is located at the end of the North Structural Axis without a continued connection to other cities (metropolitan area). This fragment is characterized by expressive residential use, low density, large land plots, and urban voids. It contains the largest green area among all terminals in Curitiba. Like Capão Raso, Santa Cândida could be more compact in density and diversified in use because it is located in a Structural Sector. An improvement in the metropolitan connection with the terminals of the nearby city of Colombo would favor a denser occupation in the area.

4.2.4 *Sítio Cercado*

Located in a connection axis between structural axes, this fragment shows an avenue diversified in

Table 1. Spatial, socioeconomic, and locational characteristics of the central and peripheral fragments. Source: Authors (based on URBS, 2020).

	Capão Raso	CIC	Santa Cândida	Sítio Cercado	Tiradentes	Rui Barbosa	Cabral	Portão
Location	Structural Sector	Amid housing complexes	End of Structural Sector	Surrounded by housing complexes	Center	Center	Structural Sector	Structural Sector
Distance of geographical center (km)	7.29	10.57	7.52	11.65	0.00	0.74	3.18	5.61
Plots of land average area (sqm)	1,274.38	219.56	889.00	410.08	494.95	732.61	1,532.35	1,574.79
Density (people/hectare)	246.11	349.86	193.08	262.99	1,303.97	1,132.65	500.65	462.54
Three main uses (per area)	RES 34% MIX 30% INS 15%	RES 63% MIX 17% INS 12%	RES 60% COM 22% SER 9%	RES 47% INS 17% COM 15%	MIX 66% COM 13% SER 9%	MIX 59% RES 16% SER 10%	RES 61% MIX 12% INS 11%	MIX 68% RES 10% INS 10%
Green areas (sqm/people)	0.69	0.77	7.77	0.57	0.21	0.14	3.24	1.03
Land cost (R$)	1,000 - 1,500	500 - 1,000	1,000 - 1,500	500 - 1,000	Over 2,500	Over 2,500	2,000 - 2,500	1,500 - 2,000
Transport connections	Crossing connections on the South axis	District feeding and access to Araucária	End of structural axis with access to Colombo	District feeding/connection between S and SE axes	Connections with the entire city	Connections with the entire city	Crossing connections on the N axis	Crossing connections on the S axis
Number of users	124,928	59,912	39,540	57,330	No data	No data	85,536	67,348
Number of lines	36	18	14	20	41	48	17	14
Terminal area (sqm)	7,400	6,700	13,200	6,500	14,300	48,100	12,200	5,800.00

uses. It is surrounded by residential areas, with low cost and late occupation plots of land. The connection to the transport axes and the district attendance consolidates the users' movement. Like CIC, this fragment could receive more diverse uses to combine with the flow of people.

4.3 *Strategic axes: A study*

Here a synthesized study about mobility alternatives is presented. It is based on characteristics evidenced in the total urban fabric, the fragments, and other elaborated data, considering transport demands.

Some strategic axes were defined through the analysis of daily frequencies (working days) of the main lines that connect the terminals, called structuring lines: BRTs, *Ligeirinhos* (direct lines with few stops), and Direct Feeders (direct lines with no intermediary stops). The frequencies in each stretch between terminals were classified into five levels, considering each terminal as an intersection of stretches. Figure 3 presents a cartogram with the levels of structuring lines quantity and evidences differences between terminals. Some terminals are marked by good connection within the metropolis, and some represent transitions to mobility inside districts or at the end of structural axes. For example, in the Center as a whole (Tiradentes, Rui Barbosa and other minor stops), there is the highest number of levels added, 30 in total. Based on the cartogram, we can imply some conclusions concerning the connections of the studied terminals:

a) Capão Raso reaches the second place after the Center, adding 13 levels in five different connections;
b) Santa Cândida has only one connection of level 3, since the terminal is at the end of the BRT corridor, counting with simple minor lines directed to the nearby city of Colombo;
c) CIC has two connections, totalizing three levels, and is characterized as an intermediary stop between Curitiba and Araucária;
d) Sítio Cercado embraces three connections in a total of five levels, which implies the terminal is used as connection between the South and Boqueirão structural axes, without a link to another municipality.

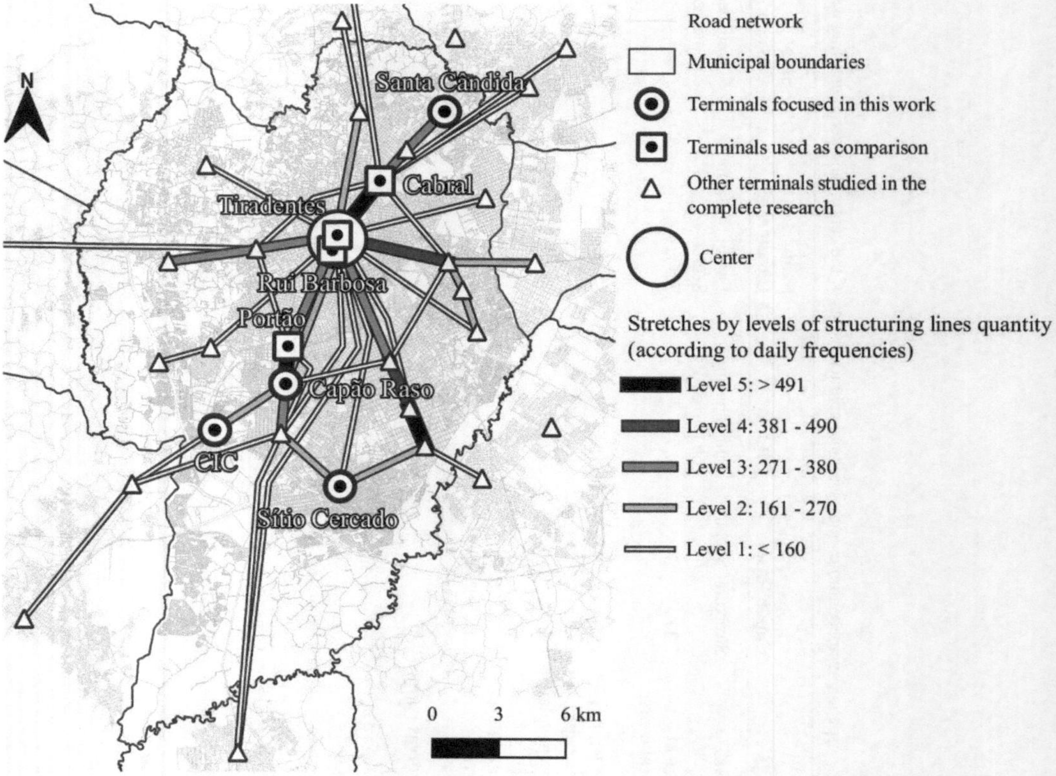

Figure 3. Levels of mobility connection in Curitiba. Source: Authors.

5 RESULTS AND DISCUSSION: POSSIBILITIES OF A NEW MODE, THE MAGLEV

There are possibilities in the situation analyzed. The TOD concept (Calthorpe, 1993) opens a range of options for implementing public policies in terminals' surroundings, combining specific urbanistic tools (Brasil, 2001). Different transport modes can be used in this process. One of them is MagLev Cobra: the magnetic levitation train developed at the Federal University of Rio de Janeiro (UFRJ) (Stephan, 2015). Figure 4 shows the composition of the MagLev vehicle, which was conceived to work in cities already consolidated. Table 2 presents the advantages of speed, costs, and impacts of the mode. To sum up, the conjugation of urbanistic tools with the TOD concept and the implementation of MagLev would constitute a way to intensify the occupation in the areas around the terminals (Nichele, 2018).

Figure 4. MagLev Cobra system showing the levitation. Source: Stephan (2015).

Based on this potential, we developed a prospective for 30 years from now, counting with denser and more varied occupation in the blocks inside the fragments. The foundation is the land use legislation (Curitiba, 2000), and we considered the typologies proportion consisting of 75% of mixed/ residential uses, and the remaining 25% of other types. Figure 5 shows a cartogram with the expansion of the connections and new axes proposed for structuring modes and also exhibits the graphic comparison between the present and the planned densities.

6 FINAL CONSIDERATIONS

Three points stand out from our study that require optimizations. Firstly, the accessibility through time spent in commuting, the connections of the transport system in the peripheral fringes. The MagLev Cobra could be an alternative combined with other policies and urban design tools related to the right to the city. Higher densities in the southeast and southwest directions could be improved based on sustainable urban design and the present mobility connections. Secondly, the necessary discussion with inhabitants to decide, plan, and manage the urban sprawl for the fundamental preservation of water springs and life in the metropolitan area of Curitiba. Finally, an emphasis on the articulation of several modes through urban design and infrastructures capable of combining terminals, cycle paths, bicycle-sharing systems, etc. All these improvements would guarantee different options for users to access destinations, with an adjustment to individual accessibility, travel time, routes, and costs.

Table 2. Modes comparison. Source: Nichele (2018).

System	BRT (Direct)	Light Rail	MagLev Cobra	Train	Metro
Vehicle	Bi-articulated	7 articulated modules	30 articulated modules	Train with 8 cars	Train with 6 cars
Passengers/vehicle	250	420	440	2.264	2.082
Minimum interval at rush hour (minutes)	2.5	3.0	1.5	4.0	2.0
Frequency (vehicles/hour)	24	20	40	15	30
Passengers/hour/direction	6,000	8,400	17,600	33,960	62,460
Average speed (commercial) (km/h)	27.5	24.0	40.0	36.0	40.0
Time of a travel in a 10 km journey (minutes)	55.1	58.8	50.0	54.2	53.5
Execution time (years)	2.5	5.0	5.0	9.0	9.0
Average implementation costs per km (R$)	33.3 million	41.6 million	33 million	148 million	600 million

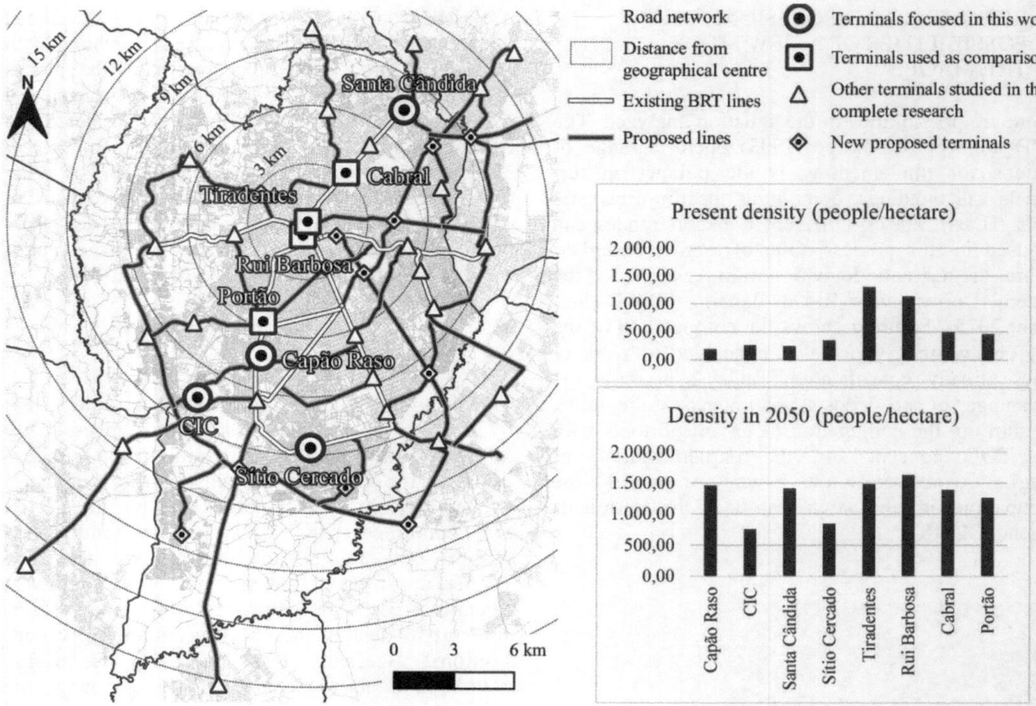

Figure 5. Prospective of transport systems and density of the fragments with the implementation of the MagLev - 2050. Source: Authors.

REFERENCES

Albuquerque, A.F. 2007. A questão habitacional em Curitiba: o enigma da “cidade modelo” (Master’s dissertation, Architecture and Urbanism School, University of São Paulo, São Paulo, Brasil).

Banister, D. 2011. Cities, mobility and climate change. JournalTransport Geography 19: 1538–1546.

Banister, D. 2007. The sustainable mobility paradigm. Transport Policy 15: 73–80.

Brasil 2001. Law n. 10,257, of July 10th 2001. Regulates articles 182 and 183 of Federal Constitution, establishes general guidelines to the urban politics and other providences. Diário Oficial da União, Brasília, Section 1, p. 1.

Brasil 2012. Law n. 12,587, of January 3rd 2012. Institutes the guidelines of the National Urban Moniblity Politics. Diário Oficial da União, Brasília.

Calthorpe, P. 1993. The next American metropolis: ecology, community and the American dream. New York: Princeton Architectural Press.

Carlos, A.F. 2007. O espaço urbano. São Paulo: FFLCH.

Cervero, R., Sarmiento, O.L., Jacoby, E., Gomez, L.F. & Neiman, A. 2009. Influences of Built Environments on Walking and Cycling: Lessons from Bogotá. International Journal of Sustainable Transportation 3(4): 203–226.

Cervero, R. & Dai, D. 2014. BRT TOD: Leveraging Transit Oriented Development with Bus Rapid Transit Investments. Transport Policy 36: 127–138.

Cheng, Y. & Chen, S-Y. 2015. Perceived accessibility, mobility, and connectivity of public transportation systems. Transportation Research Part A 77: 386–403.

Coalition for urban transitions 2019. Climate Emergency, Urban Opportunity: World Resources Institute (WRI) Ross Center for Sustainable Cities and C40 Cities Climate Leadership Group. Retrieved from https://urban transitions.global/urban-opportunity/

COMEC (Coordenação Região Metropolitana Curitiba) 2006. Plano Desenvolvimento Integrado Região Metropolitana Curitiba: PDI. Curitiba: COMEC.

Corrêa, R.L. 1995. Espaço Urbano. São Paulo: Ática.

Curitiba 2000. Law n. 9,800, of January 3rd 2000. Provides on Zoning, Land Use and Occupation in the Municipality of Curitiba. Curitiba: Curitiba City Hall.

Del Rio, V. 1990. Introdução Desenho Urbano no Planejamento Urbano. São Paulo: Pini.

DENATRAN (Departamento Nacional de Trânsito) 2020. Frota Nacional veículos. Estatística. Retrieved from https://infraestrutura.gov.br./component/content/article/115-portal-denatran

Garcez, L.A. 2006. Curitiba: Evolução urbana. Curitiba: Imprensa universitária UFPR.

Harvey, D. 2005. A produção capitalista do espaço. São Paulo: Annablume.

Harvey, D. 2012. O direito à cidade. Lutas Sociais 29 (jul-dec): 73–89.

IPPUC (Instituto Pesquisa Planejamento Curitiba) 2004. Plano Diretor de Curitiba. Curitiba: IPPUC.

IBGE (Instituto Brasileiro Geografia e Estatística) 2019. Contagem Populacional 2019. Retrieved from www.ibge.gov.br

Kohlsdorf, M.E. 1996. A apreensão da forma da cidade. Brasília: UnB.

Lima, C.A. 2012. Ignis Mutat res: cidade, mobilidade, energia (Research project, Department of Architecture and Urbanism, Federal University of Paraná, Curitiba).

Lima, C.A. 2015. Configuração urbana e sistema BRT de Curitiba – Brasil. Investigando a qualidade espacial do entorno de terminais: uma metodologia em construção. In: Seminário Internacional de Investigación em Urbanismo (7), Barcelona. Electronic annals. Retrieved from https://upcommons.upc.edu/handle/2117/80278

Lima, C.A. & Kuzma, R.O. 2017. Curitiba, uma cidade inclusiva? In Congreso Iberoamericano de Suelo Urbano: el suelo en la nova agenda urbana (3), Curitiba. Annals: 1–20.

Lima, C.A., Silva, M.N., Texeira, A.G. & Nichele, H.E. 2018. Incongruências entre moradia e mobilidade. Conjuntos de habitação popular e o sistema BRT em Curitiba. In Congresso Observatório das metrópoles 20 anos (14). Rio de Janeiro. Analls: Eixo 2 Gestão e Governança Urbana – Sessão 2.2 Mobilidade urbana.

Lynch, K. 1997. A imagem da cidade. São Paulo: Martins Fontes.

Moura, R. & Magalhães, M.V. 1996. Leitura do padrão urbanização Paraná nas últimas décadas. Revista Paranaense Desenvolvimento 88: 3–21.

Nichele, H.E. 2018. Plano de Sistema Estrutural de Mobilidade na Metrópole de Curitiba – PR. (Undergraduate Work, Department of Architecture and Urbanism School, Federal University of Paraná, Curitiba, Brasil).

Ojima, R. & Hogan, D. 2009. Mobility, urban sprawl and environmental risks in Brazilian urban agglomerations: challenges for the urban sustainability in a developing country. In: A. de Sherbinin, A. Rahman, A. Barbieri, J. C. Fotso & Y. Zhu (eds.). Urban Population and Environment Dynamics in the Developing World: Case Studies and Lessons Learned. Paris: CICRED.

Pinderhughes, R. 2004. Alternative urban futures. Lanham (USA): Rowman & Littlefield.

Renner, M. 2016. Apoyar um transporte sostenible. In: G. Gardner, T. Prugh & M. Renner. Ciudades sostenibles. Worldwatch Institute: 201–221. Washington, DC: Island Press.

Ruano, M. 1999. Ecourbanismo. Barcelona: Gustavo Gili.

Santos, A., Polidori, M., Peres, O.M. & Saraiva, M. 2017. O lugar dos pobres nas cidades. URBE. Revista Brasileira Gestão Urbana 9(3 sep-dec): 430–442.

Stephan, R.M. 2015. MagLev Cobra: Tecnologia de levitação magnética no Brasil. Ciência Hoje 55 (325): 20–25.

URBS (Urbanização de Curitiba S/A) 2020. Histórico Transporte Coletivo. Retrieved from http://http://urbs.curitiba.pr.gov.br/transporte/historia-transporte

Vasconcellos, E.A. 2010. Transporte Meio Ambiente. São Paulo: Annablume.

Wallace-Wells, D. 2019. El planeta inhóspito. Barcelona: Debate.

Public space and sustainability: Researching urban streets as citizen coexistence spaces

C.A. Duarte & L.B. Albuquerque
Architecture and Urbanism Course, Universidade Católica de Pernambuco, Brazil

ABSTRACT: Currently, large urban centres and, in particular, their historic centres, daily challenge planners and designers in their search for new alternatives and redesigns that bring along solutions that can minimize problems such as insecurity and discomfort in public spaces. Thus, the objective of this research is to analyse indicators that show the socio-spatial qualities of urban streets and the relationship of these urban streets with private spaces in a strategic axis in Recife downtown (PE, Brazil); it aims at developing analytical and integrated planning tools, "capable of recognizing and activating the potential attractor and humanizer of urban streets as levers of a viable, environmentally harmonious and socially just development" (DUARTE et al, 2018). The result of this work also means an integration of teaching, research and extension activities, enabling a broader understanding of the public space as an essential stage for the exercise of citizenship, respect for diversity, and encouragement for human coexistence, even in times of social distancing.

1 INTRODUCTION AND CONCEPTUALIZATION: FOR AN INTEGRATED PLANNING

The whole set of public spaces of a city can be considered the urban life's platform, as they are strategic places of encounter, dialogue, and social exchange. According to Gehl (2015), life between buildings encompasses different activities, involving people exploring a city's common space. These activities are inseparable from private social relations. The process of production of these spaces must be planned to foster collective ownership and human relations, creating safe and comfortable places that positively impact the physical and mental health of citizens. Their design should promote improvements in the relationship between people and the environment, incorporating the needs and identities of the place and considering the various urban elements in an integrated manner, always attentive to social, environmental and economic impacts.

The reality of large urban centres today, and especially their historic centres, daily challenges planners and designers in their search for new redesign alternatives that bring along solutions that minimize the emptying of downtown spaces, and the lack of security and comfort in public spaces, among other evils that the accelerated growth of cities and the overvaluation of the car have catalysed. The reality of the streets in Recife (PE-Brazil) today, as a reflection of these urban problems, makes us question about the link between growing spatial segregation and the feeling of insecurity and discomfort in public spaces. The social representation of "street" still is, for many, just a "circulation space" for motorised vehicles. Why are streets rarely considered as spaces of permanence, of encounter, or of leisure? Do the existing characteristics of these spaces encourage the exercise of citizenship there? How to transform their challenges into opportunities? With the purpose of giving an answer to such questions, this work is especially takes as starting point the integrated analysis methodology proposed by the Plano Centro Cidadão ("Citizen Centre Plan") (Duarte et al, 2018), which, in the light of guiding concepts, uncovers indicators and develops direct tools for a more integrated and applicable urban analysis.

Figure 1. Example of Recife's streets.
Source: The authors, 2019.

DOI: 10.1201/9781003220855-8

1.1 *The Urban Citizen Space*

The concept of Urban Citizen Space is the result of an "integrated planning, where public and private spaces are planned in a joint and complementary way" (Duarte, et al, 2018:20). In such a space, citizenship is taken into account for a human-centred approach. The concept has been built from a research which is based on some reference texts. On one hand, works such as those of Davis (1997) and Caldeira (2000) help to understand the phenomenon of socio-spatial segregation in urban environments. On the other hand, Jacobs (2000) and Gehl (2015) serve as a complement for the research on social relations in the street space. Furthermore, it is important to highlight that the concept of Urban Citizen Space integrates two other concepts: "Citizen Street" and "Urban Architecture" (Duarte, 2014), which also allow to discover tools of urban analysis in an integrated way, at the local scale. The concept of "Citizen Street" is more closely related to the public road space and its elements. The concept of Urban Architecture is especially related to the private space and the relationship of its interfaces and buildings with the street space. These two notions highlight the link between public and private spaces as primordial for the vitality of the city. So, the Urban, Citizen Space can be understood as the union or integration of these two concepts.

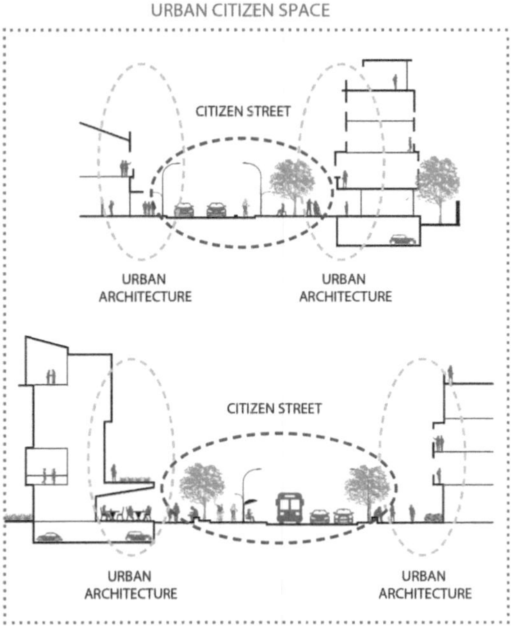

Figure 2. Citizen street + Urban Architecture diagram. Source: (Arcanjo,2018, apud Duarte, et al, 2018)

Supporting the methodology derived from the analysis of the concept of Urban, Citizen Space, we need to clarify what "Urban architecture" is. The term means a building or a set of buildings with the capacity to transmit to the citizen in the public space the feeling of interaction and integration with private spaces and the existing context". (Duarte, 2014, apud Duarte et al, 2018)

This concept, in turn, is referred to in the concept of the "Open Quadra", conceived by the architect Christian de Portzamparc in the 1970s. It is based on the simultaneous consideration of characteristics, such as permeability (physical and visual); proximity (between buildings and streets and between the buildings themselves - density); proportionality (of the buildings with the surroundings and of these with the human scale); variety (of forms and use/user); and viability (urban and economic).

The other guiding concept that supports the Urban Citizen Space is the Citizen Street. It should be understood as "the public street space that articulates places and people, and conceived in an integral way, where buildings (with their forms and uses); mobility (with all means and surfaces of displacement), vegetation (or urban nature) and urban furniture (including infrastructures) are all shown to be planned in a joint and complementary way, aiming above all at the safety and comfort of citizens" (Duarte, 2014, apud Duarte et al, 2018).

In "Rua Cidadã" (Citizen Street), the characterisation of public road spaces is studied by identifying first the "design elements of urban coexistence" (Duarte, 2005 a). These elements are taken into consideration to define urban axes and are deemed to be capable of interfering in the choice of citizens to use or not these spaces. Thus, to be planned in a joint and complementary way, aiming at the safety and comfort of citizens, these elements demand at least 12 essential attributes to be promoted on the roads (Figure 04). The buildings or architectural interfaces facing the street must present visual permeability, variety of uses, and proportionality with the human scale. Mobility must allow for multimodality (or reception of the various modals), areas of permanence and rest and, especially, characteristics that stimulate the deceleration of motorised vehicle traffic. Vegetation or urban nature (its whole variety included) should, whenever possible, be implanted in an expressive and coherent way, in order to allow for a greater shading of public foot pavements, without, however, coming into conflict with the accessibility or lighting for foot pavements. Finally, urban furniture should be thought and planned prioritizing attributes such as ergonomics, unblocked foot pavements, and lighting for pedestrians.

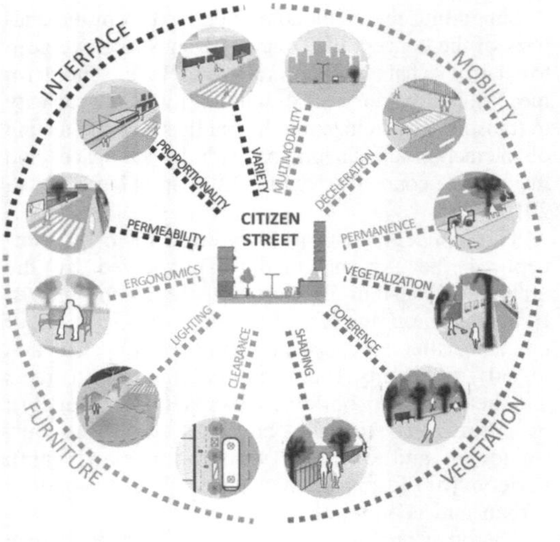

Figure 3. Citizen street diagram.
Source: (Duarte, 2014 b)

1.2 *The Sustainable Urban Space*

The Danish architect and urban planner Jan Gehl reinforces the need to take the human dimension in consideration in city planning, emphasizing the social function of public space as a meeting place, a vital issue in achieving a sustainable, open and democratic society. In his view, urban space should promote safety, comfort, and pleasure for people. Similarly, Jeff Speck, author of "Walking Cities" (2016), reinforces the need for local planning focused on an urban space that promotes safe, comfortable, profitable, and interesting walks.

The desire for a sustainable, humane, safe, and comfortable city is now a contemporary society's unanimous desire. This way, it is certain that implementing an urban space where people's well-being come first is a fundamental step.

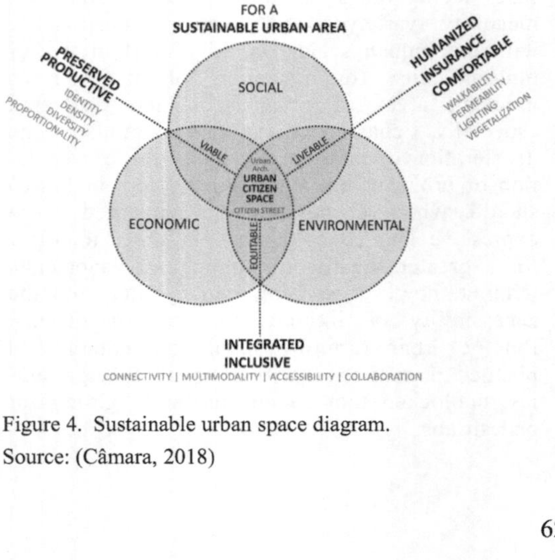

Figure 4. Sustainable urban space diagram.
Source: (Câmara, 2018)

The infographic above clearly sums up how sustainable the implementation of a "Citizen, Urban Space" can be. It is perceived that the integrated consideration of the attributes or properties of "Citizen Street" and "Urban Architecture" can catalyse an equitable, viable and habitable urban development. They may foster the production of urban spaces that can be more preserved and productive, more integrated and inclusive, and even more humanized, safe, and comfortable.

2 METHODOLOGY AND DEVELOPMENT: INQUIRING INTO THE "DEGREE OF CITIZENSHIP".

The methodology proposed for this study is based on results obtained by the author of this article in a previous, collective research (Duarte, et al, 2018) and is the result of an integrated method, where public and private spaces are studied and planned in a complementary way.

As previously mentioned, the Urban Citizen Space has 12 attributes or indicators. Together, they can promote space which are more preserved and productive (with identity, proportionality, density, diversity); more humanized, safe and comfortable (with permeability, walkability, vegetalization, lighting); more integrated; and inclusive (with accessibility, connectivity, multimodality, and collaboration).

Thus, the adopted methodology to evaluate the "degree of citizenship" of a given urban space is carried out with the aid of a diagram (Figure 05) that indicates the degree of intensity that each urban quality has in the territory. It is illustrated as a kind of daisy which petals represent the qualities under analysis, indicating a greater or lesser degree of intensity of each of these qualities in different communities (Duarte et al, 2018, p. 24). Thus, a whole petal would indicate a high or more intense degree, while a medium petal would indicate a medium degree of intensity in the area, while a short petal would indicate a low or weak degree. The analysis of these 12 indicators then makes it possible to establish the necessary factors for a sustainable urban development. Such analysis can support the management and planning of cities that aim at combining their growth and productivity with habitability, preservation of the environment (natural and built), social inclusion, and public health.

The proposed methodology for this study is based on results obtained by the author of this article in previous collective research (Duarte, et al, 2018) and is the result of an integrated method, where public and private spaces are studied and planned in a complementary way.

That said, for a better understanding of the socio-spatial qualities, the research applied the method of analysis on a road axis, taking as references previous studies developed by the Citizen Center Plan (2014-2018), the Model Office of the Architecture and Urbanism Course in 2018, and a Scientific Initiation

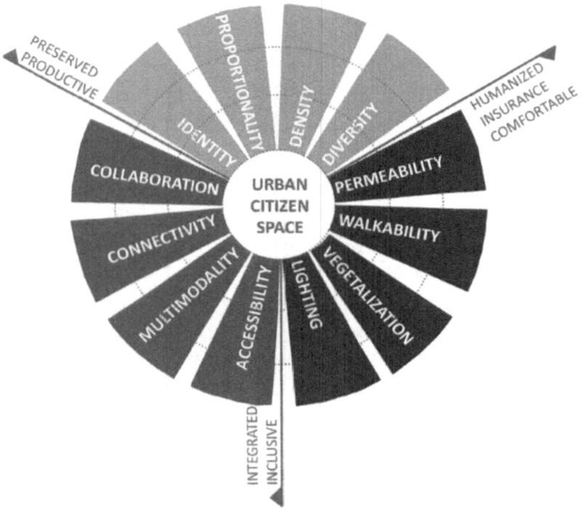

Figure 5. Urban Citizen Space diagram.
Source: (Duarte et al, 2018)

Project carried out jointly by the authors of this article in 2019, all in the same institution (UNICAP). These studies allowed the identification of a strategic route in the downtown area for the displacement on foot by students, from the mapping of educational institutions distributed along this axis.

The so-called Rota educacional do Recife ("Recife Educational Route") (Figure 06) runs through 5 city districts: Bairro do Recife, Santo Antônio, Santo Amaro, Soledade, and Boa Vista, impacting, within a 500m radius to the north and south, thousands of students in 90 educational institutions, subdivided into basic education, technical education, higher education, and preparatory courses. The methodological contour aims at providing an integrated characterization of the studied site. It is presumed that, from the example applied to the route in question, it could be used for the analysis of different locations. From the diagnosis of the attributes of this territory it is possible to establish guidelines that could feed more coherent, complete, and necessary policies and projects.

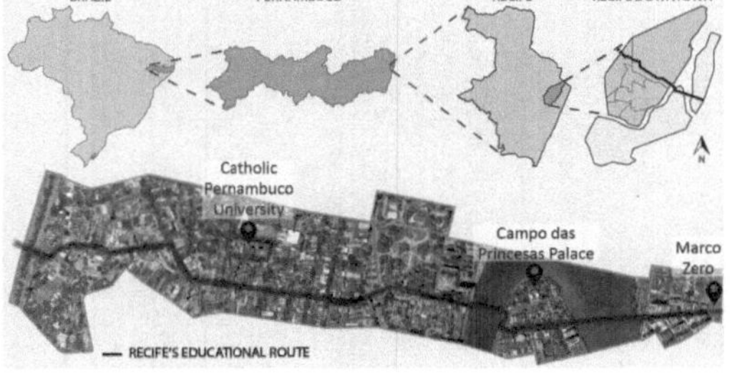

Figure 6. Location of the studied area: Recife's Educational Route.
Source: (Albuquerque,L., Duarte, C., 2019)

What follows will be an explanation of each of the 12 socio-spatial attributes of the Urban Citizen Space, highlighting its concept, its methodology of analysis, and its graphic representation, measuring qualitatively the various indicators analysed in the "Recife Educational Route". The evaluation of the socio-spatial quality of this urban axis and its attributes aims at identifying, as already highlighted, a possible "degree of citizenship" of the space, revealing susceptible parameters (or socio-spatial properties) to be included in urban policies and norms, highlighting the importance of creating specific regulations for urban roads and their elements, in different contexts.

2.1 *For a preserved and productive urban space*

To conquer an urban space so it can be preserved and productive, it is required to promote 4 specific attributes, which are defined and explained below:

The **identity** of a place may have its strength associated with the visual harmony of its built space and with the ease of recognition or identification of this built landscape by the place's users and visitors. "In the analysis, the presence of similar architectural types is adopted as defining the identity of the area. The greater the recurrence of the same type or set of types, the greater the degree of typological identity of an area" (Duarte et. al., 2018: 24). As an example, in figure 07, the studied area identity is strong, since the area is predominantly represented by "sobrados" (ancient houses), and "casas de porta e janela" (door-and-window houses), as well as (in the occupied area) by large equipment, with larger buildings.

"The **proportionality** of an area can be measured from the balance between the buildings templates. Thus, the more the built masses compose a skyline marked by a harmonic relationship, the greater can be the proportionality of the site" (Duarte et. al., 2018: 26). In figure 07 it is noted that 47.59% of its buildings have just one floor, with little differentiation in their altimetry profiles when we find that 82.42% of the buildings in the area have at most two floors, characterizing the area with a high proportionality.

Population and constructive density are attributes that must be carefully planned and controlled to promote flows of people, social dynamism, and sustainable growth. A balance must always be sought between population and constructive density to achieve day-and-night liveliness and security in a given territory. In this methodology, the focus is on population density, which will be considered high when it is greater than 250 inhabitants per hectare (Leite, 2012). In the territory analysed, there is a low population density, for 88% of its territory show a density between 0.00 - 109.43 inhab./ha (Figure 08).

The diversity of uses in cities promotes a variety of choices and contributes to greater social inclusion. In this research, the diversity of a place is measured by considering a quantitative balance between housing and other complementary activities. As mentioned above, housing has a great importance for the

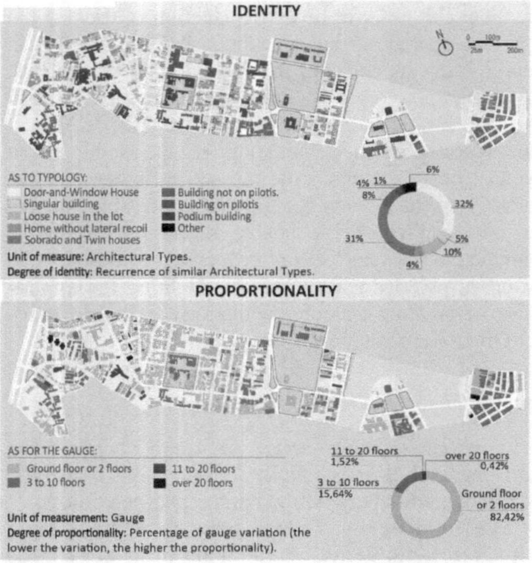

Figure 7. Identity analysis and Proportionality of the Educational Route .

Source: (Albuquerque,L., Duarte, C., 2019)

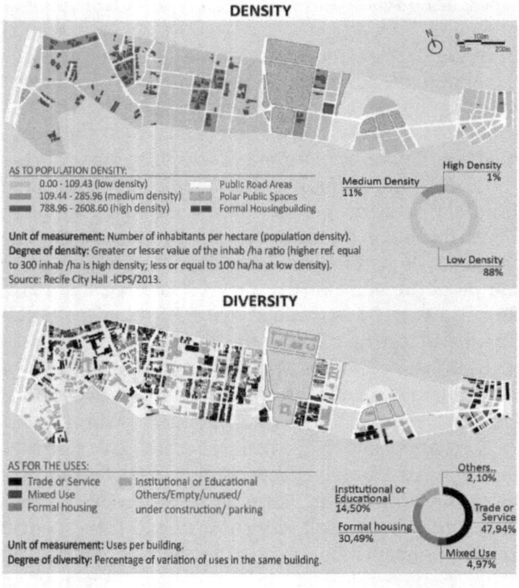

Figure 8. Density analysis and Diversity of the Educational Route.

Source: (Albuquerque,L., Duarte, C., 2019)

sustainability of the area, contributing to security, and economic, social and cultural development. In the research, the area analysed can be considered reasonably diverse since it has about 35% of buildings with housing or mixed use, almost 50% of commercial and service uses, and the other uses have similar values among themselves (Figure 08).

2.2 *For a humanized, safe and comfortable urban space*

In the search for a humanized, safe and comfortable urban space, it is necessary to promote four other socio-spatial attributes:

In this study, permeability consists of the visual interaction between public and private spaces; it is therefore the visual relationship between the pedestrian and the interior of the lot (or building). In this way, the attribute is gauged according to the relationship between the road and the neighbouring lot, that is, the visibility relationship between them. The greater the visual permeability of the architectural interfaces, the greater the degree of permeability of the area. In the example analysed in figure 09, the area presents a medium degree of permeability, where 64% of the ground floor is productive, and 65% of the interfaces show a high visual permeability.

To stimulate the walkability attribute, Speck (2016) highlights 4 main conditions for walking: to be beneficial, safe, comfortable, and interesting. Walkability refers to the urban quality of walking in an accessible, comfortable and safe way. In the research, walkability is evaluated by the "quantity of linear meters of pavements with simultaneously adequate conservation, continuity, and width, as well as with high visual permeability of its architectural

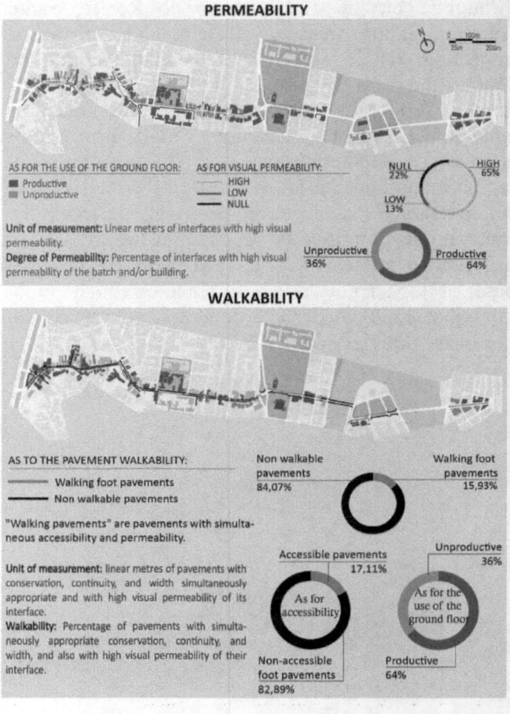

Figure 9. _ Permeability and walkability analysis of Educational Route.

Source: (Albuquerque,L., Duarte, C., 2019)

interface" (Duarte, et. al., 2018:34). In figure 09, it is shown that only 15.93% of the foot pavements are considered walkable.

Vegetalization is understood as the presence of nature and vegetation, making spaces more attractive and comfortable for people, and increasing the interest for more active and healthy modes of displacement. "The probability of walking is three times higher on routes for pedestrians with vegetation. The coverage that adult trees provide can further stimulate everyday outdoor activities" (Farr, 2013, p.37). The shading of streets and pavements, generated by treetops, is the object of this study. In this way, the degree of vegetalization of the area is measured by the linear metres of shaded pavements (10m is counted per canopy of medium or large trees). Figure 10 shows how to apply the methodology. This way, it was found that the area has a medium degree of vegetalization, with 51.41% of shaded foot pavements.

Public **lighting,** especially when considering the human scale, is of paramount importance for walking, as it conveys a sense of security to the pedestrian and encourages activities and commerce at night, bringing vitality to the area. This notion includes both security and the enhancement of the built heritage, as well as the promotion of activities and social exchanges at night. The unit used to measure this characteristic is the percentage of linear metres of directly lit foot pavements (DUARTE, et. al., 2018). It must be assessed whether the lighting meets the pedestrian's scale and whether it has physical barriers that prevent it from functioning properly, as is the case with intersections with treetops. In the area given as an example there are no posts at the pedestrian scale, all of them are turned into the motoring roads, which results in a low degree of humanized lighting, with only 31.69% of the foot pavements directly illuminated (Figure 10).

2.3 *For an integrated and inclusive urban space*

The properties described below make it possible to achieve an integrated and inclusive urban space:

Accessibility is presented in multiple dimensions, including those of attitudinal, technological, communicational and physical nature, among others. The methodology presented here "studies the physical accessibility on the foot pavements as an attribute that enables people to move around in an easy, safe and comfortable way" (Duarte, et al, 2018: 28). To assess its degree, the continuity of the foot pavements, their state of conservation and width are simultaneously analysed. The accessibility of the area under study is visibly precarious, presenting a percentage of 82.89% of non-accessible foot pavements (Figure 11).

Multimodality is understood as the capacity of a road to receive, safely and comfortably, as many modes of transportation as possible, in coherence with its road hierarchy. In the research, the units of measurement are the linear meters of multimodal streets. In

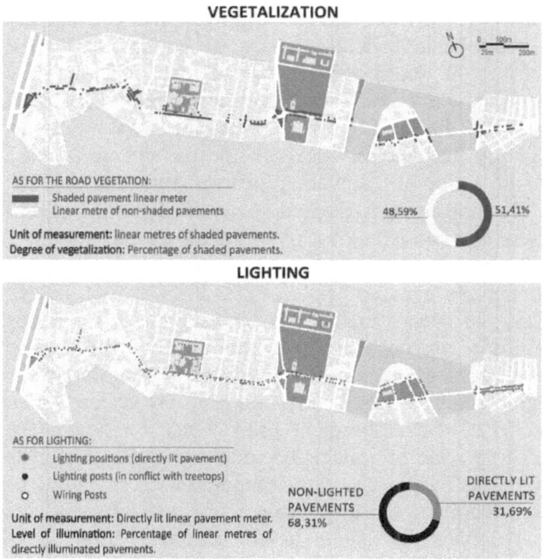

Figure 10. Vegetalization and Lighting analysis of Educational Route.

Source: (Albuquerque,L., Duarte, C., 2019)

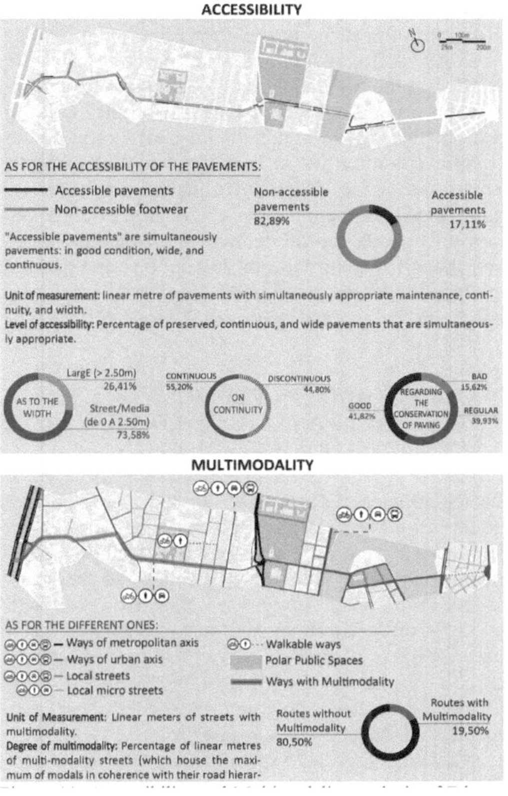

Figure 11. Accessibility and Multimodality analysis of Educational Route.

Source: (Albuquerque,L., Duarte, C., 2019)

this way, the more linear metres of street consistently accommodate different means of transportation, the greater the multimodality of the area. The territory analysed has a low degree of multimodality because it has only 19.5% of its roads with this attribute, highlighting the precariousness of the pavements and the insufficiency of the cycling infrastructure (Figure 11)

"Connectivity is defined as the attribute that allows users and residents of a given urban area to articulate or connect with other areas around them through streets and avenues" (DUARTE, 2018, p. 54). In this way, the proper connection between neighbourhoods within a city helps in urban mobility and the ease of movement of people and products. For the analysis, the hierarchy of road connections at city scales is then the unit of measurement used. Thus, the wider the existing scale of connection in the area, the greater its connectivity, since it is the breadth of the hierarchy of connection of existing roads with their surroundings that defines the degree of connectivity of the place. Figure 12 shows the example of the area studied, with the presence of large metropolitan and urban axes, which produces a high degree of connectivity of the area.

Stimulating the **collaboration** of the community and those involved is one of the 10 steps towards sustainable urbanism (Farr, 2012). Such collaboration is of utmost importance for urban planning, as it enables different visions and preferences to be included in the project, so that local needs are understood on a neighbourhood, block or single building scale. In this research, the spatial distribution of the different communities in the area is initially identified to assess the degree of collaboration (Duarte, et. al., 2018). In the example, the degree of collaboration of the area can be defined as reasonable, since even though it has important institutions and actions developed by them, there is the absence of local residents' and traders' associations (Figure 12).

3 RESULTS AND CONCLUSION: INTEGRATED ANALYSIS AND FORWARD-LOOKING VISION

The methodological procedure for the diagnosis presented in this work, as already mentioned, is based on the concept of the Citizen, Urban Space, allowing the integrated analysis of the urban qualities that characterize the space under study. The main axis analysed, called "Recife's Educational Route ", has great relevance for the promotion of active mobility and humanization of the central area of Recife (PE, Brazil), for it attracts and irradiates thousands of pedestrians daily, especially students of various age groups. In summary, by evaluating the socio-spatial quality by means of thwte petal diagram, it is possible to analyse each aspect of the area studied, both individually and in an integrated manner, classifying them with different intensities from the qualitative and quantitative interpretation of the data collected (Figure 13). The tool allows, simultaneously, a specific and systemic look at the urban space.

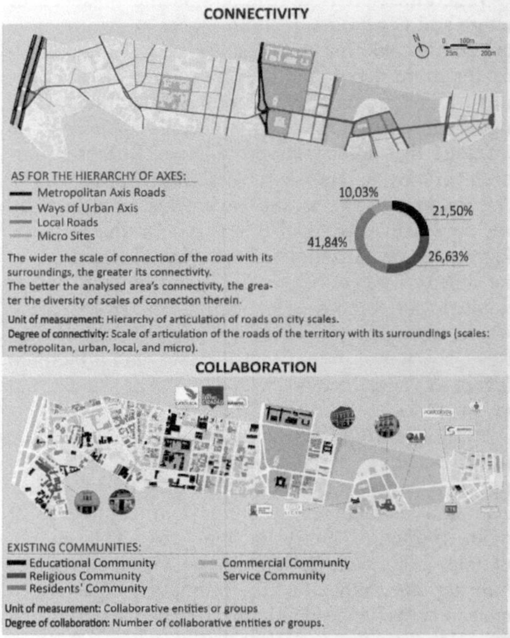

Figure 12. Connectivity and Collaboration analysis of Educational Route.

Source: (Albuquerque,L., Duarte, C., 2019)

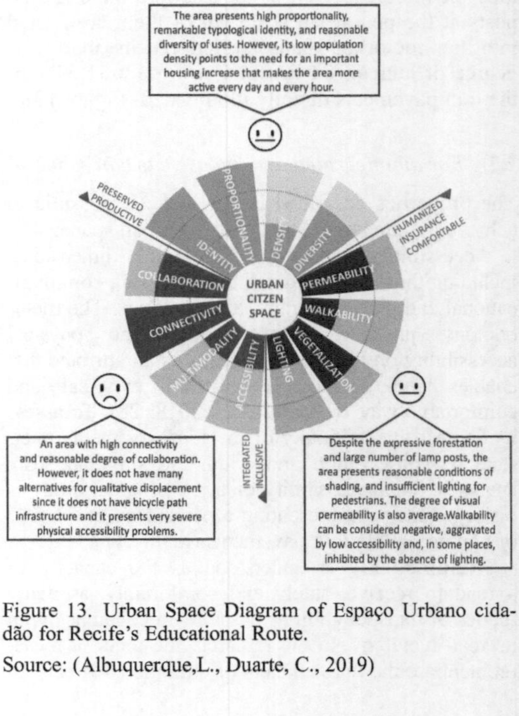

Figure 13. Urban Space Diagram of Espaço Urbano cidadão for Recife's Educational Route.

Source: (Albuquerque,L., Duarte, C., 2019)

As demonstrated, many qualities are still lacking to conquer a "Citizen, Urban Space" in Recife's Educational Route. Among the most urgent of these qualities, we can highlight population density, walkability, accessibility, multimodality, and humanized lighting. Once all the indicators of the route have been analysed, its potential for transformation can be perceived, and possible urban design strategies that could promote greater appropriation and citizen coexistence are more easily identified.

Finally, the hypothesis of this research can be validated when it is observed that, from an integrated perspective, and analysing attributes related to public and private space, it is possible to offer urban management a range of indicators that can allow a more efficient, humane, healthy and sustainable decision-making. The urgency of urban life in the open air and of healthy social distancing further reinforces the importance of stimulating citizen coexistence in contemporary streets.

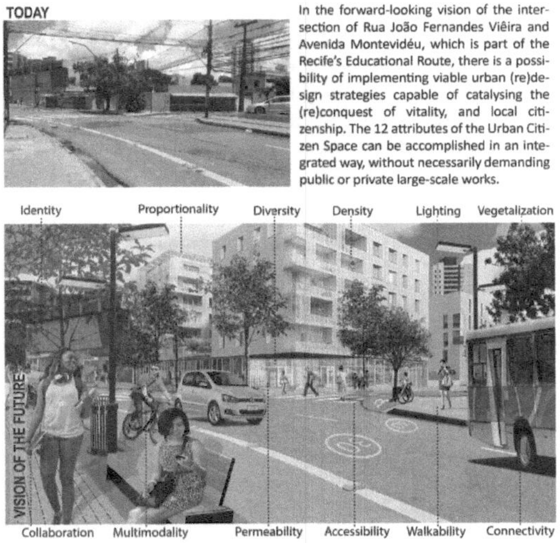

Figure 14. Forward-looking view of a street corner of Recife's Educational Route.

Source: The authors, 2019.

REFERENCES

Albuquerque, L., Duarte, C. (2019). Rua e Coexistência Cidadã: Análise do desenho e uso dos espaços públicos viários em rotas estratégicas do centro do Recife. 21ª Jornada de Iniciação Científica, UNICAP, 18-21 novembro (em papel).

Câmara, A. et al (2018). Plano Centro Cidadão: Diretrizes Urbano-Arquitetônicas para o Centro Expandido Continental do Recife. Vol. 03. Recife: Fasa

Duarte, C. (2005 a). Le dessin de la coexistence: L'impact des politiques d'aménagement urbain sur l'usage de l'espace public - Axes Urbains d'une ville brésilienne - Recife. Master Paris: Université Paris 1 Panthéon - Sorbonne. — (2005 b): A Rua Cidadã e os elementos da coexistência. Metodologia de Urbanismo. Recife: Curso de Arquitetura e Urbanismo da Universidade Católica de Pernambuco.

Duarte, C. et. al. (2018). Plano Centro Cidadão: Estudo preliminar de desenho urbano do setor de ensino e conhecimento do Centro Continental do Recife. Vol. 04. Recife: Fasa.

Farr, D. (2013). Urbanismo Sustentável: Desenho Urbano com a Natureza. Porto Alegre: Bookman Editora LTDA.

GDCI/NACTO. (2018). Guia global de desenho de ruas. 2ª Ed. São Paulo: Senac.

Gehl, J. (2015). Cidades para pessoas. São Paulo: Perspectiva S.A.

Portzamparc, C. (1996) L'Ilôt Ouvert- The Open Block. Paris: Semapa-Ante Prima-AAM Editions. 2004.

Research Tracks in Urbanism: Dynamics, Planning and Design in Contemporary Urban Territories – Allegri et al (eds)

Public space and social innovation – The project "Reinventar a Roda"

J. Coimbra
Master degree in Regional and Urban Planning, Department of Social, Political and Territorial Sciences, University of Aveiro, Portugal

P. Silva & B. Bittencourt
Department of Social, Political and Territorial Sciences, University of Aveiro, Portugal

ABSTRACT: This paper approaches the project "Reinventar a Roda" (Reinventing the Wheel), which involves the concepts of public space and social innovation and aims to build a digital platform that allows the sharing of wheelchairs across services spread throughout the city, so that they can be requested in an easy and accessible way by citizens with mobility limitations. The central objective of this project is to generate conditions of urban mobility for short distances, facilitating displacements in the public space. The wheelchairs used for this purpose are the result of a recycling process of old or damaged wheelchairs from institutions or individuals who are willing to collaborate with this project. In this way, the products will have a circular use.

1 INTRODUCTION

This paper presents an innovative concept of accessibility and sustainable mobility – the creation of a digital platform that allows the sharing of wheelchairs in different services in the city, with the purpose of serving people with reduced mobility. This concept is innovative for three reasons. First, due to the idea of creating a wheelchair sharing platform, based on the recycling of disused material. Second, for generating conditions of urban mobility for short distances. Third, for the interface and complementarity it generates between the public space, social innovation and circular economy with the aim of solving a problem of accessibility and sustainable mobility.

Data from the latest census in Portugal (INE, 2011) reveal that the lack of mobility is a problem that affects people of all ages. Namely, walking is the main difficulty for the elderly, reaching about 35%. The National Statistics Institute (INE, 2017) states that the population is very aged and in the city of Aveiro the aging rate, in 2016, was 137.6%. Other studies corroborate these figures and, according to an OECD/OESPS report (2019), in developed countries such as Portugal, the average life expectancy has been increasing and, with it, the number of elderly people and the chances of having mobility limitations.

Although it is known that the aging of the population combined with the country's budgetary reality and the low purchasing power of most Portuguese (OECD/OESPS, 2019) has made the practice of reusing support products necessary, little is known about the reuse and the destination of wheelchairs. What happens to wheelchairs when they no longer fulfill their intended purposes? How would a shared wheelchair system benefit the public space? Is the disabled person interested in using a shared wheelchair? The project "Reinventar a Roda" (Reinventing the Wheel) aims to develop a digital platform to promote access to wheelchairs in a shared and easy way in Aveiro, avoiding the constant acquisition of new products. It is assumed that there are disused wheelchairs in institutions and private homes, which no longer fulfill the purpose for which they are intended. Therefore, it is possible to generate an environment of donation, of this material in disuse, so that people who find it difficult to walk can use the wheelchairs available in the city.

In the public space, "Reinventar a Roda" is a social experience that triggers social innovation and establishes a direct relationship with the concept of circular economy, where resources are used in a more sustainable way. The project proposal will contribute to "closing the life cycle" of a wheelchair through the recycling and reuse of the disused product. Public space, social innovation, and circular economy are three concepts that are interconnected and complement each other in order to present a solution of accessibility and sustainable mobility.

DOI: 10.1201/9781003220855-9

2 STATE OF THE ART

2.1 *Public space*

For many reasons, the public space has been at the center of worldwide attention, as far as urbanism is concerned, rehearsing incremental forms (Kamalipour & Dovey, 2020) or discussing their production contexts (Rossini & Bianchi, 2020). The focus has been as much in quantitative terms, because many settlements have their formation both before and after modernist standards, as in qualitative terms, because many of the existing spaces, regardless of their quantity, are the target of criticism as to the experience and conviviality they encourage (Piyapong et al., 2019; Xu et al., 2019).

Even in those cases in which the physical form corresponds to the parameterized form, there are gaps and insufficiencies in the relationship between the object (the public space) and its user, the citizen (Torabi et al., 2020). One of these reasons is that the design needs to be adapted to changes in terms of demographic dynamics (Pots & Yee, 2019). In fact, a rapid rate of urbanization, marginal construction processes, and the inadequacy of old areas generally constitute barriers to the enjoyment of an increasingly diverse population (Suminski et al., 2012), who are older, living even later with any type of disability, generating loss of autonomy (Haupt & Skalna, 2020).

The simple idea of "leaving no one behind" was the way UN-HABITAT took on the challenge of designing an urban agenda for 2030 (United Nations, 2017). This agenda is embodied in 17 objectives for sustainable development, in which urban spaces play an important role, given the weight that the world population living in urban settlements already has (more than half) and is expected to have in the future (exceeding 70%). The objective for sustainable development number 11 – sustainable cities and communities – explicitly addresses the context of inclusion, establishing the need to create more inclusive cities.

In a more specific line, the European Urban Agenda reinforces the issue of inclusive public spaces by relating urban mobility to soft modes. In this topic of the European Urban Agenda, accessibility appears as a key element to introduce the issue of inclusion. It is explicitly stated that accessibility should cover people with disabilities, the elderly and children (European Union, 2016).

The National Program for Spatial Planning Policies integrates the concerns that these two agendas (the New Urban Agenda, accompanied by sustainable development objectives and the European Urban Agenda) contain regarding the issues of inclusion and sustainable mobility. It advocates the polarization of the territory, in compact urban areas and a better public transport service. It also looks at public space from the perspective of promoting urban rehabilitation and qualifying the urban environment (Act 99/2019).

Many interventions for the rehabilitation of public spaces have taken these aspects into account, covering from some spaces with centuries of existence to others of very recent genesis. Funds from the European Union have focused on these actions, either through the constitution of Urban Rehabilitation Areas (provided for in the Legal Regime for Urban Rehabilitation, in its most recent version published in Decree Law no. 66/2019), or through the design of Sustainable Urban Mobility Action Plans.

In this context, there is a general improvement in public spaces, with the abolition of architectural barriers and impact on accessibility and maintenance of mobility of the elderly (Haupt, Skalna, 2020), dynamization of the local economy and involvement of the populations (in a context where the concept of placemaking is familiar) corresponding to dominant groups in society (Silva, 2016; Short & Martinez, 2020). However, these continue to exclude those who depend on equipment such as the wheelchair (Bonehill, von Benzon, & Shaw, 2020) and even other people to move around and enjoy the "new" public space that is being created and which is accompanied by multiple other forms of mobility (Jiménez, De La Fuente, & Hernández-Galán, 2018).

"Reinventar a Roda" has the ambition to respond to this aim: to increase inclusion at the level of the public space and, simultaneously, generate dynamics of social innovation that respond to this immediate need, but that also respond to other long-term needs. By bringing together agents with this objective, it will be possible to contribute to the development of networks that, in the future, can respond to other needs, with an impact in the identification of new values in the public space. The long term of this process, therefore, may have a circularity component, which can be evaluated/identified in a more advanced moment. It is about social innovation and the role it can play in this context, which the next section of this paper deals with.

2.2 *Social innovation*

Definitions of social innovation are abundant, and because there is no distinct concept, we easily engage in a debate about their meanings and variations. We adopt a simple definition here: social innovation can be understood as a rupture in the way of doing things, a new element in each context that is sufficient to resolve an uncomfortable situation (Bittencourt, 2016). Genuine social innovation is changing the societal system – it permanently changes the behaviors and designs of structures that previously gave rise to social, cultural, economic, and environmental challenges. In a simple way, we can say that social innovation is an idea that works well for the public. It refers to new ideas that solve real problems for the benefit of people and the planet.

This definition highlights a central feature of social innovation: it represents continuous doing, usually presented as a creative response to problems not solved by the market or the State. As Hulgar and Ferreira (2019: 26) refer, "social innovation offers a space for interaction between government and society".

Taylor (1970: 70) notes that the central idea of social innovation is revealed through "new ways of doing things". In the author's understanding, actions are improved to respond to the explicit social needs through new and better ways of human integration, resulting in unprecedented social inventions that corroborate the quality of life of society. In this line of reasoning, social innovation can therefore be dealing with the wellbeing of individuals in the public space. The elements of this wellbeing are linked to their quality of life and social activity (Nyseth & Hamdouch, 2019).

Deep approaches to social innovation require individuals with decisive ideas that bring about social change (Mulgan et al., 2007). In this context, we understand that one of the goals of social innovation is to change the way the actor interacts with the public space. For social innovation to exist in a public space, we need to combine ingenuity and a diversified initiative of innovative actors from the different sectors that make up society (public, private and third sectors) alongside users and communities, to find new solutions to emerging economic and social problems. In other words, regardless of whether they are ideas, performance models or products and services, social innovation must be the purpose of satisfying the needs of a population. Furthermore, it should help solve some of its problems through a dynamic of transformation of social relations (Moulaert et al., 2013), which implies the revocation of previous practices and relationships of power and/or economic, social, and environmental dependence, from the actors of the community in question.

In the continuation that social innovation directed to the public space is the innovation that causes social change for social and public wellbeing, this can be understood as a broader innovation for presenting solutions to problems not perceived by the public sector. In other words, social innovation brings about social change and/or transformation of social relations to the extent that it achieves not only the objectives of the actors involved, but also of the collective, achieving that the results become new social practices generally assumed and reproduced. Bearing in mind the referred content, we can cite as an example of social change and/or transformation of social relations the creation of a digital platform that allows the sharing of wheelchairs in different services spread around the city. The character of this innovation expands the social component by proposing the recycling of wheelchairs in disuse, that is, the reuse of the product with the purpose of generating urban mobility for short distances, combining this idea with innovative practices of accessibility and sustainable mobility. In this context, social innovation establishes a close relationship with the concept of circular economy, which is a strategic concept based on the reduction, reuse, recovering and recycling of materials and energy (European Commission, 2015).

However, we must be aware that there is an important relationship between the actors involved and the public space, social innovation, and the circular economy, which could be explored more effectively to respond to the main societal challenges, namely accessibility and sustainable mobility. With the project "Reinventar a Roda", the intention and desire to promote social innovation involves social change and the transformation of social relationships in the public space. It makes it clear that the sense of desire for social change may not be univocal; otherwise, not all actors in the community will benefit equally from the transformation of the proposed and/or achieved social relationships. This is an aspect that we should also consider when we approach the conceptual commitment of the circular economy.

Social innovation has been an object of interest, namely by public intervention policies. For at least two decades, in the European Union, several programs have included social innovation as a strategic policy to facilitate interventions in the public space, urban regeneration, urban planning, public procurement, etc. (Hulgar & Ferreira, 2019). Despite this finding, reporting to the public space and social innovation, it inspires the strengthening of society affected by the promotion of new social relationships strengthened in solidarity, participation, and civic cooperation. "Reinventar a Roda" is a social innovation inspired by the desire to meet the social needs of a community neglected by traditional ways of using the public space. The increase in participation of social actors and the collaborative relations between them are understood as the resources capable of leading to the desired social change.

In "Reinventar a Roda", we can say that the relationship of the public space, social innovation and circular economy occurs through innovative initiatives that lead to greater civic participation and cooperation between social actors and focus on valuing the people who use the public space. Social actors have participated in the process since the initial phase: donation of damaged wheelchairs to be recovered and reused; encouraging the circular economy cycle: reduce, reuse, recover and recycle (Geissdoerfer et al., 2017). It is in this sense that social innovation must be a way of creating social value in a concrete community. The results of its initiatives extend beyond the objective of generating wellbeing for a specific part of society's actors and must ensure the inclusion of society.

It is considered that, to face societal challenges in the scope of accessibility and sustainable mobility, a social experience is needed – the sharing of wheelchairs – that originates social innovation and that links actions across the field and supports full exploitation potential, both locally and globally. While there remains a growing interest and enthusiasm for social innovation and the circular economy as paths that fight societal challenges globally, considerable variation also persists as different countries and regions have adopted social innovation and the circular economy in the public space. Thus, a political and social manifest is in place to build an inclusive and interconnected society that helps create a more favorable environment for social innovations to emerge in the public space, to grow and develop globally, across all cities. It is this interconnection between social innovation and the circular economy that will be addressed in the following section.

2.3 *Circular economy*

The European Commission (2015) defines circular economy as a model of production and consumption that involves the reuse, repair, reform and recycling of existing materials and products to keep these in the economy whenever is possible. A circular economy implies that the waste itself becomes a resource, thereby minimizing the actual amount of waste. In general, it opposes a traditional linear economic model, based on a take-make-consume or consume-throw away pattern (European Commission, 2014). What was considered waste can be turned into a resource. This is one of the foundations of the circular economy.

Circular economy, according to McDonough and Braungart (2010), in practice, implies reducing waste to a minimum. When a product reaches the end of its useful life, its materials are kept economical whenever is possible. These can be used productively repeatedly, thus, creating more value (Hawken, Lovins, & Lovins, 1999; Lifset & Graedel, 2002). This is a departure from the economic mainstream, based on a take-make-consume or consume-throw away pattern (European Parliament, 2014). This model depends on large quantities of cheap and easily accessible materials and energy. In addition, part of this model is planned obsolescence, when a product is designed to have a limited shelf life to encourage consumers to buy it again (Lifset & Graedel, 2002).

The interconnection established between social innovation and the circular economy is based on a similar principle between them, since the circular economy propagates social innovation while still conceiving the idea of a product. Therefore, since social innovation is an innovation that is social from the beginning of the action to the end (Bepa, 2011), we note that the circular economy describes a world in which reuse through repair, reconditioning, and reform refers predominantly to a social and economic model.

3 METHODOLOGY

The methodology applied is exploratory and empirical. Based on a master's project with the same name, practical advances were made in the application of the approached concept that talks about generating a shared system of wheelchairs, fed by disused materials from institutions. Legislation related to the theme of accessibility was also a target of reflection, as well as empirical material associated with the themes of public space and social innovation.

The process was guided by the construction of research questions based on hypotheses. Two of the hypotheses were validated through contact with institutions belonging to the Local Council for Social Action of Aveiro (CLASA network), a group of people with motor disabilities, their families and an institution in Valladolid, Spain.

We found that there are effectively disused wheelchairs in the CLASA network's institutions, however, in some cases, they cannot be donated because they do not belong directly to them. Other institutions claim that the project may have their wheelchairs when the equipment is no longer in use. Initially, telephone contacts with institutions were analysed, and later questionnaires were applied, which allowed for the collection of institutional feedback.

A campaign for the collection of wheelchairs was disseminated through social networks and the local newspaper of Aveiro, speeding up the donation of the first two. These did not come from the CLASA network institutions, but from two nearby institutions. The first was a higher education institution that, in the need to provide better equipment to its audience, felt the need to dispose of a chair that it had been using for about 10 years, to support students, teachers and researchers. The second was a social institution, which donated a vacant wheelchair and some parts of others, including wheels, brakes and footrests, as one of the employees considered the equipment relevant and useful in some future context.

With the reception of wheelchairs and the application of questionnaires, the first two hypotheses were validated. These hypotheses were reinforced by a contact with the ASPAYM, an institution that welcomes people with mobility problems located in Valladolid, Spain. According to reports, when a person with motor limitations dies, wheelchairs are left there by their family members.

The last hypothesis related to the interest of the disabled citizen, was validated through some informal conversations with people with mobility difficulties, their families, and permanent users of wheelchairs. We already had the validation of the hypotheses when we received the collaboration of the university to study solutions for the chairs that were collected.

In general, all groups consider the existence of this form of shared mobility relevant because they share the opinion that, without the proper resources, the most frequent option of the person with mobility problems is not to participate in many of the activities that occur in their social circles. The analysis of the theoretical material allowed to reinforce the relevance of this idea and contributed to its development.

The project under study is involved with the concept of public space since it will be implemented therein, and with the concept of social innovation, because there is a mobility solution at the city level that is not yet available in the market or by the State to solve the problem of lack of mobility for people with mobility limitations.

4 FINAL NOTES

In addressing the public space in connection with social innovation in this paper, we intended to highlight the unique characteristics of the current project – "Reinventar a Roda" (Coimbra, 2019) –, which consists in creating a system for reusing and sharing wheelchairs, so that in cities this support product that helps mobility becomes easy and accessible. The study revealed that it is possible to implement a shared wheelchair system to benefit from the public space.

The project, under development, intends to operationalize the recycling and reuse of wheelchairs, with the immediate objective of extending the life of this equipment. In addition to this immediate objective, there are several other possible results of this project. In terms of social innovation, encouraging interaction between a group of stakeholders who do not have the habit of working together (institutions that provide support to people with mobility limitations, users, companies, and individuals with know-how in repair). This conquered habit of interacting and cooperating is a resource with enormous potential to make the project successful, but not only that, it may be the key to developing other levels of cooperation in other contexts.

In terms of technological issues related to wheelchairs, we based the equipment recycling process on the reuse of parts and other materials, taking advantage of technologies and practices like bicycle repair. It is expected that innovative products will emerge from this process. By now, the project counts with the development of solutions by Mechanical Engineering students of the University of Aveiro.

As this project has the consequence of putting more disabled citizens on the street, there is a huge reflection on the necessary improvements in the public space. For their action and use of these spaces, these citizens will be in a privileged position to evaluate it and become indispensable partners in future mobility projects carried out by institutions with responsibility in the area.

The "Reinventar a Roda" proposal joins the public space and becomes significant in modern society by including the needs of people with disabilities in the plan of the built environment, facilities and urban services, to improve the quality of life of all and, thus, support community integration. The public space can have a positive impact on the individual's quality of life, attributed to the increase in wellbeing opportunities (Besters, Marrades, & Kahne, 2019). However, the built public space and the access to essential facilities and services, including mobility, are not always compatible with the needs of vulnerable groups, including the elderly and individuals with disabilities. The awareness of the guidelines in the strategic planning of the public space must also be improved to achieve the sustainable development of an accessible mobility system.

REFERENCES

BEPA, Bureau of European Policy Advisers (2011). Empowering people, driving change: Social innovation in the European Union. Luxemburg: Publications Office of the European Union.

BONEHILL, J., VON BENZON, N., SHAW, J. (2020). 'The shops were only made for people who could walk': impairment, barriers and autonomy in the mobility of adults with Cerebral Palsy in urban England. Mobilities, 15(3), 341–361.

BESTERS, M., MARRADES, R. & KAHNE, J. (eds.) (2019). Our city? countering exclusion in public space. STIPO, A Placemaking Europe Publication.

BITTENCOURT, B. (2016), A inovação social como propulsora do desenvolvimento local sustentável. In J. M. Carvalho Ferreira (Org.). Perspetivas de Desenvolvimento Sustentável. Lisboa: Clássica Editora, pp. 135–164.

COIMBRA, J., (2019). Reinventar a Roda, Final Project Report, Master degree in Regional and Urban Planning, University of Aveiro.

COMISSÃO EUROPEIA (2015). Fechar o ciclo – plano de ação da UE para a economia circular, COM (2015) 614 final. Bruxelas.

COMISSÃO EUROPEIA (2014). Towards a circular economy: A zero waste programme for Europe, COM/2014/0398 final. Brussels.

DEVILE, E. C. P. G. L. (2014). Dinâmicas de envolvimento das pessoas com incapacidade nas atividades turísticas. Tese de doutoramento. Universidade de Aveiro.

GEISSDOERFER, M., SAVAGET, P., BOCKEN, N.M.P., HULTINK, E.J. (2017). The Circular Economy – A new sustainability paradigm? Journal of Cleaner Production, 143, 757–768.

HAWKEN, P. LOVINS, A. B. & LOVINS L. H. (1999). Natural Capitalism: The Next Industrial Revolution. London: Earthscan.

HULGARD, L. & FERREIRA, S. (2019). Social Innovation and Public Policy. In, J. HOWALDT, et. al., (org.). Atlas of Social Innovation. New Practices for a Better Future, (26–29). Dortmund: Sozialforschungsstelle, TU Dortmund Universit.

HANSON, J. (2004). The Inclusive City: delivering a more accessible urban environment through inclusive design. In: (Proceedings) RICS Cobra 2004 International Construction Conference: responding to change. York, UK.

HAUPT, P. & SKALNA, B. (2020). The design of public space augmentations for encouraging activities conducive to social inclusion, prevention and rehabilitation. International Journal of Psychosocial Rehabilitation, 24(1), 9–17.

JIMÉNEZ, D., DE LA FUENTE, Y., HERNÁNDEZ-GALÁN, J. (2018). Diversity of "pedestrians on wheels", new challenges for cities in 21st century. Studies in Health Technology and Informatics, 256, 357–366.

KOVACIC, Z., STRAND, R. & VÖLKER, T. (2019). The Circular Economy in Europe Critical Perspectives on Policies and Imaginaries. London: Routledge.

KAMALIPOUR, H. & DOVEY, K. (2020). Incremental production of urban space: A typology of informal design. Habitat International, 98, 102133.

LIFSET, R. & GRAEDEL, T. E. (2002). Industrial ecology: goals and definitions. In Robert U. AYRES and Leslie W. AYRES (eds.). A Handbook of Industrial Ecology. Cap. 1 (3–15). ElgarOnline.

MACHADO, M. H. & LIMA, J. P. (2015). Avaliação multicritério da acessibilidade de pessoas com mobilidade reduzida: Um estudo na região central de Itajubá (MG). Urbe, 7 (3), 368–382.

MACHADO, M. H. & LIMA, J. P. (2017). Avaliação da acessibilidade pela perspectiva da pessoa com mobilidade reduzida. R. Tecnol. Soc., 13 (29), 1–21.

MCDONOUGH, W. & BRAUNGART, M. (2010). Cradle to Cradle: Remaking the Way We Make Things. New York: Farrar, Straus and Giroux.

MULGAN, G. et. al., (2007). Social innovation: what it is, why it matters and how it can be accelerated in Working Paper, Oxford Said Business School, pp. 1–52.

MOULAERT et. al., (Ed.) (2013). The international handbook on social innovation: collective action, social learning and transdisciplinary research. Cheltenham: Edward Elgar Publishing.

NaÇÕES UNIDAS, (2017). New Urban Agenda, United Nations.

NYSETH, T. & HAMDOUCH, A. (2019). The Transformative Power of Social Innovation in Urban Planning and Local Development. Urban Planning, 4(1), 1–6.

OCDE/Observatório Europeu dos Sistemas e Políticas de Saúde (2019), Portugal: Perfil de Saúde do País 2019, Estado da Saúde na UE, OCDE, Paris/Observatório Europeu dos Sistemas e Políticas de Saúde, Bruxelas.

PIYAPONG, J. et al. (2019). Empirical evidence of the roles of public spaces and public activities in the promotion of community relations and sense of place in a coastal community. Journal of Place Management and Development, 12(4), 485–507.

POTTS, R. & YEE, L. (2019). Pokémon Go-ing or staying exploring the effect of age and gender on augmented reality game player experiences in public spaces. Journal of Urban Design, 24(6), pp. 878–895.

ROSSINI, L. & BIANCHI, I. (2020). Negotiating (re) appropriation practices amid crisis and austerity. International Planning Studies, 25(1), 100–121.

SILVA, P., (2016). Tactical urbanism: Towards an evolutionary cities' approach? Environment and Planning B: Planning and Design, 43(6), pp. 1040–1051.

SHORT, J. R., MARTINEZ, L., (2020), The urban effects of the emerging middle class in the global south, Geography Compass, e12484.

SUMINSKI, R. R. et al. (2012). Park quality in racial/ethnic minority neighborhoods. Environmental Justice, 5(6), 271–278.

TAYLOR, J. B. (1970). Introducing Social Innovation. Journal of Applied Behavioral Scienc, 6(1), 69–77.

TORABI, N., et al. (2020). Widening the lens: Understanding urban parks as a network. Cities, 98, 102527.

UNIÃO EUROPEIA, (2016). Establishing the Urban Agenda for the EU: 'Pact of Amsterdam'. Agreed at the Informal Meeting of EU Ministers Responsible for Urban Matters on 30 May 2016 in Amsterdam, the Netherlands.

XU, N., WANG, J. & WANG, W. (2019). Revealing urban public space patterns through quantitative comparison between the old city of Nanjing and Zurich. Sustainability, MDPI, Open Access Journal, vol. 11(13), pp. 1–20.

Research Tracks in Urbanism: Dynamics, Planning and Design in Contemporary Urban Territories – Allegri et al (eds)

City sense. Legibility from sensorial experiences in three squares of Barcelona

L.K. Vila Solier
Departament d'Urbanisme i Ordenació del Territori, Barcelona School of Architecture, Universitat Politècnica de Catalunya, Barcelona, Spain

ABSTRACT: This paper explores the idea of legibility from the analysis of sensorial experiences in three public spaces in Barcelona. For this purpose, the research stands out the influence of perceptual systems in the configuration of the idea of place as a consequence of everyday practices in public space. The methodology proposed for this study brings together both physical and social factors and includes a diagnosis of sensory dynamics through ethnographic and survey work. The objective is to show the relationship between the city sense, in terms of being perceived actively in a sensory experience, with a sense of the city, in terms of the configuration of a coherent and memorable mental image. The analyzed cases are the Plaza del Sol in the neighbourhood La Vila de Gracia; Cathedral Avenue in El Barrio Gótico and Plaza de Mar in La Barceloneta.

1 INTRODUCTION

Can characterizing urban multisensory parameters help to develop a better version of cities? When Kevin Lynch travelled to Italy in the decade of 1950 to start the bases for his theory on the image of the city, he documented his entire sensory experience (sounds, smells, textures, etc.). Although later he would focus on studying the visual dimension, in the beginning, he approached what he called the *sensuous form*, worrying about sensory perception. This paper aims to recover this holistic vision.

Although Barcelona is a benchmark in the resurgence of public space since the decade of 1980, its sensory dimension has not been studied deeply. The parks, squares and streets are constantly crowded because of its benign Mediterranean climate. These spaces become the stage of constant sensory interaction. This paper is focused on the sensory stimuli that make memorable public spaces.

2 HYPOTHESIS AND OBJECTIVES

The hypothesis proposes that sensory experience in public spaces in Barcelona influences the formation of the *legibility* of the city, creating symbolic places and connected structures. The article introduces four sub-hypotheses. In the first place, that a pedestrian environment influences the perception of sensory stimuli. In the second place, that nature in public space produces memorable sensory microclimates. In the third place, that the diversity in zoning has an impact on the daily sensorial dynamics. And finally, that the variety of human activities feeds the sensory experience of different social groups.

The main objective is to study the multisensory urban experience in Barcelona's public spaces. Furthermore, this article seeks to identify what types of urban elements are identifiable in the citizens' memory from a multisensory approach.

3 METHODOLOGY

The work proposes a socio-spatial methodology that includes a diagnosis of sensory dynamics through ethnographic work and perceptual surveys. The analysis is made by two different blocks of factors. On the one hand, the *permanent factors* are the elements or connections with a fixed presence and a sensory impact. These are divided into three sub-factors with indicators: Pedestrian characteristics (the type of pavement, connection with the urban fabric, urban furniture and accessibility), the presence of nature (connection with a geographic element, type and density of trees, shrubs or ground covers), and the diversity of zoning around the public space (restaurants, commerce, culture, etc.).

On the other hand, the *non-permanent factors*, are the dynamics that have alternate repetition processes. These are registered in two parts: daily activities with sensory potential (food-related, play-related, recreational, sports, performative, civic and commercial)

DOI: 10.1201/9781003220855-10

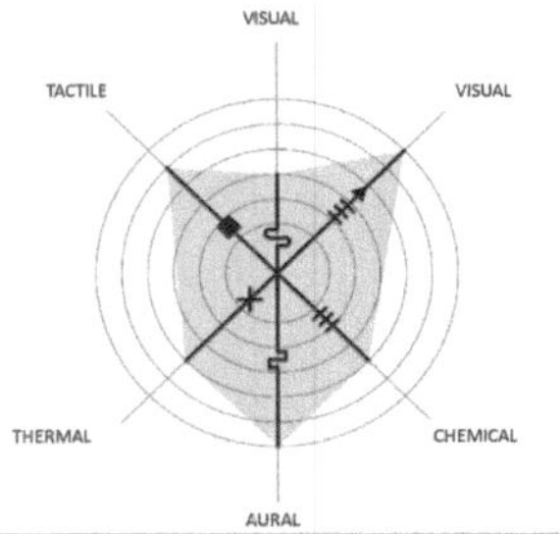

DESCRIPTORS OF MULTISENSORY RADAR

TACTILE	VISUAL	KINETIC	CHEMICAL	AURAL	THERMAL
Static	Dark	Strong	Weak	High pitch	Hot
Mobile	Bright	Light	Intense	Low pitch	Cold
Rough	Saturated	Free	Stagnant	Quiet	Dry
Smooth	Neutral	Bound	Fresh	Loud	Wet
Light	Perspectival	Indirect	Musky	Clear	Natural
Heavy	Flat	Direct	Putrid	Reverberant	Artificial
Porous	Intimate	Level	Floral	Vocal	Ambient
Resistant	Vast	Graded	Fruit	Non-vocal	Source
Hard	Solid	Sustained	Spice	Natural	Radiant
Soft	Void	Quick	Resin	Artificial	Convective
Warm	Detailed	Crowded	Meaty	Attack	Constant
Cold	Blank	Empty	Oily	Decay	Responsive

TEMPORALITY INDICATORS

Singular	Constant	Situated	Ambient	Directional	Repetitive	Switching	Varying
Sensory mark	Experienced over a period of time	Located in a specific place	Source is hidden	Sense of direction indicated	Pattern, regular or irregular, fast or slow	Change between two poles immediately	Change between two poles gradually

Figure 1. Multisensory radar and its descriptors and indicators. Source: Own elaboration based on Lucas, R. (2014).

and sensory dynamics documented by multisensory radar (Figure 1). This is a methodological instrument proposed by Lucas Raymond (2014) that assesses six sensory aspects simultaneously: auditory, chemical (olfactory and gustatory), haptic, thermal, kinetic and visual in a range of intensity from 1 to 6. It incorporates descriptors and indicators for temporal and spatial information.

4 THEORETICAL FRAMEWORK

Understanding the role of the senses in the urban experience is essential to understand its sensory dimension. On the classic five senses, Gehl (2014) states that they are classified into two groups: the senses of distance (sight, hearing and smell); and the senses of closeness (touch and taste). Touch and taste complete the image with particularities building a spatial memory. "Regarding perception (…), there are few variations between 100 and 25 meters. After this threshold, the visualization of details (…) dramatically increases (…) between zero and seven meters, all the senses are used (…) (Gehl, 2014: 35). Therefor urban spaces can be perceived in a multisensory way in different intensities according to their spatial configuration and scale.

Furthermore, "(…) both our senses and our perception system are adapted to pedestrian movement. (…) with a speed between 4 and 5 kilometers per hour, we have time (…) to observe what is happening around us" (Gehl, 2014: 42). In this sense, an environment with a high index of pedestrian areas can create a responsive and sensed narrative. "Every moving experience of architecture is multi-sensory; the qualities of space, matter and scale are measured in equal parts by the eye, ear, nose, skin, tongue, skeleton and muscle "(Pallasmaa, 2006: 43).

As Gibson (1966) states the *perceptual systems* are how the body's perceptual apparatus is oriented; listening, touching, smelling, tasting and looking. In this sense, Pallasmaa (2006) refers that architecture does not appeal to the five classic and separate senses, but to the sensory experience in which there are interaction and encounter between sensory perceptions. Therefore, multisensory perception is an active experience in which there is a simultaneity of perceptions that produce complex and variable ways of understanding urban environment.

On the other hand, the *legibility* of the urban landscape refers to "the ability with its parts can be recognized and organized in a coherent pattern" (Lynch, [1960] 2015): 12). This visual value is essential to understand the importance of public spaces as articulating elements. This identification exists thanks to the *environmental image* that is defined as: "the generalized mental portrait of the external physical world that an individual possesses (…) is the product of both the immediate sensation and the memory of previous experiences" (Lynch, [1960] 2015: 13). However, if we incorporate the theory of *perceptual systems*, this portrait comes from a multisensory perception. "All our body constitution and our senses," think "in the fundamental sense of identifying and processing information about our situation in the world and mediating reasonable behavioral responses" (Pallasmaa, 2012; 131).

The *mental image* demands from the observer an active role. As well in the perception of the information, as in the creativity with which he will do it; so that he can be able to adapt it according to his needs (Lynch, [1960] 2015). This dynamic role combined with the *perceptual systems* can configure the citizen as a person who experiences in a complex form the urban landscape. "We contemplate, touch, listen and measure the world with our entire bodily existence, and the experiential world begins to organize and articulate around the centre of the body" (Pallasmaa, 2006: 66).

To structure the *mental images*, Lynch raised five elements: paths, edges, districts, nodes and landmarks; and to analyze them he proposed three parts. The *identity* refers to the distinction between objects by its singularity and individuality, the *structure*, to the spatial relationship between this object with others and its observer; and the *meaning*, to its practical or emotional value (Lynch, [1960] 2015). He also introduced the concept of *imaginability*, which is defined as "that quality of a physical object that gives it the ability to bring forward, (…), a strong image in any observer" (Lynch, [1960] 2015: 19) and the *visibility* "when it is not only possible to see objects, but they are presented to the senses acutely and intensely" (Lynch, [1960] 2015: 19).

In this sense, it is vital to mention sensory landscapes. On the one hand, Murray Schafer ([1977] 2013) indicates that the *soundscape* is an environment with sounds observed as a field of study that focuses on the way it is perceived and understood by society. This idea is aligned with *perceptual systems* since "a soundscape consists of heard events, not seen objects" (Schafer, [1977] 2013: 25). The soundscape presents different components according to Schafer. In the first place, *keynote sounds* or not consciously heard background sounds, that are created by geography and climate. In the second place, *sound signals* or consciously heard foreground sounds. Finally, *sound marks*, or unique sounds of a community perceived with special attention.

On the other hand, in the case of urban *smellscapes*, Paul Rodaway (1994) affirms that these are not continuous and for this reason, it is impossible to detect the smellscape of an area as a whole at a specific moment. To document the types of perceived olfactory stimuli, Quercia et al. (2015) proposed the term *smell notes*, which are defined as the different levels of intensity that urban smells can have. These are classified as: *base notes* or macro-level smells perceived by visitors of a city; the *middle notes*, or mixed smells between the dominant scents and the *base notes*; and *high notes* or smells at the micro-level, of short duration and perceived in very localized spatial and temporal points.

4.1 *Case selection criteria*

There are three selection criteria for the cases: chronological, typological, and referring to the factors of the methodology. In the first place, it was decided to choose three cases between the 80s and the present, since urban renewal policies in Barcelona emerged at that time. In the second place, to study only squares because they constituted a benchmark in these urban policies. Also, because they have dimensions that allow the simultaneous study of several dynamics. A final filter has been made from the analysis factors set out in the methodology, seeking to choose spaces where there are pedestrian features, various activities, the presence of nature and zoning that give a certain dynamic.

For these reasons, the following cases were chosen: Plaza del Sol, remodeled in 1983, located in Vila de Gràcia, Avenida de la Catedral, inaugurated in 1992 located in the Gothic Quarter, Plaza del Mar, opened in 2009, in La Barceloneta. The zoning established around the public space is diverse (Figure 2.). In Plaza del Sol, the uses are related to food such as restaurants, takeaway bars or cafes. Avenida de la Catedral has a cultural dimension with museum, archive and Cathedral, while in Plaza de Mar there are temporary places related to food and drinks.

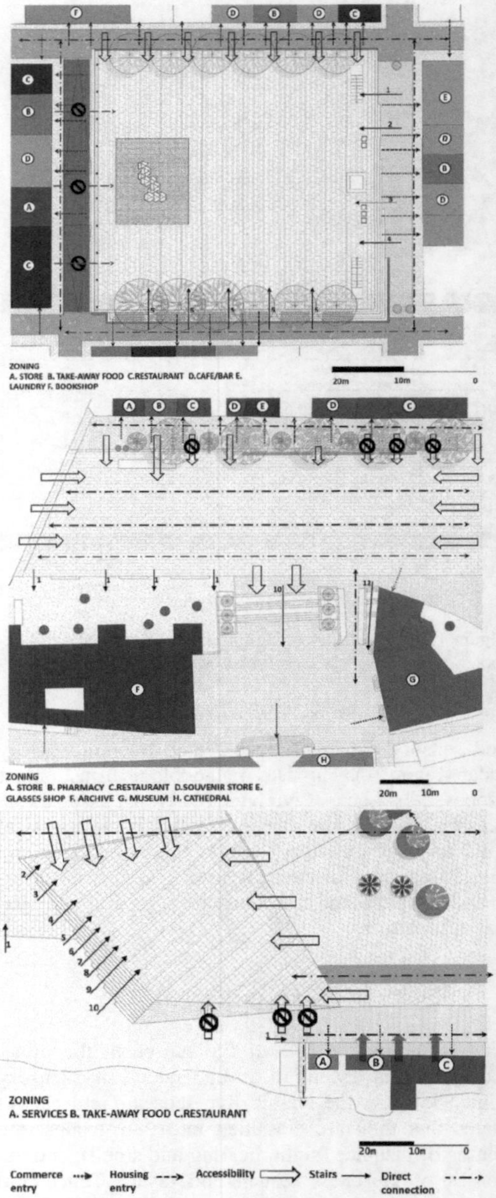

Figure 2. (From top to bottom) Permanent factors in Plaza del Sol, Avenida de la Catedral and Plaza Mar. Source: Own elaboration, 2018.

5 RESULTS

5.1 *Plaza del Sol*

In the first case, Plaza del Sol, it can be concluded that the most important sensory aspects are auditory and tactile. Both show a growth correlation reaching their peak at night. The first begins to grow in the afternoon with family activities in the playground and esplanade. It has its greatest intensity at night with the recreational activities of young people groups who sit on all the surfaces (esplanade, playground, furniture and stairs). This direct contact with the pavements represents a large area of tactile sensibility. Simultaneously, a deep sound perception comes from conversations between people, both in the plaza and on the restaurant terraces. (Figure 3).

In this square, there are two *sound marks*: the human voice and the chimes of the clock on the Plaza de la Vila tower (a very near square). Since the zoning is mainly related to restaurants, the vocal *sound mark* on terraces is intensified, as well as with play-related, recreational and performative activities. The other *sound mark* is a particular sound shared by the neighbourhood. Also, as a consequence of the extension and number of terraces the olfactory *base note* is food. The combination of tobacco and beer with the *base note* is the *middle note*, especially at night. The perception of these sensory details is possible above all due to the pedestrianized nature of the square's surroundings.

The square is remembered in the *mental image* due to the high occupation of its esplanade and stairs (Figure 4). This spatial relationship encompasses sensory aspects such as tactile, auditory, olfactory or thermal, and this perception is influenced above all by its pedestrian characteristics and the recreational (by young people) and play-related activities (by children at afternoons).

On the other hand, in terms of its relationship with other spaces, it is given in two modes: as a central and arrival space. The first occurs when the people use different types of linear references such as kinetic and sound (streets or avenues); specific points such as equipment associated with the visual and auditory; or neighbourhood squares as multisensory nodes. The second case places the square in a corner of mental map and the paths are differentiated in more detail with clues such as advertisements, or olfactory corners such as bakeries or rotisseries. In both cases, the vehicular flow through the busiest roads is taken as a guide since it contrasts with the pedestrian nature of the neighbourhood. Besides, this perception is influenced by the diversity of shops and activities around. (Figure 5).

Likewise, the streets associated with the square are distinguished by their auditory, olfactory or kinetic characteristics (Carrer Gran de Gràcia, Travessera de Gràcia and Torrent de l'Olla) thanks to the variety of shops. The main nearby node is Plaza de la Vila, a very important and memorable meeting space because of its everyday use. It presents multisensory dynamics thanks to the variety of shops, the diversity of activities and social groups, which interact in a full pedestrian area. Also, the northeast building of the square is a tactile and visual landmark due to its saturated colours, height and relief.

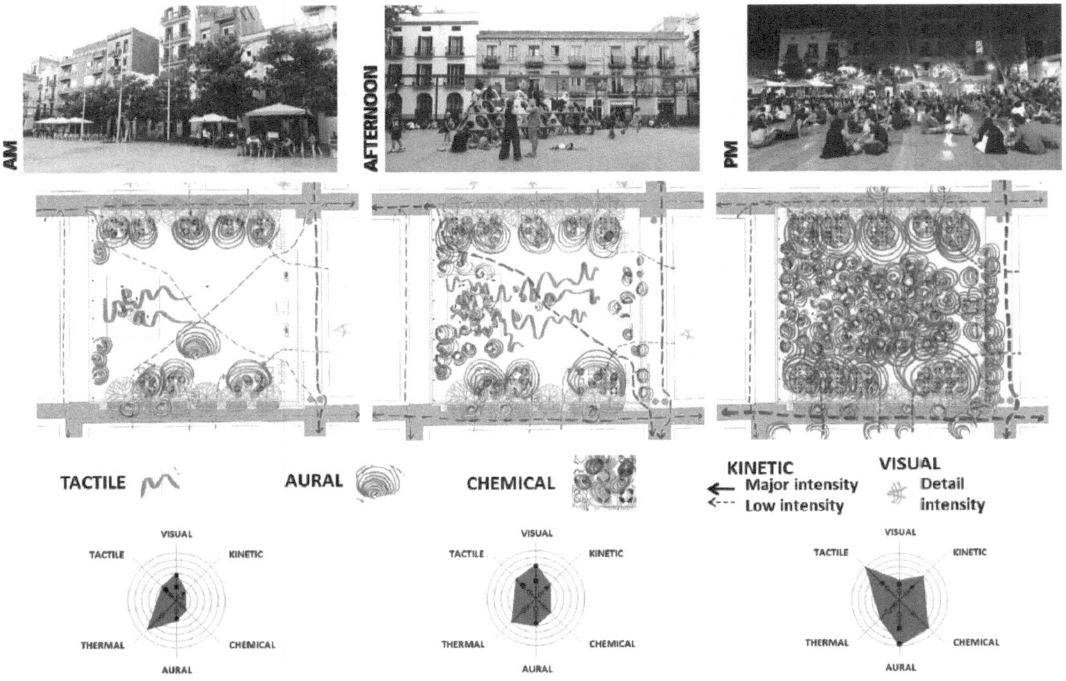

Figure 3. (From left to right) Daily evolution of sensory dynamics in Plaza del Sol. Source: Own elaboration, 2018.

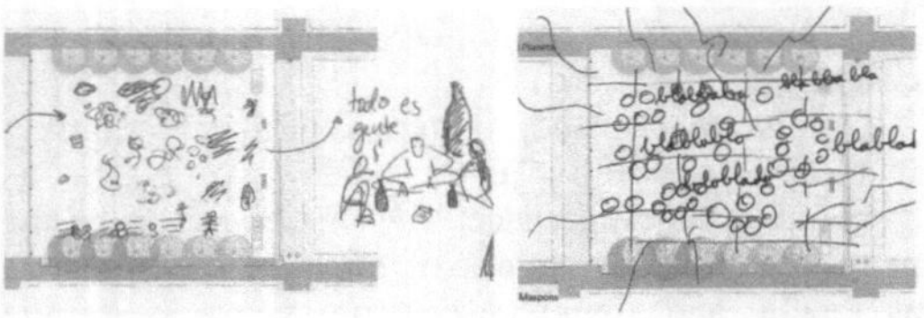

Figure 4. Mental maps of the Plaza del Sol. Source: Own elaboration, 2018.

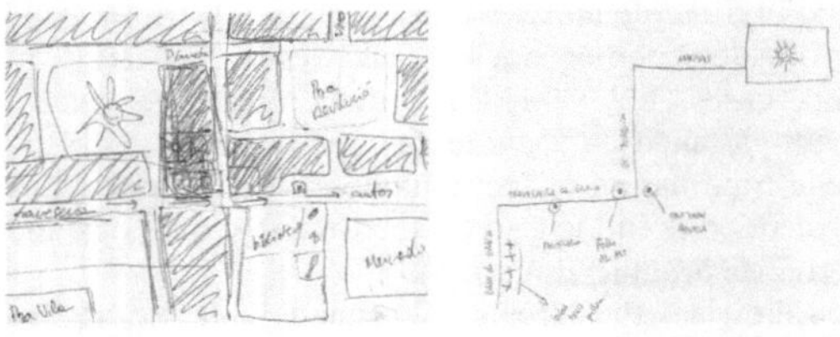

Figure 5. Mental maps of the neighborhood that locate the Plaza del Sol. Source: Own elaboration, 2018.

Finally, it is important to highlight the perception of Plaza del Sol as a tactile and auditory node due to the direct contact of the body with pavements, exerting a singular haptic perception, and the voice present in various forms. Both dimensions make it attractive, constituting it as an intensive focus of sensory interactions.

In conclusion, legibility is given thanks to the high level of pedestrian characteristics and recreational and play-related activities, that influence the sensory experience of the place.

5.2 *Avenida de la Catedral*

The second case, Avenida de la Catedral, is a quite popular square due to Barcelona Cathedral. In this place, the most important sensory aspects are kinetic, tactile, and auditory. These have a correlative evolution, due to their occupation on benches and stairs, the simultaneous presence of musicians and also people circulating (Figure 6).

The kinetic aspect reaches a peak of intensity at night due to the increased movement of people in both directions. Friends, work and family groups with young people, adults and older people enter or leave the Gothic neighborhood. Also, this index can rise if there are performative activities (juggling, hula hoop) or civic activities (choreographic dances) that involve movement. The auditory also presents its peak of perception at the same time due to musicians in performative activities. The tactile aspect intensifies according to the level of occupation on the marble benches or the stone stairs and parapets. Both allow the body to be directly in contact with the matter configurating preferential seats to chat or perceive the sound and kinetic movements.

In Avenida de la Catedral two *sound marks* were identified: the musical sound and the bells of the

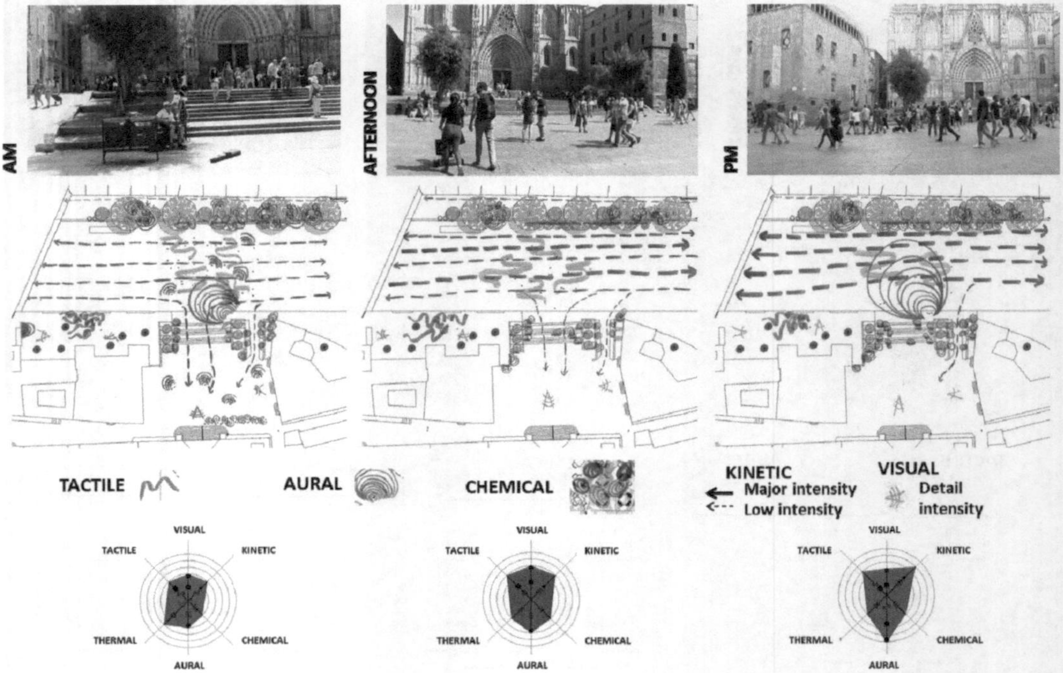

Figure 6. Evolution of activities and sensory dynamics in Avenida de la Catedral. Source: Own elaboration, 2018.

Barcelona Cathedral. The first occurs at different times and types, configuring an active and constant dynamic. The other is a sound which is related to the history and symbolism of the square. Although the type of shops around the square does not affect the soundscape, civic activities do because they increase typical music sounds. Besides, in its smellscape, the *base note* is mineral because of the stony character of its configuration with materials such as stone, granite, etc. The *high notes* are tobacco or floral aromas that come from the trees. In both cases, the complete pedestrian character of the square allows them to be perceived very clearly. Also, the furniture and pavement protected by tree shadows influence the number of people who experience them.

On the other hand, the square is remembered due to the cathedral which is associated with visual and tactile dimensions. This as a result of the complexity of its facade and height (Figure 7). Likewise, the remains of the Roman wall also provide a certain haptic pattern. This set of perceptions is influenced by its high pedestrian level and performative activities.

As for its relationship with other spaces, this occurs in two forms. As a crossing space where the references to nearby streets and squares are associated with kinetic, visual and auditory aspects. This perception is influenced by the high level of the pedestrian area of the neighbourhood and the commercial activity that produces a lot of movement. Also as an arrival space where streets that lead to it are associated with a mobile perception given by the flow of people. It should be noted that none of the respondents was a resident of the Gothic Quarter, which is why their mental map of the neighbourhood does not have a detailed urban plot (Figure 8).

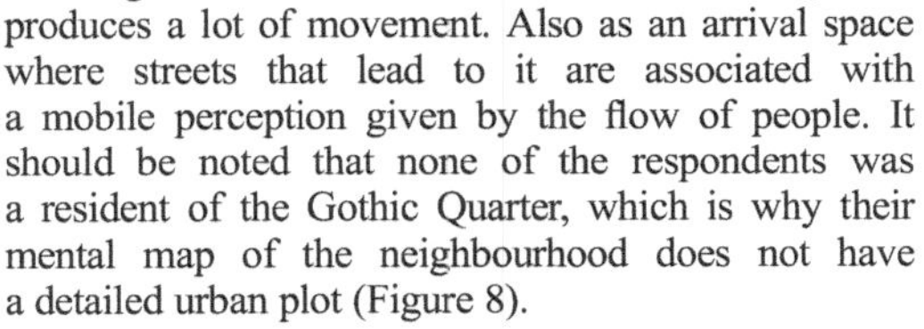

Figure 7. Mental maps of Avenida de la Catedral. Source: Own elaboration, 2018.

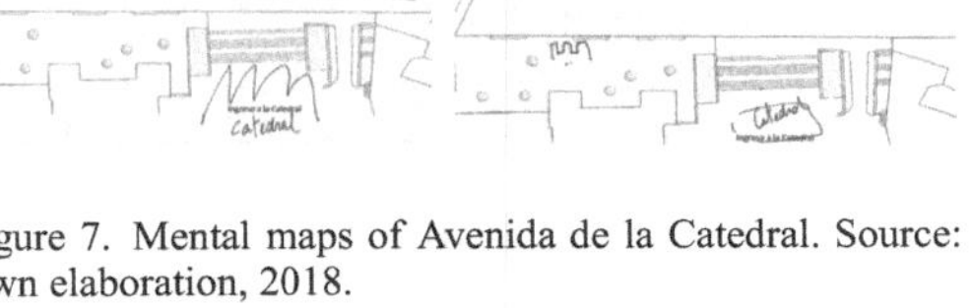

Figure 8. Mental maps of the neighborhood that locate Avenida de la Catedral. Source: Own elaboration, 2018.

Concerning urban elements, there are kinetic paths (Avenida del Portal de l'Angel and Carrer del Bisbe) and a multisensory node (Plaza Sant Jaume). Likewise, the cathedral was identified as a visual, auditory and tactile landmark that is distinguished by its material complexity and its distinctive sound. These sensory dimensions are perceived thanks to its calm environment and the level of pedestrian characteristics of the square and the neighbourhood.

Lastly, it is important to highlight the roles of Avenida de la Catedral as an auditory and tactile node, and as a sound and kinetic path. The first role is configurated by the simultaneity of its musical soundscape and constant occupation of its urban skin, while the second because it is a musicalized scenario at the moment of greatest movement. In these roles, pedestrianism and performative and recreational activities play a decisive role. Therefore, the conclusion relates the urban *legibility* to the sensory dimension.

5.3 *Plaza del Mar*

In Plaza del Mar, inaugurated in 2009, the most important sensory aspects are thermal, tactile, kinetic and auditory with an inverse correlation between the first and the others (Figure 9). On the one hand, the tactile aspect reaches its peak in the afternoon when the granite pavement stabilizes its temperature and it is comfortable to sit on the stairs and the esplanade. Also, at this time there are still many people who are on the beach, in direct contact with the sea and the sand, emphasizing haptic dimension. Kinetic dimension reaches its maximum at night when it is cooler for sports and recreational activities; that provides a variety of speeds and routes around.

The auditory perception is remarkable, reaching its highest level also in the afternoon. Is produced by performative activities, with musicians, or recreational activities both in the stairs and on the beach. Also, its intensification is related to the street trade that offers drinks or beach items. The thermal aspect has a development that increases until it reaches its highest intensity at noon. Also, it appears as very warm with a component of intense humidity and there are very few places that offer shadows. The peak of its intensity contrasts with the other sensory aspects and it is an example of how this perception affects occupation and therefore sensory dynamics.

On the other hand, the *keynote sound* is the sound of the sea that highlights the geographical location and the maritime character of the city. This geophony presents a low volume, due to the calm nature of the Mediterranean. *Sound signals* are the warning sounds about the state of the sea from speakers and helicopter that supervises the beaches. The *sound mark* is the musical sound of the performative activities that take place at the end of the afternoon. In this square, the smell of the sea is the *base note*; while *middle note* is given by the combination of it with body smells, beer or garbage; and the *high notes* are smells of tobacco or alcohol. The smell of the sea highlights the importance

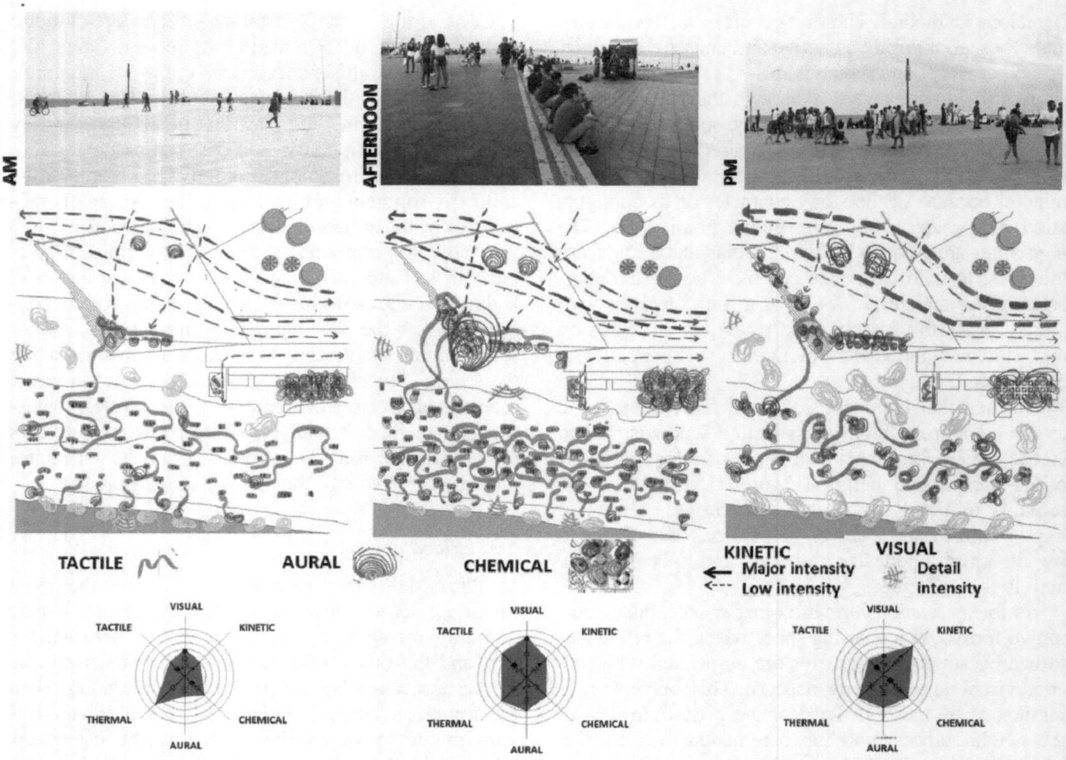

Figure 9. Evolution of activities and sensory dynamics in Plaza del Mar. Source: Own elaboration, 2018.

of the urban connection with nature. Recreational activities and street trade increase the perception of certain smells. Both the olfactory and the sound are perceived very clearly thanks to the pedestrian level of the place.

The square is remembered in the *mental image* by three elements: the sea, which has a visual, olfactory and auditory perception; the musicians with their sonic dimension; the beach that combinates the tactile, visual and thermal dimensions (Figure 10). This group of perceptions is influenced by its pedestrian indicators, the geographic element present, and performative and recreational activities.

The square's relationship with other spaces occurs in three modes. As a central space in which olfactory, thermal and visual references converge thanks to the pedestrian character of the maritime border. As an arrival space for recreational activities with multisensory perception influenced by shops and the large flow of people. Finally, without any reference other than the beach, which denotes the lack of daily routes (Figure 11). Within this diagnosis, it should be noted that none of the people who did the survey was a resident of La Barceloneta neighbourhood.

The results characterize the Paseo Joan de Borbó and Paseo Marítimo as kinetic, olfactory and sound paths. Shops influence the perception of the first avenue; while the pedestrian character influences the multisensory dimension of the second one. Likewise, three milestones were identified: Hotel W; that stands out visually for its great height and haptic character; the two towers of the Paseo Marítimo; and

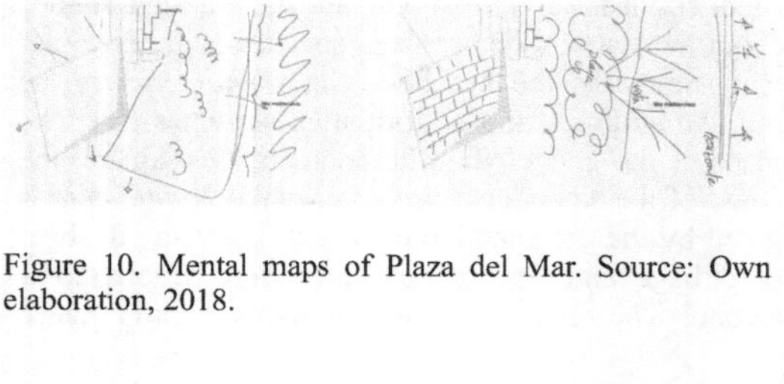

Figure 10. Mental maps of Plaza del Mar. Source: Own elaboration, 2018.

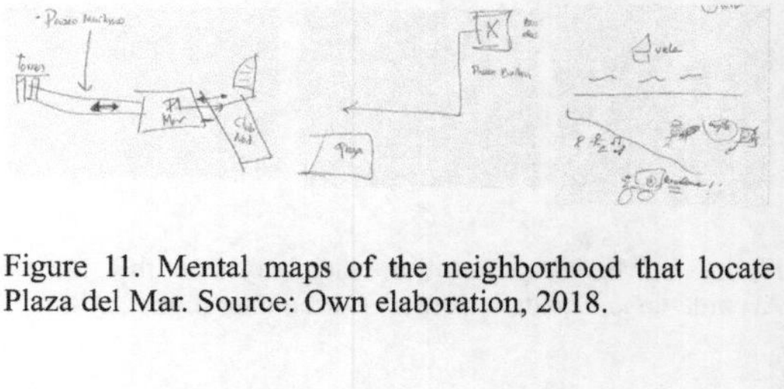

Figure 11. Mental maps of the neighborhood that locate Plaza del Mar. Source: Own elaboration, 2018.

the Peix Olímpic, which are associated with visual, olfactory and thermal perceptions. For these perceptions, the pedestrian characteristics of the Maritim boulevard are crucial. There is also a large edge with high permeability, made up of the beach and the sea. The porosity and roughness of the beach, as well as the strong smell and calm sound of the sea, configure a multisensory edge.

Finally, the Plaza del Mar plays three roles in the environmental image. As an olfactory, auditory and kinetic path because of sports, street trade, performative and recreational activities and the sea. Also, as an auditory and tactile node due to the occupation by performative and recreational activities. Lastly, as part of the multisensory edge formed by the coastline. In summary, in this case *legibility* is mainly associated with the high level of pedestrian configuration, the presence of the sea, and the performative and recreational activities.

6 CONCLUSIONS

In conclusion, the multisensory dimension in the urban experience does affect the *legibility* of the urban environment. The methodology showed that sensory dynamics is constantly given by the interrelation between the physical characteristics of space and activities that take place in it.

The pedestrian characteristics are very important since an accessible, calmed and connected with the urban fabric public space is an environment in which attention is paid to sensory experience and becomes meaningful. Stairs and level changes are important as areas of contact with the urban skin. The presence of natural elements is also a factor that feeds sensory *legibility*. Trees usually connect with the thermal and tactile aspects of the human being, while geographic elements do so with the olfactory and tactile dimension. Both create atmospheres with sensory perceptions that generate *identity*. The diversity in the zoning also has repercussions since it produces continuous sensory stimuli towards the public space. However, the monofunctional orientation produces a sensory monotony that does not make the spaces particularly memorable, even if they have high occupancy rates.

One of the most important factors is the presence of activities since if there is a simultaneous diversity of these, the place is enriched and allows different social and generational groups to enjoy and interact. The sound of a musician while children play on the pavement and groups eating and chatting on the restaurant terraces makes the place more complex and allows an urban experience with a complex sensory mixture. Also, activities such as performative or recreational, create strong and memorable environmental images, with memories of melodies and textures.

Finally, it is important to conclude that each of these factors cannot influence *legibility* alone. They act under a constant feedback dynamic and are interrelated with each other under quotidian practices. Sensory appropriation comes from this interaction. The daily sensory experience is linked in their quotidian biography with each of these components, building a practical memory. Therefore, it can be concluded that the multisensory experience of public spaces contributes to making them symbolic and memorable places.

REFERENCES

Gehl, J. (2014). Ciudades para la gente. Buenos Aires: Ediciones Infinito.

Gibson, J. (1966). The senses considered as perceptual systems. Boston: Houghton Mifflin.

Henshaw, V. (2014). Urban smellscapes: understanding and designing city smell environments. New York: Routledge/Taylor & Francis Group.

Lynch, K. ([1960]2015). La imagen de la ciudad. (E. Revol, Trans.) (Tercera ed). Barcelona: Gustavo Gili.

Lucas, R. (2014). Sensory Notation Handbook. Reino Unido: Flâneur Press.

Pallasmaa, J. (2006). Los ojos de la piel: la arquitectura y los sentidos. Barcelona: Gustavo Gili.

Pallasmaa, J. (2012). La mano que piensa. Sabiduría existencial y corporal en la arquitectura. Barcelona: Gustavo Gili.

Pink, S. (2015). Doing sensory ethnography. Londres: Thousand Oaks.

Porteous, J. D. (1990). Landscapes of the mind: worlds of sense and metaphor. Toronto: University of Toronto Press.

Quercia, D., et al. (2015). Smelly Maps: The Digital Life of Urban Smellscapes. Proceedings of 9th International AAAI Conference on Web and Social Media. Retrieved from http://arxiv.org/abs/1505.06851

Rodaway, P. (1994). Sensuous geographies: body, sense, and place. Londres: Routledge.

Schafer, R. M. ([1977]2013). El paisaje sonoroy la afinación del mundo. (V. Cazorla, Trans.). Barcelona: Intermedio.

ARQUITETURA magazine | IAB Brasil _1961 to 1968

M.L.G. Lima
PROURB, Faculdade de Arquitetura e Urbanismo, Universidade Federal do Rio de Janeiro, Brazil

ABSTRACT: ARQUITETURA was the name of the journal edited by the Institute of Architects of Brazil - IAB. In this article, we seek to fit it into a global network of specialized publications, which arose in that period and was responsible for the reorganization of the modern architecture movement after WWII. Our goal was to introduce and contextualize the content of the editions, focusing on texts on urbanism and urban policy. The results indicate that ARCHITECTURE was a vehicle for the dissemination of texts by countless important Brazilian and foreign authors, revealing an intense intellectual exchange between IAB architects and other scholars. In addition, it was a radically transdisciplinary publication, connected with the political and cultural events of the period, representing a professional class committed to social justice and the social function of architecture.

1 INTRODUCTION

"ARQUITETURA" was the title picked for a specialized journal on architecture and urbanism published by the Brazilian Institute of Architects (IAB) in the 1960s. The periodical was formatted as a journal, an extension of the Monthly Newsletters that had been published by IAB in Rio de Janeiro since 1958, which was initially organized by the IAB – Guanabara State center with the name GUANABARA, whose main goal was to be IAB's official voice to the society. A little later, the journal became the voice of IAB nationally, and its name was changed to ARQUITETURA. This continued until issue #67, when the journal went back to being published "under the exclusive responsibility and full control" of IAB – Department of Guanabara (ARQUITETURA #66, 1967: 10).

The journal was published irregularly: between August 1961 and November 1962, while using the name GUANABARA, only five issues were published. From December 1962, now named ARQUITETURA, it became monthly with almost no interruptions up to issue #78, when it stopped being publish. The publication was resumed later in 1977 and continued up to 1998. In our study, only the first phase of the journal will be addressed.

Our hypothesis is that ARQUITETURA is part of a global phenomenon that took place in the 1960s in which various publications specialized in the field of architecture were published. New technologies made the printing process more accessible and democratic, which allowed the reorganization of the modern architecture movement in the post-war period. The phenomenon was recognized by Denise Scott Brown (1968) and Beatriz Colomina (2010), which dubbed it the "explosion of small architecture journals". Despite the aspects explored by Colomina – experimentation, collages and provocations –, we believe that ARQUITETURA was part of a network of dialogues that sought to expand possibilities of doing and thinking architecture and urbanism.

According to Torrent (2018), specialized architecture publications in the 1960s in Latin America took on the "double task" of being partly a professional representation and partly a critical voice regarding sociospatial development issues. After consolidating as authorities in theoretical discussion – addressing economic, social and political issues and using interpretational frameworks from sociology, structuralism and Marxism –, these journals expanded their range from architecture and urbanism to a predominantly interdisciplinary landscape. This broader perspective of the professional made architects capable of addressing problems that are more complex than simply drawing project, such as the development of plans, programs and projects dedicated to guiding the economic growth of the territory. (Torrent, 2018: 4734).

Our study will focus on articles on urbanism and urban policy, to the detriment of other projects. The methodology consisted of the indexation of all content in the summaries of the 78 issues of the journals, which produced a total of 1,139 items. Each item was classified regarding issue, year, month, title of section, title of article and author. Based on the initial classification, the indexation was submitted to other more subjective evaluations, which ranged the following categorization: type of content, general theme, specific theme, tags and content summary. This indexation was organized in a single spreadsheet, which allowed us to visualized groups in the content.

DOI: 10.1201/9781003220855-11

Figure 1. Cover of issue #6 (1963).

Based on the document, we sought to respond to questions such as: Which were the major issues in urbanism addressed in the journals and why? What other disciplines to which they are connected? Who are their interlocutors? What are the opinions of these architects/authors/editors concerning the then current political and economic events? What is their stance on the urban reality experienced? Which role do they assign themselves as professionals concerning that reality? What answers did they provide in the field of urbanism to the socio-political problems they identified?

The most important discussion in the journals was the consolidation of the profession of architect in Brazil, based on class awareness, which can be perceived already in the first editorial written as a manifest, in the #1 issue of GUANABARA (1961: 8):

> *There's a struggle.*
> *No Brazilian architect aware of their duties as architect and Brazilian can abstain from the basic struggle of our profession.*
> *All of us, within our means, must engage in the matter, less for class interest than for our duty to the society, which emerges the moment we choose the profession.*
> *Work, for every citizen, is both a means to provide their needs and their families' needs and a way to contribute with their share to the joint effort of collective progress and wellbeing; the first is a right, the second, a duty.*
> *(#1, 1961:8)*

2 AUTHORS AND NETWORK OF DIALOGUES

The journals were produced to be the voice of idealism for Brazilian architects, engaged in the fight for the assertion and consolidation of the profession in Brazil, fulfilling their social role, a professional and cultural entity. This commitment to the trade, often expressed in their editorials, was conceived and written by architect Maurício Nogueira Batista, considered the "soul of the journal" (Konder Netto, 2019). Not only he was responsible for editorials, but also, mainly, he chose, coordinated and produced the material that would be published, also signing many articles and translations.

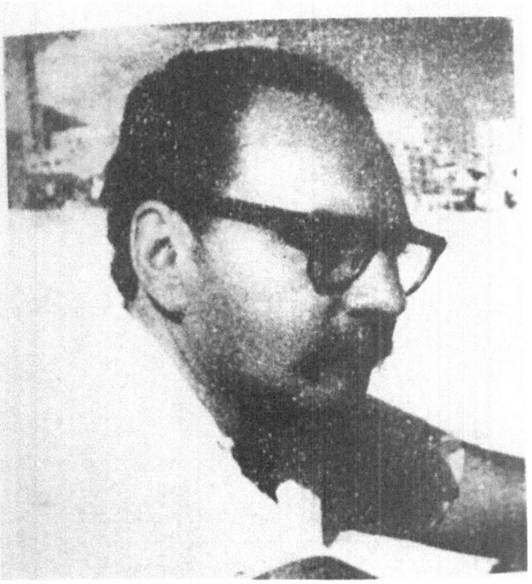

Figure 2. Maurício Nogueira Batista. Source: Pinheiro, 2001.

ARQUITETURA had the privilege of gathering a generation of young talents, later recognized in the cultural sphere of the country, such as journalist Zuenir Ventura – hired to teach writing to architects. Filmmaker Cacá Diegues, poet and art critic Ferreira Gullar, literary critic and essayist José Guilherme Merquior, playwright Cléber Ribeiro Fernandes, and composer Nelson Lins de Barros kept a permanent section – "Galeria" –, dedicated to arts in general. It should be mentioned that, in that period, they were all between 21 and 38 years of age, except for Nelson Lins de Barros, who was older.

Among the more than 240 authors that wrote articles, news pieces, translations, transcriptions and other texts for the journal, we should highlight, by number of contributions, architects Maurício Nogueira Batista (20 texts), Alberto Vieira de Azevedo (12 articles on acoustics), and Jorge Wilhein and Maurício Roberto (both with 8 contributions). We should also point out the work of architect Cláudio Ceccon – cartoonist Claudius –, responsible for a series of editorial cartoons exclusive of the journal.

Figure 3. Claudius's cartoon for ARQUITETURA #17 (1963).

The journal also gave space for the work of young women like architect and landscape artist Rosa Grena Kliass, author of an article entitled *Joinville and Curitiba: Two landscape studies*, in issue #55 (January 1967), and architect Philomena Chagas Ferreira, who wrote *Data on the climate of region of Brasília*, in issue #30 (December 1964). We should also mention a transcription of a lecture by Argentinean-Brazilian urbanist Adina Mera, at the *National School of Architecture* (FNA), in April 1961, about definitions for the terms "urban planning" and "urbanism", published in issue #1, in August, 1961.

We also highlight the publication of an article by architect and urbanist Hélio Modesto, titled *The Problem of Urban Planning*, in a double edition (#72 and #73), in 1968. Hélio Modesto was an important urbanist from Rio de Janeiro. The recovery of his work in the field of urban planning is still "very timid" (Oliveira, 2018), and this article published in ARQUITETURA helps filling that gap.

ARQUITETURA was an important venue for architects and other professional in areas related to exchange, especially concerning the housing problem in Brazil. In that field, the contribution of Brazilian sociologists José Arthur Rios and Maurício Vinha de Queiroz in issue #8 (1963) was also important. José Arthur Rios's article *Favelas* had been published originally in the *Aspectos da Geografia Carioca* journal and introduced the situation and aspects of the *favelas* in Rio de Janeiro. Maurício Vinha de Queiroz's article *Architecture and Development* had been published prior in the *Revista do Instituto de Ciências Sociais.* The rationale behind these publications was raising social awareness on the housing problem in Brazil, a first step toward a solution, which should later be substantiated in propositions.

Figure 4. Picture that illustrated José Arthur Rios's article Favelas. Photo by Flávio Damm (#8, 1963: 19).

ARQUITETURA was, still, a venue for the dissemination of pieces by numerous important foreign authors, which shows an intense intellectual exchange between various nationalities, stimulated via translations to English and French of the summary of the journal's contents for two years straight – between issues #36 (June 1965) and #59 (May 1967).

Among other contributions by international authors, we highlight that of North-American historian Lewis Munford, whose works were published in threefig_ch10_209_5 occasions: the translation *Regional Planning*, an article from the classic book *A Cultura das Cidades* (1956), published in issue #2 (1961); a review on a conference in Rome, transcribed by the *Central Architects Society Bulletin* in September, 1962, and published in issue #7 (1963); and a translation of the article *The Promise of a New World,* published originally in the *Journal of the American Institute of Architects*, in issue #45 (1966).

Furthermore, we highlight the dissemination of the work of German architect Otto Koeningsberger, who created the idea of *Action Planning* – through which he sought to review the basic tenants of urban planning conceived in Great Britain in the 1930s, given its applicability to booming societies, like those of Asia and Africa. His conclusions concurred with the experiences of Brazilian architects in planning actions in the country and, therefore, ARQUITETURA issue #39 (September 1965) included a translation of one of his conferences, at the *Architectural Association*, in London, in 1964, published in the *Architectural Association Journal*, volume 79, #882. The same issue was addressed in a translation of an article of that same author in ARQUITETURA #55 (1967).

The Journal also published the work of English architect Christopher Alexander in issue #65 (1967), through a translation of an article entitled *Street Patterns and Geometry,* only two years after the publication of the essay *A City is not a Tree* (1965) and ten years prior the launch of his seminal book *A Pattern Language,* of 1977. Just like the article *Semiology and urbanism*, of French historian Françoise Choay, published by *l'Architecture d'Aujurd'hui* in June/July 1967, which was translated and published in ARQUITETURA's issue #78, on December 1968. This publication happened only three years after the publication of the book *L'Urbanisme, utopies et réalités* (1965), which introduces an analysis of the evolution of urbanistic thinking and an anthology on urbanism. In the article translated, the author discusses the possibility of a semiology of the city and the legitimacy of a study of that nature in the field of urbanism.

ARQUITETURA #68 included two translations of pieces by English architect and planner John C. Turner, a member of the *Urban Studies Center* of the MIT and of Harvard University. Turner had been in Brazil at the time, invited by the *Federal Housing and Urbanism Service* (SERFHAU), for a series of lectures on "subnormal housing agglomerates" (#68, 1968: 17) in which he discussed aspects of urban development and housing with Brazilian technicians. The first piece translated is one of Turner's statements about his experience in Brazil. The second is an article entitled *Barriers and Channels for Housing Development in Developing Countries*, which had been previously published in the *Journal of the American Institute of Planners*, on May 1967, and translated for ARQUITETURA by architect Cláudio Ceccon.

The translation of two other articles written by architect Pierre Mas are worthy of mention. The pieces are on urbanism and urbanization in Morocco, an attempt to expand the debate between the so-called "third-world countries". Although this was the term used in the piece (#34, 1965: 5.), since the 1950s the Economic Commission for Latin America and the Caribbean (CEPAL) suggested the idea of "underdevelopment" to express the productive, social, institutional and international specificities, as well as other characteristics concerning the international presence of Latin-American countries (Fridman, 2018).

In the first article (*Urbanism, Architecture and Development*), the author refers to an ongoing experience of development in rural agglomerates in Morocco (#34, 1965).

The second article, published in issue #47 (1966) of ARQUITETURA is a translation from a piece published in the Moroccan journal *A+U, African Journal of Architecture and Urbanism,* issue #3 (1965), entitled *Urbanism and the Organization of the Territory in Underdeveloped Countries.* Beyond those, ARQUITETURA also published a translation of an article originally published at the ATU – *African Journal of Architecture and Urbanism,* written by French landscape artist Jean Mallet, *Urbanism and Landscape* (#52, 1966: 17), as well as a translation of a piece by French architect Henri Chomette on housing in Africa (n° 63, 1967: 3). Chomette was responsible for a lot of the modern architecture in Sub-Saharan Africa, developed in the 1960s. It should be mentioned that, despite the space provided by ARQUITETURA to Africa, the articles published in the journal about the continent were all written by Europeans.

3 THEMES

Themes included architectural and urbanism projects, legislation, heritage, design, urban policy and construction materials, as well as film, visual arts, theater and music, among others. However, the sections that addressed the role of architects in the housing and urbanization policy of cities were more frequent and emphasized in the publication, as well as the editorials. Generally, the editorials introduced the main piece in the issue, and were written as a public statement, in which IAB's architects declared their points of view concerning a specific theme.

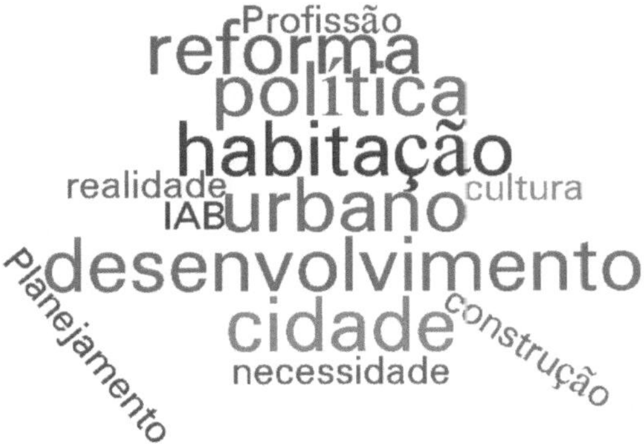

Figure 5. Word cloud generated with the titles of editorials.

The titles of all editorials in the 78 editions analyzed in the study were submitted to word counting tools (excluding the terms "architecture" and "architects" from the list since they would always be there and should not be determining for the counting). The titles of the editorials produced a total of 307 words. The most present words in all editorials concerned urban matters, with 14 instances of "urban" + "urban" + "city". The issue of "housing" – terms "housing" + "habitational" – comes in second with 13 instances. In third place, it is worth mentioning the terms "development", "policy" and "reform", with four instances each.

The housing issue also stands out in a section intitled *The problem of housing and the architects*, published in the 1962 and 1963 issues. The goal of the section was to "decisively state the stance of architects concerning the serious housing problem in Brazil" (#6, 1962: 4). It published various types of documents.

We should highlight the publication of the results of the *1st National Housing Round*, organized by the IAB in São Paulo in 1961 (#6, 1962). It was a discussion of the consequences of large housing complexes to physical and mental health (#12, 1963); a statement of IAB's architects on issues regarding laws affecting the Brazilian housing policy, such as the Tenancy Law (#17 1963) and Decree 1,281, which changed the National Housing Commission into the Federal Housing Council and took it out of the Ministry of Labor and Social Security and put it under direct subordination to the President of the Council of Ministers (ARQUITETURA #6, 1962).

Figure 6. Cover of issue #15 (1963).

Still in the field of housing, ARQUITETURA was the main source of dissemination for the famous *Seminar on Housing and Urban Reform (People, their Homes, their Cities)*, which was later known as the *Quitandinha Seminar*. Issue #15 of ARQUITETURA (September 1963), published the entirety of the results of the Seminar.

Another important issue in ARQUITETURA were the critiques to the *National Housing Bank* (BNH), a government agency created during the military regime to implement a national housing policy. BNH's policies were criticized especially regarding a lack of integration of the new housing policy with a true urban and social planning scheme. The editorial from January 1966, #44, *The Favelas and the City* stated the following: "it is inadmissible that we continue thinking only in terms of building new homes, isolated homes, in a vain attempt to resolve the situation of a population that lives in infra-human conditions" (1966:4). Emphasizing the stance, editorial *Architects and Housing*, from May, 1966, stated that:

> *Therefore, architects are concerned about the development of housing programs that do account for the deeper, multiple roots of the problem and could turn into elements that aggravate the situation, not only the residential crisis, but also the likely lockdown of the harmonic development of our urban areas (# 47, 1966: 4).*

Since the early 1950s, the ideas of "planning" and "development" were linked and included into the ideology that the State, the main agent of modernization, would implement planned policies as a national development strategy. There was a belief that a body of rules, universally valid and independent from local political structures, if implemented correctly, would have the power of serving the public interest (Fridman, 2014). The ideas on urbanism advocated by ARQUITETURA fit into that planning paradigm: city planning should obey regional and national guidelines, and its physical elements should be integrated to a planning in which economic and social aspects are necessarily included.

For such, architects defended the creation of specific agencies to address urban planning issues, both at the regional and federal levels. This agenda was advocated by IAB's architects since 1953, and had been discussed and complexified in congresses, reports and study commissions (França & Leite, 2017). They asserted the creation of legal mechanisms that allowed for greater municipal intervention capacity, concerning the use and distribution of urban land, given the social interest in that case, since the debate over the issue of urbanization should not be restricted to its physical limitations, but also include economic and social factors. In this context, various articles, editorials, translations and other documents addressing the issue were published. "Planning" was, actually, the title of a special section in the journal in 16 editions between #22 (1964) and #58 (1967).

The debates organized by architect and urbanist Harry James Cole on urbanization and development should be mentioned here. The background of that architect is still little known by Brazilian researchers (Lucchese, 2009), despite his field work on urban planning, such as the implementation of SERF-HAU's integrated development plants. The pieces published at ARQUITETURA are crucial to fill that historiographic gap.

Similarly, pieces by Brazilian architect Francisco Whitaker Ferreira *A Roadmap for Studies on Urban Planning* (#41, 1965: 27), *A Roadmap for Studies on Urban Planning II* (#42, 1965: 30) and *Development Research and Planning* (#70, 1968, p. 31) are worth mentioning. Francisco Whitaker Ferreira had been a direct collaborator of Joseph-Louis Lebret in Brazil, director of planning in the Oversight Authority for Land Reform during the João Goulart administration (1961-1964). Whitaker had been exiled in France since 1966, escaping political repression in the country.

Another important piece is the translation of an article of North-American urbanist Catherine Bauer Wurster's *Economic and Urban Development, Social Implications* (#10, 1963: 26), in which she addresses and discusses problems related to economic development, urban growth and the social implications included. As well as the pieces of Lewis Munford, John C. Turner, Otto Koenigsberger, Helio Modesto and Adina Mera, which have already been mentioned in this article.

Figure 7. Claudiu's cartoon for Quitandinha Seminar (1963).

Another relevant issue on planning in the journal was the reaction of Rio de Janeiro architects to the Doxiadis Plan. This was a plan created by the State of Guanabara in the 1960s to guide the growth of the city until the year 2000, produced by the office of Greek architect Constantino Doxiadis, who had been invited by then governor Carlos Lacerda (1960-1965). Despite its qualities and/or defects, the Doxiadis Plan was never accepted by Brazilian architects and urbanists. The reaction of architects to the hiring of Doxiadis can be recovered in pieces published in issues #20 and #21 – February and March 1964 –, in which IAB Guanabara's stance was put forth and the reasons for opposing the hiring.

Besides the central issues of housing and urban planning included in this article, ARQUITETURA also covered several other issues relevant to the universe of Brazilian architects and urbanists, such as in the special edition on the use of computers in the field of architecture and planning, #57 (March 1967). Moreover, between 1967 and 1968, ARQUITETURA produced a series of pieces on the architecture of penitentiaries on a partnership with the Secretariat of Justice of the State of Guanabara, aiming at a National Open Call for a Preliminary Draft of a New Penitentiary in the State of Guanabara.

4 FINAL CONSIDERATIONS

We conclude that ARQUITETURA was radically transdisciplinary. When the journal established a dialogue with other international publications, and not merely as a disseminator of their content, but as an active participant of the debate, it sought to respond to the motivation of the Brazilian historical context of that time. It was a media that was viscerally interlinked with the political and cultural events of the time. ARQUITETURA represented a profession committed to social justice and the social function of architecture, gathered under the Brazilian Institute of Architects (IAB), of which the journal was the voice.

Like the Little Magazines (Colomina, 2010), ARQUITETURA worked as an incubator for new ways of thinking, an important arena in which to discuss the emerging problems of the urban occupation of space, especially how they emerge in Third World countries. Many of the problems discussed in its pages are either still relevant or were aggravated with the growth of cities, which makes the journal's contributions important subsidies for reflections and the formulation of new projects dedicated to improving the lives of people in large cities.

REFERENCES

Colomina, B.; Buckley, C. Clip, Stamp, Fold: The Radical Architecture of Little Magazines, 196X to 197X. Nova York: Actar and Media and Modernity Program, Princepton University, 2010.

França e Leite, V. L. S. Caminhos que levam à cidade. [s.l.] Universidade Federal Fluminense, 2017.

Fridman, F. Notas sobre o planejamento no período João Goulart. Revista Urbana, v.6, jun. 2014.

Leme, M. C. DA S. (ED.). Urbanismo no Brasil. 1895-1965. São Paulo: FUPAM/FAUUSP, 1999.

Lucchese, M. C. (2009). Em defesa do planejamento urbano: ressonâncias britânicas e a trajetória de Harry James Cole. Tese de doutorado. São Carlos, Universidade de São Paulo.

Oliveira, L. L. Cidade e Patrimônio: o pensamento de Carlos Nelson. n. c, p. 1–4, 2018.

Rezende, V. F. O planejamento de cidades nos anos 60: uma reflexão a partir do Plano Doxiadis para o Estado da Guanabara São Paulo: III Enanparq, 2014.

Torrent, H. Disciplina y profesión frente al problema del desarrollo, ideas publicadas: 60-70, el desencanto de la arquitecturaAnais V Enanparq. Anais…Salvador: 2018.

Periodicals

Collection of journals GUANABARA/ARQUITETURA, from issue #01 (1961) through #78 (1968).

Interviews

Konder Netto, Marcos. Interview (22/04/2019). Interviewer: Marina Lage da Gama Lima. Rio de Janeiro, 2019

 ISBN 978-1-032-10788-2

Santa Fe-Paraná metropolitan area: Metropolis of second order. Mobility of people dwelling - work and their strategies

M. Soijet, V. Gramaglia, V. García, J. Santiago & B. Pennisi
Facultad de Arquitectura, Diseño y Urbanismo, Universidad Nacional del Litoral, Argentina

ABSTRACT: This work presents the outcomes of the project "Urban Observatory of the Metropolitan Area Santa Fe-Parana". This research made some progress in the treatment of the metropolitan phenomenon in a second reading of the Santa Fe-Paraná Metropolitan Area, based on social practices. It deepened the analysis of daily mobility from the evidence of the consolidation of the metropolitan scale as a significant unit of everyday experience for the inhabitants included, which constitutes relevant approximation towards the recognition and prioritization of problems in the field. That is why this research is based on the theory of mobility studies as a methodology to overcome the static view of the territory, enabling an exploratory inquiry open to relational readings that explains today's dynamics and mobility variables made invisible in traditional records that were previously applied.

1 INTRODUCTION

1.1 *Presentation, aims and hypotheses*

This work shows the progress in the Project "Urbanistic Observatory of the Metropolitan Area Santa Fe-Paraná (AMSF-P)" and leads to a connected reading from social practices, which is being carried out in the Facultad de Arquitectura, Diseño y Urbanismo, Universidad Nacional del Litoral (Argentina), with the participation of Urbanism Professors.

In this research, the metropolitan phenomenon is approached by performing an observation of the Metropolitan Area Santa Fe-Paraná, with a base on social practices and analysing daily mobility from the evidence of the consolidation of the metropolitan scale as a meaningful unit of the daily experience for city dwellers. This involves a meaningful approach for the recognition and relevance of the problematic.

As it comes to mobility, the researchers made use of traditional records (Annual Daily Mean Traffic-TMDA and frequency of public transportation), which were then mapped and analysed. Nevertheless, these records do not enable us to count with the details, which permit the segmentations of the features of these travels and travellers. Therefore, in the search for alternatives and facing the limitations of the information provided by the formal system of statistics, this research is carried out from the theorical turn of mobility studies as the methodology to proceed on an exploratory inquiry open to related readings which explain certain mobility dynamics and variables, not observed in traditional records.

In the presentation and being of territorial connectivity – a key issue for the functioning of the area and the region related –, it is highly relevant to examine thoroughly the "residence-work connection", which is influenced by the availability and costs of urban land.

In this aspect, the aims of this work are:

- To inquire about the new ways of gaining access to information about people's mobility;
- To gather information about people's intra-metropolitan travelling, considering residence and work, and observing the strategies used;
- To contribute to the definition of a metropolitan mobility map.

As a general hypothesis, the possibility arises to identify tendencies in daily mobility in metropolitan dynamics (in working and nonworking days). This poses new variables to the territorial scheme defined, recognizing integration clues of new areas to the agglomerate and consolidation of some metropolitan subcentres, with capacity of influence on the whole structure, although the AMSF-P historically shows a hierarchical organization strongly characterized by the presence of central cities (Santa Fe and Paraná). This has been shown by different indicators.

Based on this proposal, some questions arise: Which are the attractors by which mobility in the AMSF-P is produced? Which are the consequences these mobility flows have for people or the area profile? Which are mobility possibilities and impossibilities within the area, and which are the strategies people, families and organizations use to overcome them?

DOI: 10.1201/9781003220855-12

The recognition of mobility attractors enables the introduction of new aspects that stand out as a question of further inquiries within the Project. Fundamentally, two questions arise. Firstly, the conditioning factors for selecting a living site, which then cause displacement (work, land cost, the landscaping qualities or, in terms of safety given to the place of residence, among others). Secondly, the predominance in individual transportation use for displacements connected to work.

1.2 *Background and topic status in our area*

Within the framework of the promulgation of the Provincial Law on the recognition of Metropolitan Areas in the Santa Fe Province in 2016, the identification of population displacements (concentration/de-concentration), which appear in the prevailing stage of the urbanization process in urban agglomerates in the last years, not only enables the understanding of different urban matters but it is also essential to:

- Plan and arrange territorial, local and urban development (e.g., intra-metropolitan migration processes);
- Design infrastructure policies (future transportation demands);
- Project public service demands and other additional services in attraction centres.

1.2.1 *Information sources*

Residence-work pendulum displacements are not recorded directly in a questionnaire form by the formal statistics system in Argentina, unlike in other countries. There is background material, such as the 2001 Census, which recorded the relationship of residence-work among cities, some head cities and mainly "bipolarity" situations (Paraná-Santa Fe, Resistencia-Corrientes, etc.), although the results provide general information not describing age or gender groups.

In other countries such as Brazil, Mexico and Chile, the Census of the year 2000 included, for the first time, an inquiry about place of work or study. Processing of this census generated surprising results about socio economical segmentation of daily mobility in Río de Janeiro, Sao Paulo, Santiago and Mexico City.

Likewise, the methodology background of the urban agglomerate analysis from information of census daily movements is relevant to define the functional integration of municipalities of the Brazilian Urban Net. These data, published by IPEA, IBGE and UNICAMP (2000), are used for a regionalization proposal, contributing to the discussion of analytical definition criteria of the Brazilian metropolitan regions. Other countries, such as Spain, benefit from the residence-work Mobility Atlas (2017) in the Madrid Community. It is a document featuring a design on workers' displacements between their place of origin and their place or work. This is a highly relevant issue in a Community like Madrid, in which the metropolitan issue has grown to the point of crossing administrative borders.

In addition, studies from Mexico (2017) examine the relationship between urban design and travel behaviour, having census series as the source, which provides comparative data of each decade and even inter-census surveys. In this sense, and in the first place, the definition of urban area attractors of residence-work displacements connected to the concentration of industrial and/or service areas and the possibility of developing working flow diagrams must be highlighted as well as a mapping of a composite index of centrality.

There are studies from Chile (2018) and other cases from Brazil (2008, 2011) and Mexico (2017) that should also be observed, where approaching the topic is performed from big and secondary metropolis.

1.2.2 *Mobility concept*

Research has traditionally recognized that our displacements to perform daily chores involve several practices and the ways people move are closely connected to the lifestyles in cities and territories. This daily mobility is thought to play a central role in the organization of contemporary urban life. Jirón and Imillan (2018) propose the difference between mobility as the "object of study" and as "theoretical-methodological orientation" and put forward the idea of a moving dimension in "dwelling".

The concept of "mobility turn" (Urry BS Ahwllwe 2006; Sheller 2017) describes the displacement of people, goods, ideas, knowledge, practices and capital in relation to the change of place. This enables urban studies to develop study objects from multiple relationships, considering city and territory movements and rhythms.

On addressing the issue of mobility as central, it is intended to recognize that the movement itself is not the centre, but mobility is understood as a way of dwelling (Urry, 2007). The justifying principle of this observation is the confirmation that daily life is experimented in a continuous way, overcoming daily life segregation and fragmentation from fixed time-spaces.

In this sense, methodology plays an important role, whenever it allows the "follow up" and "combination" of the set of practices, which give life to the objects apart from the delimitations seen in the observation. This opens spaces and times and the relationship nets found in the basis of the phenomenon itself.

1.2.3 *Residence-work mobility*

In general terms, population displacements through geographical spaces can adopt different ways:

1. Internal and external migration, which involves a change of residence;
2. Seasonal Migration. Although there is no change of residence, displacements can last for long periods. Among these, we may find: tourism and displacements associated to harvest cycles;
3. Daily or pendulum displacement. There is no change of residence. It takes place among residence locations and working sites, study and/or access to services. It happens with certain intervals and in general they refer to round trips and are repetitive movements.

With information on pendulum displacement, it is possible to:

1. Get to know the socio-demographical features of people involved in daily displacements;
2. Describe the sites where people habitually perform their work and study tasks or their access to different services;
3. Gather information about distances to travel from residence locations to work or study centres or access to services;
4. Calculate the time employed in round trips from residence locations to work, study or access to service centres;
5. Identify round trip flows originated in certain periods of time between residence locations and the above mentioned places;
6. Identify the reasons that justify daily displacements;
7. Be an indicator to determine agglomerates.

Furthermore, for the purpose of our interest: Identify strategies and tendencies of daily mobility (in working and nonworking days) in the metropolitan dynamics, which contribute to the approach of new variables to the territorial scheme defined.

2 MATERIALS AND METHODS

2.1 *Methodology. Metropolitan mobility displacements and life stories*

The current research is carried out by relating a qualitative methodology with the spatial analysis through "life story cases and daily mobility situations", placing emphasis on metropolitan displacements. This work is based on semi-structured interviews which provided very important accounts about the participants' experiences. We also make use of a digital geospatial tool named "Google Maps Chronologies" which allows us to obtain a set of vectors displaying the different movements in the AMSF-P. This spatial analysis tool permits the use of communication technologies and geographical references for methodological development. As a condition, the person interviewed should have a cell phone, a google account and an active GPS. Then, by using Google Maps, it is possible to have access to the route taken in the day, which is seen through geographical reference vectors.

In this way, such net enables to detect different mobility attractors; it allows the analysis of the reasons why displacements reach the metropolitan scale together with other related variables such as timetables and transportation times, as well as the recognition of the potentialities and limitations of those analytical procedures. Vectors also display the temporal travel information, which allows us to maintain a chronology – if desired – of displacements, compare periods of years or different years from future systematic records. To all this, life stories are added.

This is a technical "nonconventional" alternative for traditional transportation studies. In this case, it is used in the "mobility turn" context, which permits (very cheaply) to gather primary and qualitative

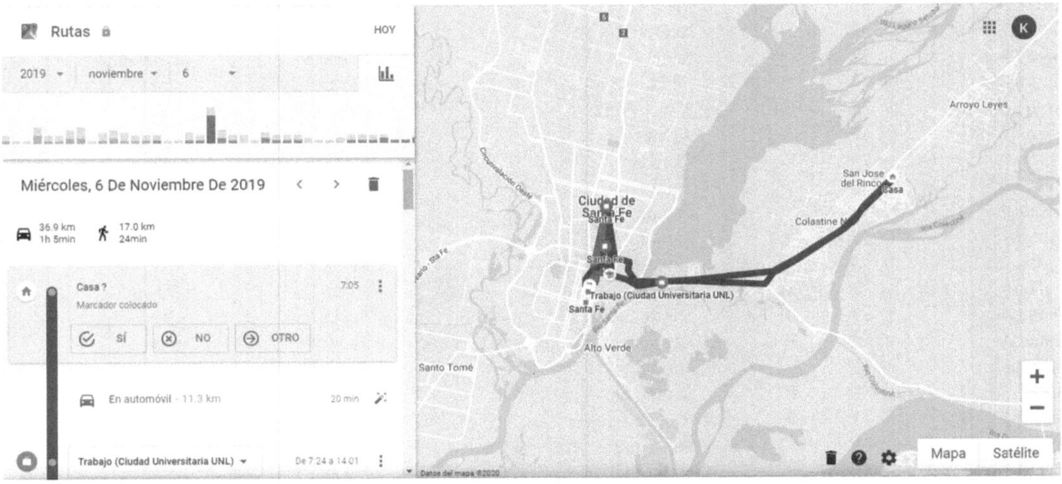

Figure 1. Traditional route vector of a person interviewed. Source: Google Maps Chronology, 2019.

"satisfaction" data, which traditional departure-arrival surveys do not consider, or at least, neglect since their priority is departure and arrival points from the displacements.

This tool does not have (it could not) the aim of quantifying data or justifying a statistical representation. On the contrary, it collects narrative records that come from the participants' "experience".

In this way, qualitative (life stories) and spatial methodological modalities (Google Maps) are related and they must cross and complement each other in order to perform this AMPSF-P relational study.

2.2 *Study area and selected cases*

At this stage, the working area is the metropolitan region, which will then resolve situations of interest deriving from this instance. It refers to a second order metropolis or a metropolis in the heart of the country with a population of approximately a million inhabitants. It has the peculiarity of resembling a mirror-system separated by the Parana River: each of them with a head city, Santa Fe (Santa Fe Province) and Parana (Entre Ríos Province), located in the central-east river coast of Argentina.

As mentioned in the introduction, this work has recorded daily metropolitan mobility conditions. For this, the participants were randomly selected and had to fulfil the condition of completing daily residence-work pendulum displacement within the metropolitan area.

Approximately 60 sample cases were collected. It is worth mentioning that being this a research stage, the starting point was the selection of people close to the researchers' circle to then continue with a more heterogeneous sample.

The result is a set of vectors, which show different displacements in the AMSF-P, and comprise "working day displacements" and "weekend displacements".

This mapped record allows the recognition of pendulum displacement and the inquiry, through the stories, about the strategies they are based on.

Finally, certain constants are separated and compared with the situations synthetized in other research papers.

In order to analyse daily mobility practices, a matrix is drawn, being the most important axes: age, gender, place of residence, place of study or work, means of transportation used, amount of travels in one day, duration of displacement, time living in the residence, reasons for choosing the place of residence, and difficulties faced, among others.

The sample features do not permit their projection to the whole population in order to make data generalization of quantity and percentage of population that fulfils this condition.

Table 1. Data base fragment. Source: Soijet M., Gramaglia V., Garcia V., Santiago J. & Pennisi B. (2019).

Nickname	Age	Gender	Place residence	Place study/work	Means of transport
María Laura	43	F	Oro Verde ER	Santa Fe	Car
Miguel/esposa	58	M	Esperanza SF	Santa Fe	Car
Mauro/esposa	46	M	Franck	Santa Fe	Car
Graciela/Norberto	61	M	Santa Fe	Laguna Paiva	Car
Gastón	30	M	San Benito ER	Paraná	Car
Irina	39	F	Paraná	Santa Fe	Bike
Ode	32	F	Recreo	Santa Fe	Car
Nico Orazi	28	M	Santa Fe	Sauce Viejo	Carpooling
Andrés N	38	M	Santo Tomé	Santa Fe	Car
Andrés G	32	F/M	Monte Vera	Santa Fe	Car
Anto	32	F	Arroyo Leyes	Santa Fe	Car
Pochi	70	F	Rincón	Santa Fe	Bus
Ema W.	30	F	Paraná	San Benito	Carpooling/bicicleta
Gisela	30	F	Paraná	Oro Verde	Bus
Miriam	49	F	Paraná	Oro Verde	Car
Paula	33	F	Ángel Gallardo	Santa Fe	Car
Nacho	35	M	Sauce Viejo	Santa Fe	Bus
Diego	39	M	Santo Tome	Santa Fe	Car
Claudia	59	F	Rincón Limite con Arroyo Leyes	Santa Fe	Car
Miriam	39	F	Paraná	Santa Fe	Bus
Aleja	32	F	Santa Fe	Rincón	Car
Flor	35	F	Rincón Pueblo	Santa Fe	Car/bicicleta
Carla	41	F	Santo tome (aires del Llano)	Santa Fe	Car
Pia	35	F	Humboldt	Santa Fe	Bus/car
Sil	36	F	Santa Fe	Monte Vera	Bus

3 RESULTS

3.1 *General description of daily mobility in the AMSF-P*

As described in previous works (Soijet, 2010, 2017), the 2010 Census show that the area has exceeded the million inhabitants and as a "peculiarity" it records the "duplication" of its central city in two urban cores – Paraná and Santa Fe – each with crowded towns around, perimeter cities to the scheme defined, and a series of small towns in the inner ring between the agglomerate and those terminal cities.

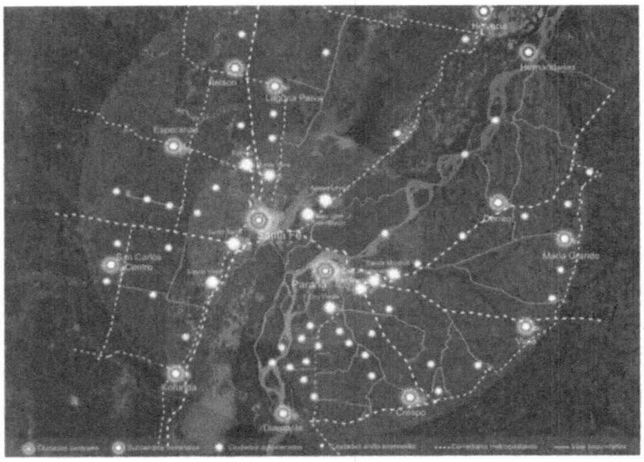

Figure 2. Satellite image. City system of AMSF-P. Source: OAMSF-P 2017.

From the observation of the cases studied (and all the TMDA previous information and transportation frequency), we can see that Paraná and Santa Fe (in their condition of head cities) show the highest agglomeration of daily displacements.

Although it is not possible to describe the population included in such displacements, sample results highlight the high schooling level as an attribute, mean ages (young adults) and employment situation related to tertiary activities. This mobility experience analysis, in which gender differences can be unveiled, is important for further studies.

As for means of transport used, we observe a generalized use of private cars; this is strongly related to the deficiencies of public transportation that can be verified in the life stories.

Although the reason for travelling is work (forced mobility) when central cities are the destination, to this motivation we may add displacements for shopping, formalities and other reasons defined by the location of education and health facilities.

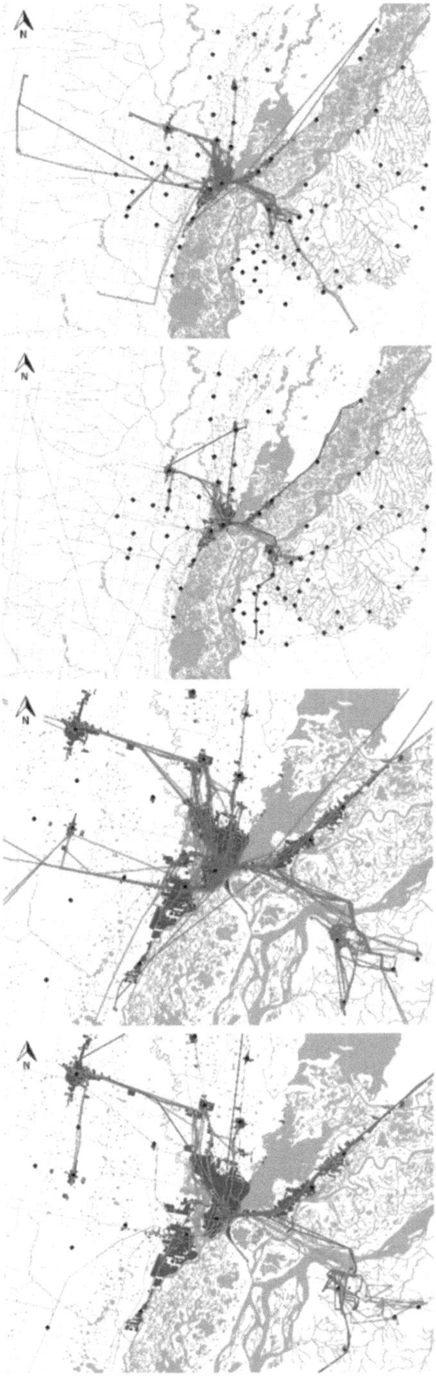

Figure 3. Working days route; picture 2: Weekend route; picture 3: Zoom. Working days route; picture 4: Zoom. Zoom. Weekend route. Source: Soijet et al. (2019).

In this sense, life stories show that pendulum displacements from agglomerated cities include the family (children to school, husband or wife to university or to his/her job).

When the destination is other cities from the areas, the reasons for travelling are visit to relatives and friends as well as leisure and recreation. These displacements occur mainly at the weekends.

According to this, the need to address daily mobility from time variability becomes relevant, since some of the functions generate a changing mobility during the week or the year: for example, functions connected to leisure show a wider territorial extension for relaxing periods than for working activities.

At the same time, according to the relationship between place of residence and place of work, that is to say, according to the place of origin of displacements, from these data we can infer that the agglomerate municipalities have a smaller autonomy degree as work markets for their own residents or they show a higher degree of interrelation for work displacements with central cities. If, on the one hand, it is possible to prove a tendency towards a no concentration pattern in the demographical scheme, this tendency decreases when it comes to employment.

Another aspect of the same issue is population dwelling in central cities and working out of the municipality. Displacements are observed to happen in the agglomerate and outer ring cities and not in intermediate ring sites.

Regarding the residents of the municipality in the area, who travel to work to other municipalities which are not a central city, displacements take place towards terminal cities (Esperanza, Crespo).

On this aspect, the focus is on the internal relationship of the areas from the emergency of dynamic urban cores. These cores are set in the ring, whose functional support in the last years has had a direct effect on the increase of mobility in the nearby municipalities.

As for mobility periods, records show that work displacements of people interviewed – who make up the universe of analysis – is predominantly a short and a medium distance; being long distance displacements only a few (less than half an hour, an hour, less than two hours).

In addition, no perceptions are recorded about the time devoted to transportation; there are no references to "wasted time" due to transport ineffectiveness but there are allusions to the stress it causes. This matter is evident among women's stories, where "risk" of transportation adds to the feeling of distress they suffer from going away from home and leaving the children. The stories also allow the observation of other matters such as:

- The connection with other participants at the moment of planning day-to-day mobility, a fact that is more common in families with children;
- The economic variable of transportation, which means the coordination of sustainable alternatives, such as carpooling;
- The reduction of "unnecessary" displacements. Everyone agrees on the advantage of doing formalities and shopping online.

Even though the sample does not attempt to be representative – for such reason statistically formatted percentage data are not given – some of the matters which are relevant and repetitive can be mentioned: There were several answers regarding the matter of leisure; nevertheless, Esperanza stands out for the repetition in the recorded accounts. It is also convenient to include leisure sites in the area, considering mobility, which correspond to weekends and holidays, to get to the places where leisure facilities or second residencies are found; these are locations connected to the riverside landscape.

Regarding the choice of place where people live, the answer "where it was possible according to budget", was the most usual and in some cases, these residences were inherited properties.

This is a process by which a population previously living in a central city move residence out of this city permanently but continues being connected to it through jobs and other daily activities. Therefore, we do not face a true rupture as for the space where the life of this population unfolds. This refers only to a change of house, a residential displacement mainly directed to the centre of the suburbs.

In these residence changes, which are much related to the need of creating new family cores, the following features are observed when choosing a place to live: job location, cost and availability of urban land, landscape features as well as safety, among others – all of these being an "excuse" for the distance to places where other activities are performed. In the case of women, some stress is added when approaching their couples' working place.

There could be other reasons not related to the ones mentioned in the justification for residential changes. In this respect, Ascher (1995) suggests the habitat style and typology, the symbolic social space where the new residence is located, the reproduction of a cultural and familiar model, etc.

These residential expansion processes generate a very relevant type of mobility – residential mobility –, which constitutes a phenomenon that deserves much interest, although it does not represent a type of daily mobility.

When examining these elements, it is possible to recognize the consolidation and growth of the metropolitan area. In this, mobility increase is not related to the spatial expansion but it is connected to the reinforcement and difficulties of the space organization of such mobility, to the development of residential mobility and to a usage and activity guideline more diversified in the metropolitan territory.

3.2 *Sources of information*

The results confirm the restrictions in information because census records do not include any inquiries related to daily displacements. It is important to study the relevance of inquiries in census' records, which justify this situation, and to check the same in other countries.

With this information, the capacity of statistical data would be greater, and perform a follow up of a 2030 Agenda, since elements will be provided for the evaluation of Objectives 8 and 11.

In the same way, information will be given for the evaluation of "Consenso de Montevideo" priority measures such as number 10, 28 and 54, which relate to the working conditions of different population groups, and those measures from number 76 to number 84, which consider the territorial resolution, development and planning, as well as the surveillance of environmental vulnerability.

4 FINAL CONSIDERATIONS

The current case of the Santa Fe-Paraná Metropolitan area is of relevance as it embodies dynamics that take place in other interior metropolis with either a higher or lower intensity.

We highlight the importance of getting Census records on residence-work daily mobility and the capability of the methodology applied for the study of territorial organization of the urban systems, given the high significance of mobility problems in the organization of this territory.

As for these contributions, we should mention that the Province of Santa Fe is under a process of creating guidelines in the metropolitan location, from the release of the Law for the Creation of Metropolitan Areas. These guidelines set the definition of mobility strategies and land use as a priority, among others.

With recent transformations happening so quickly, it is essential and convenient not only to study each one of the variables analysed, with a strong purposeful analytical base from the discipline field and with the aim of gaining from interdisciplinary but also to carry out an interinstitutional operative scheme. All this leaves an open working field which will be the purpose of our particular attention in the pursue of inquiries.

REFERENCES

Ascher, F. (1995). Métapolis ou l'avenir des villes. París: Editions Odile Jacob.

Herrera, A. Y Razmilic, S. (2018). De la casa al trabajo: Análisis de un tiempo perdido. Debates de Política Pública. Centro de Estudios Públicos (CEP), 29, 2–30.

INSTITUTO DE ESTADÍSTICA COMUNIDAD DE MADRID. (2017). Atlas de la movilidad residencia-trabajo en la Comunidad de Madrid. Madrid: Consejería de Economía, Empleo y Hacienda.

INSTITUTO DE PESQUISA ECONÔMICA APLICADA. (2000). Caracterização e tendências da Rede Urbana no Brasil. Campinas: Unicamp/IE, Convênio IPEA/IBGE/UNICAMP-IE-NESUR. 2v.

Jirón, P. Y Imilán, W. (2018). Moviendo los estudios urbanos. La movilidad como objeto de estudio o como enfoque para comprender la ciudad contemporánea. Quid 16, 10, 17–36.

Ojima, R. (2011). Fronteiras metropolitanas: um olhar a partir dos movimentos pendulares. Revista paranaense de desenvolvimento, 121, 115–132.

Sheller, M. (2017). From spatial turn to mobilities turn. Current Sociology, vol. 65, N 4, pp. 623–639

Sheller, M. Y Urry J. (2006). The new mobilities paradigm. Environment and Planning A: Economy and Space, vol. 38, N 2, pp. 207–226.

Soijet, M., Gramaglia, V., García, V., Santiago, J., Poeta, P. Y Aranovich, E. (2017). La complementariedad de procesos opuestos. Santa Fe–Paraná en el inicio de la consolidación metropolitana. Pampa, 15, 83–106.

Soijet, M., Santiago, J. Y García, V. (2018). Gran Santa Fe (Argentina): metrópoli de segundo orden. Problemáticas y dinámicas recurrentes. Limaq, 4, 81–100.

Soijet, M (2010). Las localidades del Área Metropolitana Santa Fe-Paraná: caracterización y estrategias de desarrollo. Observatorio Urbanístico Área Metropolitana Santa Fe – Paraná, 4, 6–21.

Stamm, C. Y Staduto, J. A. R. (2008). Movimentos pendulares das cidades interioranas de porte médio de Cascavel e Toledo, no Paraná. Revista Brasileira de Estudos de População, 25, 131–149.

Urry, J. (2007). Mobilities. Cambridge: London, Polity Press.

GOBIERNO DE SANTA FE. (2016). Ley Provincial 13.532 de creación de Áreas Metropolitanas de Santa Fe. https://www.santafe.gov.ar/index.php/web/content/download/228816/1198134/file/Ley%20de%20Areas%20Metropolitanas.pdf

Granados-Alcantar, J. A. Y Franco-Sánchez, L. M. (2017). Migración y movilidad laboral entre las zonas metropolitanas de la región centro de México. DOI: http://dx.doi.org/10.22185/24487147.2017.91.006.

Research Tracks in Urbanism: Dynamics, Planning and Design in Contemporary Urban Territories – Allegri et al (eds)

Threats and evictions: Dispossession and displacement in the center of São Paulo, Brazil

R. Abramowicz Santos
FFLCH, Faculty of Philosophy, Languages and Human Sciences, Universidade de São Paulo, Brazil

ABSTRACT: This article discusses forced urban evictions, and uses eviction threats and its executions to analyze the process of urban space production and the way of life led by those communities affected by continuous dispossession and displacement experiences. The research took place in the center of the city of São Paulo, State of São Paulo, Brazil, a region marked by successive threats, demolitions, evictions and violence, which transform and affect spaces and ways of life. The article aims to describe how these processes are territorialized and how this displacement-settlement movement – even when a precarious, unstable and temporary settlement – produces spaces, individuals and the city.

1 INTRODUCTION

This article discusses forced urban evictions. According to the United Nations Committee on Economic, Social and Cultural Rights, forced eviction is the "permanent or temporary removal against the will of individuals, families or communities from their homes or land, which they occupy, without the provision of, and access to, appropriate forms of legal or other protection"[1]. Amnesty International defines evictions similarly, and adds that it happens "often with the threat or use of violence" (Brickell et al. 2017, p.1). As reported by the Evictions Observatory[2], between 2017 and 2018, there were at least 808 eviction threats or cases in the metropolitan area of São Paulo; 75% of them due to repossession (Rolnik & Lins, 2018, p.21). Those numbers are under-calculated; moreover, they hide the dynamics and the dimension of the human tragedies related to each event. The United Nations Human Settlements Program (UN-Habitat) recognizes forced evictions as a global phenomenon and a global crisis (cf. Brickell et al. 2017, p.1).

Saskia Sassen (2014) analyzes this global dimension by using and developing the concept of "accumulation by dispossession", created by David Harvey, and proposes the notion of expulsion as an analytic and descriptive category of ongoing processes in contemporary capitalism. The author focuses her analysis on the diverse mechanisms and modes of population expulsion and forced evictions in different contexts and regions around the world. Even though expulsions are not a new phenomenon in the history of capitalism, Sassen affirms that, over the last decades, they have happened based on new configurations, with significant changes in scale, range and intensity.

To describe and analyze the system dynamics taking place behind the expulsions, in their various forms, Sassen claims that it is necessary to combine empirical research and conceptual elaboration and formulation work (2014, p.215). This research can be done by analyzing the political instruments, tools and agencies working in the contexts in which they operate: increasing employment and precarious work; downsizing social protection systems; expanding mining projects, and other transnational corporations that expel small farmers from their land; and the loss of housing resulting, for example, from the bursting of the "housing bubble" in the United States and other European countries (Sassen, 2014). It means to investigate the "emergence of new logics of expulsion" (Sassen, 2014, p.1) that accompanies these movements of market frontiers expansion under the auspices of the financial

1 "Committee on Economic, Social and Cultural Rights, general comment No. 7, item 3", available at: https://tbinternet.ohchr.org/_layouts/15/treatybodyexternal/Download.aspx?symbolno=INT/CESCR/GEC/6430&Lang=en. Accessed on December 11, 2020.

2 The Evictions Observatory is a project that involves researchers and research laboratories from different Brazilian universities (from São Paulo, Rio de Janeiro, Belo Horizonte and Fortaleza), whose aim is to monitor, analyze and reflect upon the collective processes of ongoing evictions. To know more, access http://www.labcidade.fau.usp.br/observatorio-de-remocoes/.

DOI: 10.1201/9781003220855-13

capital. It is an expansive logic that operates through instruments that allow expanding what can be financialized, in a way that finances end up "to invade – that is, securitize – nonfinancial sectors" (Sassen, 2014, p.9).

Raquel Rolnik (2019) addresses the advance in financialization and its effects on people's lives. In her book, she analyzes the "colonization" of housing and urban land by finance in different contexts and cities around the world. According to Rolnik (2019), the financial colonization process that has happened in the last decades, at a global scale with local peculiarities, has transformed housing and urban policies in a mechanism for profit extraction, wealth accumulation and financial gain. With financialization, urban and rural land became "highly disputed assets" which increased "market pressure on territories controlled by low-income communities" and put those communities "under the constant risk of territorial assets being despoiled" (Rolnik, 2015, p. 114).

One of the ways in which "colonization" operates and makes this "threat of spoliation" effective is by occupying territory "by means of evictions and demolitions and replacement of previous ways of life" (Rolnik, 2019, p. 196). The consequence is "mass displacement" of land and people – especially the poorest and most vulnerable, more exposed to the "global crisis of tenure insecurity [due to] the advance of real-estate financial complex and its direct impact on housing rights" (2019, p.6). As a result of threat and violence situations, conflicts and insurgencies explode in the urban space, to which the author refers as "war of places" and "for places", the urban warfare (Rolnik, 2019).

Using these "systemic tendencies" as reference, researchers who study the "forced eviction" issue have analyzed how those global processes are localized and territorialized. To that extent, ethnographic research projects have been carried out to analyze and criticize the ongoing processes and used as a research tool to monitor and develop translocal and multiscale dynamics. In a recent issue of the *Année Sociologique* journal, the dossier coordinators raise important questions about the relation between "recent capitalism transformations, the government of territories and populations, and poverty and outskirts management" (Aguilera et al. 2018, p.16). For the authors, the challenge is to formulate a concept of expulsion that is operative and grants a comparative perspective that fills the space between the different scales – the global and the local (Aguilera et al. 2018, p.16).

As they recognize the diversity and plurality of forms, causes, effects and actors (Aguilera et al. 2018, p.31). Furthermore, the authors identify three big common traits concerning evictions: it is always a matter of expelling people, either from a country or from their homes; it is always a forced movement, coercive; and displacement is always both spatial and social (Aguilera et al. 2018, p.29). Evictions can also be a chronic experience for individuals and families (Aguilera et al. 2018, p.21), resulting in worse living conditions, which mark and shape their social trajectories.

Another collection, *Geographies of forced eviction: Dispossession, violence, resistance*, published in the United States, addresses this local scale from an empirically grounded perspective, connected with ethnography, in order to retrace the cartography of these evictions. According to their coordinators, it is a "range of experiences and multi-scalar expressions of forced evictions that are unique but also common to these diverse local environments" (Brickell et al. 2017, p.2), also understood as an "embodied, located and grounded phenomenon" (2017, p.3).

Based on empirical research, the authors aim to, on one hand, overcome definitions and analysis that deal with evictions as phenomena that happen exclusively in the Global South and rural zones. On the other hand, the authors aim to highlight "the distinct urban logics of precarious housing or involuntary displacements" (2017, p.2), which have in common "the convergence of changing market forces, deepening housing insecurity and intensifying logics of dispossession" (2017, p.4). The authors call the attention to the "creative destruction of urban landscapes and the cardinal significance of cities as engines of dispossession" (2017, p.4) and focus "on the relationship between neoliberal urbanization and the historical production of specific 'regimes of dispossession'" (2017, p.5).

Sassen (2014) mentions the importance of spaces built by expelled and displaced people. Those spaces have multiplied in many different ways and, according to the author, they need to be recognized by means of a descriptive, analytical and conceptual work. In this sense, it is important to describe and analyze the dynamics and spaces in which evictions happen, as well as the territorialities built where the displaced people are settled, taking into consideration: the urban materialities (material conditions, infrastructure and urban networks) that affect ways of life; the social and affective bonds that are forged (and perhaps, broken); and the resistance and support networks built during such critical events (Brickel et al. 2017). In short, it is about analyzing the dynamics and practices that produce both displacement and emplacement (Schiller & Çaglar, 2015).

In Michel Angier's terms (2013), the construction of these places must be understood as processes that produce forms of urbanism, spaces that interact and affect local urban dynamics, create political facts and build relationships with public agencies, social groups and political movements. It is a live and dynamic process of "making the city", which, according to Agier (2015) can, therefore, be ethnographic.

2 BUILDING SPACES AND LIVES ON THE IMMINENT, RECURRING AND UNDETERMINED RISK OF EVICTION AND DISPLACEMENT

Harvey (2012), when analyzing what he calls "accumulation by dispossession", explains the gears that move the urban, economic and political dynamics behind mass dispossessions. Evictions are only one side of these, connected with the constant movement of promoting new capital accumulation. Harvey investigates how urban restructuring projects, for example, fulfill the task of showing how urbanization has historically played the role of absorbing capital surplus and, thus, leading to a "creative destruction" whose inherent and intrinsic effects are violence, outbreak of conflicts and "the dispossession of urban masses" (Harvey 2012, p. 22).

If the practice of "accumulation by dispossession" is recurrent in urban history, it makes one wonder if it is different in the contemporary dynamics. As Sassen (2014) shows in her book, under the Fordist production system there were processes of social integration through labor market, consumer market and urban policies. The question now arises concerning what place conflict holds in urban spaces where, while dispossession processes are accelerated and amplified, we witness the erosion of conventional conflict management and negotiation bodies and mechanisms. At the same time, the process of integration through labor market, consumer market and urban policies is interrupted and reduced. Those integration forms have been historically established as social and political grounds for the demand and development of urban social movements, and have also designed the research and urban studies agenda (Arantes, 2009).

Yiftachel uses ongoing processes in the so called Global South – also increasingly more frequent in the Global North – as a reference to show that the increase in populations' dispossessions and displacements ends up producing "gray spaces", which, for the author, "have become a dominant feature of contemporary urbanism" (2009, p.243). The gray spaces are characterized by an ambivalent and unstable game: on one side, the legality, stability and safety; on the other, eviction, destruction and death. This situation is controlled by indefinite waiting and expectancy of a possible eviction or a possible solution, being neither of them fully integrated nor eliminated (2009, p.244).

The uncertainty of living on the verge of being evicted, on the "condition of displaceability" marked by the threat of displacement (Yiftachel, 2018), establishes a social and urban experience, a way of experiencing the "contemporary urban citizenship" (Yiftachel, 2018), lacking the promise or normative reference of future stability and security of their ways of life. The uncertainty, therefore, has effects that go beyond the territorial limits of the spaces affected by this condition of "permanent transitoriness" (Rolnik, 2015, p. 128), insofar as it constrains people's urban lives and trajectories. This circumstance of uncertainty and threat, which is part of the permanent transitoriness experience, starts to govern the lives of people and families who live there as well as their access to home, places of work and socialization, and urban mobility.

On one hand, as Schiller and Çaglar (2015) state, it is about retaining the "urban scale" as an analysis plan that enables to deal with the convergence points – from the point of view of urban materialities (urban networks and infrastructures), and the broader and shared social experience of precarity and dispossession – to put under perspective the different social groups that are subject to socio-spatial displacements, together with their trajectory – the trajectories used as a tool to analyze the mechanisms as well as the operating and situated methods of eviction and displacement.

On the other hand, due to the scale and centrality of evictions, it is important not only to understand the connection between the dynamics and the mechanisms that cause displacements, but also the process of establishment/settlement, or as Schiller & Caglar call it, "emplacement" (2015), even when temporary and precarious. Mobility practices and how they are installed and rearranged (emplacement) assume and build spaces and territorialities as well as mediations, circuits and sociability. Those elements can also be used to analyze the procedures, instruments, social relations and political groups that are mobilized and participate in moments of conflict or removals, or in processes and arrangements that produce territorializations and permanence. Providing permanence is an attempt and an "*expectancy* of stability", which is different from stability itself (Rizek et al. 2015, p.302, emphasis by the authors).

At the heart of the different territorialization forms and practices is the dispute over the status of these places, which is revealed in the distinction between formal-informal, in the right to existence and permanence, and in the threats posed. What is at stake behind these disputes is the political difficulty of State power to define what can be accepted and tolerated from what must be repressed and criminalized.

According to Ananya Roy, formality is not fundamentally separated from informality, as two distinct and opposite states (Roy, 2005). The author discusses the urban informality as "a *mode* of urbanization" and sees it as "a series of transactions that connect different economies and spaces to one another" (2005, p. 148, emphasis by the author). In her perspective, informality must not be understood as a mere object of state regulation (assuming that formalization or regulation through market can solve structural problems related to deeper inequalities), but as produced by the State itself. Roy affirms that "the planning and legal apparatus of the state has the power (...) to determine what is informal and what is not, and to determine which forms of informality will thrive and which will disappear" (p.149). Therefore, "State power is reproduced through the capacity to construct and reconstruct categories of legitimacy and illegitimacy" (p.149).

This capacity, the rules and codes that produce and define informalities and their (il)legitimacy to produce urban space, provoke disputes that comprise a variety of actors and networks, with multiple scales and repertoires. Those can involve different state agencies and institutions, housing movements (more or less organized), technical and legal assistance, civil society organizations, culture and art collectives, as well as churches, organized crime groups, political parties and labor unions, researchers and college students, etc.

All those various actors and collectives, with their different repertoires, practices and agendas, revolve around the different spatial arrangements and ways of life, reinforcing the heterogeneity and the political aspect associated with the production and the management of those spaces and people (Santos & Guerreiro, 2020). The dispute is not only about appropriation and use of the space, but also about how the population in those spaces is managed by the State and its agencies, or by those various actors and repertoires, present and active during conflicts and space configuration and arrangements, and its production and territorial management.

Because the spaces and ways of life that happen in zones of indeterminacy, in different degrees of informality, are subject to so many legal acts, orders and norms – not only the State –, they also produce modes of subjectivation. In Roy's words, that is the "civic governmentality" (2009) marked by the ambivalence that emerges from the movement between legal and illegal, the different types of subjections, accommodations and confrontations of those spaces and people – both expected and produced to be governable – as well as what escapes those instruments and regulations. According to Yiftachel, it is "the emergence of new subjectivities through the production of space" (2009, p.240) by means of the invention, appropriation and transformation of coping skills and survival strategies, in order to make those places "bases for self-organization, negotiation and empowerment" (p.243).

3 THREATS OF EVICTION, EVICTIONS AND DEMOLITION IN THE CAMPOS ELÍSEOS NEIGHBORHOOD, IN THE CENTER OF SÃO PAULO CITY

The Campos Elíseos neighborhood, in the center of São Paulo, is the heart of the so called *cracolândia* (crackland). The place is identified by an itinerant but constant presence of a crowd of crack and other drugs addicts, and open-air drug market, known as *fluxo* (flow). Because it is one of the main places to buy and sell drugs, there are frequent operations by the police and regular military forces in the region in the name of *the war on drugs*. They are violent events, marked by violation of human rights, which affect the addicts as well as those who live and work in the area.

This region is also the focus of a large-scale urban development project currently under implementation. In the last years, there have been frequent police operations either with the excuse of combating crimes associated with the *cracolândia* (organized around the war of drugs), or during repossession processes in the region and the immediate surroundings. The area is a privileged place of convergence and concentration of organized practices of space production, marked by the threat and the recurrence of dispossession and displacement processes.

In the heart of Campos Elíseos neighborhood is the *Largo Coração de Jesus* (Heart of Jesus Square) across from the *Santuário do Sagrado Coração de Jesus* (Shrine of the Sacred Heart of Jesus), a church whose construction began in the 19th century, when the neighborhood was place of the coffee aristocracy. The Square is very close to *Estação da Luz* (Luz Railway Station), and especially close to *Estação Júlio Prestes* (Júlio Prestes Train Station), two main urban nodes of the city, where train lines connect with two subway lines, making it an important urban gravity area in São Paulo (Santos, 2019). It is a densely populated area, home to a heterogeneous population and to complex dynamics, set in a well-urbanized region in terms of infrastructure, facilities, shops and services.

According to the mappings and field surveys carried out in the area in recent years, most of the population live in rented places, usually in precarious conditions due to lack of maintenance. Many tenants use their place of residence as a workplace too – some have small street-facing businesses. The mixed-use property makes it difficult to homogenize and assign a unique category to those places. The same happens with many rental buildings, small hotels and tenements in the area, whose public is varied: from those who spend a night on a bunk bed or just a mattress, to month-to-month tenants, or a room by the hour, usually used for prostitution and drug use. In those boarding houses, there are many people just passing through the region, but also individuals and families who have lived for a long time in the same place. There are also migrants, especially Latin-Americans, and gypsies, who, in a kind of nomadic pendulum, spend some months away and then return – these dynamics have been repeated for almost twenty years, alternating boarding houses in the *cracolândia*. Therefore, the socioeconomic situations and realities, origins and social arrangements vary significantly. Besides the tenements, residences, and small businesses, there is also the occupation of small and precarious places, squats that are not linked to any of the traditional and more consolidated housing movements in the central area.

Around those blocks, we have the Júlio Prestes Train Station, which is part of an urban complex comprised of the *Sala São Paulo* (São Paulo Concert Hall – headquarters of the São Paulo Symphony Orchestra) and the *Memorial da Resistência de São Paulo* (São Paulo Memorial of Resistance), housed in the former *Departamento Estadual de Ordem Política e Social de São Paulo, Deops/SP* (Department of Political and Social Order of São Paulo), active during the civil-military dictatorship. Across from the station, there is the Júlio Prestes Square, which was renovated and inaugurated in 2018, and new housing towers, built very recently and quickly under the Public-Private Partnership (PPP) model carried out by the State Department of Housing in agreement with the City Hall. The partnership established a new (and concrete) source of production and market expansion, which has already impacted and transformed the territory landscape and urban life as well as become one more example of the "new type of partnership between state and local governments that lead to processes of territorial restructuring" (Rolnik et al. 2018, p.207).

During the last decade, the region has been affected by demolitions and evictions, sometimes of whole blocks, due to a series of urban actions. The nature, dimension and effects of different project implementation (or attempts to implement, since many never left the drawing board, despite the demolitions) vary significantly from one project to the other, or one period of time to the other. They range from large cultural projects that began in the 90s to, more recently, the major housing public-private partnership (PPP) (Villela et al. 2019).

In 2005, as a result of a project called New Luz Project, which was strongly confronted by the housing movements and many others actors (Gatti, 2005), whole blocks were demolished and tenements and hotels were closed. In 2007, the old Luz Bus Station building was expropriated and, in 2010, it was demolished to give place to a cultural complex that never left the drawing board. In 2013, another whole block was demolished next to the old bus station, and their residents were removed. The result was a huge wasteland left behind until recently, when PPP Housing started being built. From May 2017 on, after an extensive state government operation in *cracolândia*, together with buildings being closed and demolished, threats and evictions restarted, more intensely and frequently, on the promise that the area would be integrated to the PPP Housing Project, next to the blocks affected by the operation (Santos, 2019; Villela et al., 2019). In 2018, the residents of a nearby block were removed and all the buildings were demolished to give place to a new PPP, this time led by the State Health Department for the construction of the Pérola Byington Hospital.

This brief report reveals how intense and frequent evictions and demolitions have become part of the urban dynamics, affecting the lives of those who live, work and move within the region, and shaping an experience based on permanent threats and risks. The evictions, their effects and the intricate web of relationships that trigger them happen in a territorialized way: the threat and eviction processes not only territorialize but also produce spaces, experiences, networks and urban circuits. The ethnographic research carried out in the territory over the years has revealed political, social and urban dynamics that can be initially and briefly organized and described as follows.

There is a first dimension regarding the reconstitution of power games that are built around the events related to the evictions. The evictions trigger a network of actors and measures, which include, besides the residents themselves, State agents, technicians and politicians, lawyers and those responsible for applying the law, media, researchers, collectives, and support and solidarity networks. Those events operate as a gravitational field (Telles, 2015), which involves subjects, spaces, practices, and resource mobilization and management, whether it is to face, avoid or carry out the evictions. The interactions and tensions implied in this network, and the power games that happen because of the different actions and disputes, establish new facts, effects, and urban and political arrangements.

A second matter refers to the experiences and the modes of subjectivation produced by living under the threat or the possibility of recurring displacement. These events produce effects that affect directly the ways of life and urban trajectories built by these populations, who experience or are on the verge of losing everything – and in many cases, losing everything again. Despite those people's suffering and trauma, their urban mobility and trajectories are led by uncertainty, and many times, involve successive displacements.

The emotional dimension related to forced evictions is of the utmost importance. In addition to connecting the global and public dynamics of the situation with the private, personal and intimate aspects of the people affected by it, this dimension also reveals how people are deeply and emotionally affected by the experience. It is a traumatic experience, since its achievement is not restricted to material and physical conditions and impacts, but it is the destruction of a house, a home, circuits and networks, a whole life – subjective, affective, domestic and community universes (Brickell et al., 2017).

Around each of the events that put the population at the risk of eviction or displacement, we have the main thrust that shapes the city's political and urban dynamics as well as the conflict situations that happen in those friction points in the "encounters with the state" (Trouillot, 2001, p. 125).

The state, its agents and its laws are present in different forms, from the most open and direct, as in confrontations and repossessions, to more subtle, ordinary and silent ones, as in the tracks and "signatures" (Das, 2004) that regulate and shape the ways of life, displacement and circulation, and the management of bodies and spaces. A paradigmatic example of "undesirable" population (Agier, 2008) meeting and management was the camp established after the *Largo do Paissandu* (Paissandu Square) fire and consequent collapse of a squat in the Wilton Paes de Almeida building, in the center of São Paulo, in May 2018 (Santos & Guerreiro, 2020).

Those forced displacements and the distress due to the threats and violence present, not only mark the life histories of those who, amid uncertain and adverse conditions, are displaced and remake or build their lives under constant insecurity and uncertainty, but these conflicting components and events also reconstruct the history of production – and dispute – of the city itself.

Moreover, the evictions – and their repetitions – that we have followed in the center of São Paulo in the last years, have shown us the common and recurrent aspects in the related and engendered dynamics: (I) the precarious, vulnerable and insecure conditions experienced by the population are, many times, worsened by the forced evictions, resulting in more precarious situations; in addition, there is the possibility of going through new evictions in a near and uncertain future. And (II) forced evictions disrupt, but also produce, amplify, rearrange, and strengthen certain circuits and networks in urban areas; they also promote new modes of circumvention, circulation and establishment/settlement around the city, often on a regional level.

There is still a third dimension related to the "geographies of resistance" (Brickell et al. 2017), built and triggered by the threats and evictions, and to the local solidarity and contestation networks as well as the translocal networks, which reveal the lines of force that frame the conflict area at the occurrence of forced evictions. It is in this sense that analytical (and political) density can be given to the forms of interaction that were (and are) present during violent operations of public authorities in the region of Campos Elíseos. In view of the extensive regional and local governments' operation in May 2017, and the series of interventions and uncertainties that followed, diffuse confrontations and resistance were intensified (Santos, 2019; Villela et al., 2019): activist collectives were mobilized, and public prosecutors and defenders interacted with local population, entities and collectives. As a result, a local Community Council was elected and established, a process which was only possible because the blocks under siege were demarcated as *Zona Especial de Interesse Social ZEIS-3* (Special Zone of Social Interest, type 3).

The Special Zone of Social Interest type 3 "stands out in relation to the other ZEIS demarcated by the local government for its geographic location – in a central area, with good public transportation system and facilities, where property interest and appreciation are strong" (Gatti, 2015, p. 38). In the beginning, according to the City Master Plan, all interventions, projects and assistance for the population who live and work there should be discussed by one local Community Council, whose members represent both the government and the civil society.

Another consequence of these events, the *Fórum Aberto Mundaréu da Luz*[3] (Open Forum *Mundaréu da Luz*) was created – a cooperation between several organizations and collectives already existing and socially legitimized that had been acting in the region or had already dealt with forced evictions and human rights violations. The members came from different areas: urbanism and social sciences, mental health, harm reduction programs and anti-prohibition groups, culture and arts (such as the theater companies in the region), etc., all of them in contact with residents and workers of the affected areas. In that manner, the collective aims to develop and organize a political, urban and cultural planning in the area, engaging with and respecting people's realities, needs, aspirations, and knowledge along with their histories, which shape and build the place.

4 FINAL REMARKS

When we take the city as a reference (Telles, 2006) or retain the urban scale (Schiller & Çaglar, 2011), we can observe a common analytical and empirical process which enables us to see and work on the points of convergence among all these people who experience displacement and dispossession processes and threats. A shared experience of instability affects all threatened or displaced populations. Despite the specificities and differences between them, there is a common share set of experiences that is singularly urban. As Feldman-Bianco (2015; 2018) proposes, it is an analysis plan that allows to consider the causes of mobility (and immobility) and inequality, by putting under perspective those who are displaced (internally and externally) and dispossessed in a global scenery marked by and made of precarious work, financialization and state violence. It is also important to remark that these processes and dynamics cannot be understood without considering the mechanisms of racialization that constitute them (Feldman-Bianco, 2015; 2018; Yiftachel, 2018).

For Schiller and Çaglar (2015), the urban materialities and the shared experiences of precarity and dispossession result both in displacement as in emplacement – settlement and permanence. For the authors, the materialities and the shaping of these common conditions allow and lead to strong sociability, which represents new forms of integration and social actions of broader urban and social collectives. Hence, they are relationships that establish, or can be established, from precarity (Butler, 2015).

3 To know more, access *Mundaréu* page on Facebook or the website https://mundareudaluz.org/.

Therefore, the analyses and connections of different interrelating and intersecting planes and dimensions of the threat and displacement experiences in the center of São Paulo, which were presented here briefly, help us put into descriptive and analytical perspective the several dimensions, scales and effects associated with forced evictions and forms of territorialization: the impact and consequences on the lives of the displaced populations and urban networks; the role of public agents in the context of evictions, and the mechanisms put into action for emergency management and population support; the outbreak of conflicts; the forms of resistance, the collectives and political networks that are established and show the political aspect of the urban settlements; and the emplacement practices and strategies produced by those people, which build spaces and lives in precarious and unstable arrangements. In conclusion, it is an urban event experienced by people whose continuous movement of displacement-settlement-threat produces subjects, spaces, State and city.

REFERENCES

Agier, M. 2008. *Gerer les indésirables*. Paris: Flammarion.

Agier, M. 2013. *Campement urbain: du refuge naît le ghetto*. Paris: Payot & Rivages.

Agier, M. 2015. *Anthropologie de la ville*. Paris: PUF.

Arantes, P. 2009. Em busca do urbano: marxistas e a cidade de São Paulo nos anos de 1970. *Novos estudos CEBRAP* (83), 103–127.

Aguilera, T. et al. 2018. Politiques de l'expulsion: acteurs, enjeux, effets. *L'Année sociologique* 68(1): 11–38.

Brickell, K. et al. 2017. *Geographies of forced evition – dispossession, violence, resistance*. London: Palgrave Macmillan.

Butler, J. 2018. Corpos em aliança e a política das ruas: notas para uma yeoria perfomativa de assembleia. Rio de Janeiro: Civilização Brasileira.

Das, V. 2004. The signature of the State: the paradox of illegibility. In Veena Das & Deborah Poole (eds.), *Anthropology in the margins of the State*. Santa Fe: chool of American Resarch Press.

Feldman-Bianco, B. 2015. Apresentação: deslocamentos, desigualdades e violência do Estado. *Ciência e Cultura*, 67(2), 20–24.

Feldman-Bianco, B. 2018. Anthropology and ethnography: the transnational perspective on migration and beyond. *Etnográfica* 22(1): 195–215.

Gatti, S. 2015. *Entre a permanência e o deslocamento: ZEIS 3 como instrumento para a manutenção da população de baixa renda em áreas centrais*. Thesis, Universidade de São Paulo.

Harvey, D. 2012. *Rebel cities: from the right to the city to the urban revolution*. London: Verso.

Rizek, C. et al. 2015. Viver na cidade, fazer cidade, esperar cidade. Inserções urbanas e o PMCMV-Entidades: incursões etnográficas. In Caio Amore et al. (eds.), *Minha casa... e a cidade?* Rio de Janeiro: Letra Capital.

Rolnik, R. et al. 2018. *Cidade Estado Capital – reestruturação urbana e resistências em Belo Horizonte, Fortaleza e São Paulo*. São Paulo: FAU-USP.

Rolnik, R & Lins, R (eds). *Observatório de Remoções – relatório 2017-2018*. São Paulo: FAU-USP.

Rolnik, R. 2015. *Guerra dos lugares*. São Paulo: Boitempo. In english: *Urban warfare: housing under the empire of finance*, 2019. London: Verso.

Roy, A. 2009. Civic Governmentality: The politics of inclusion in Beirut and Mumbai. *Antipode*, 41(1): 159–179.

Roy, A. 2005. Urban informality: toward an epistemology of planning. *Journal of the American Planning Association* 71(2): 147–158.

Sassen, S. 2014. *Expulsions – brutality and complexity in the global economy*. London: The Belknap Press of Harvard University Press.

Santos, R. 2019. Na cidade em disputa, produção de cotidiano, território e conflito por ocupações de moradia. *Cadernos Metrópole* 21(46): 783–805.

Santos, R. & Guerreiro, I. 2020. Ocupações de moradia no centro de São Paulo: trajetórias, formas de apropriação e produção populares do espaço – e sua criminalização. In Fernanda Moreira, Raquel Rolnik & Paula Santoro (eds). *Cartografias da produção, transitoriedade e despossessão dos territórios populares: relatório bianual 2019-2020*. São Paulo: Observatório de Remoções.

Schiller, N. & Çaglar, A. 2011. *Locating migration: rescaling cities and migrants*. New York: Cornell University Press.

Schiller, N. & Çaglar, A. 2015. Displacement, emplacement and migrant newcomers: rethinking urban sociabilities within multiscalar power. *Identities* 23(1),17–34.

Telles, V. 2006. Trajetórias urbans: fios de uma descrição da cidade. In Vera Telles & Robert Cabanes (eds). *Nas tramas da cidade: trajetórias urbanas e seus personagens*. São Paulo: Humanitas.

Telles, V. 2015. Fronteiras da lei como campo de disputa. In Patricia Birman et al. (eds). *Dispositivos urbanos e trama dos viventes: ordens e resistências*. Rio de Janeiro: FGV.

Troulliot, M. 2001. The anthropology of the State in the age of globalization. *Current Anthropology* 42(1), 125–138.

Villela, F. et al. 2019. Permanent transitoriness and housing policies: inside Sao Paulo's low-income private rental market. *Radical Housing Journal*, Issue 1.2, 27–43.

Yiftachel, O. 2009. Critical theory and 'gray space' – mobilization of the colonized. *City* 13(2-3): 240–256.

Yiftachel, O. 2018. Displacebility – a southeastern perspective. Displacement Research & Action Network, http://mitdisplacement.org/symposium-oren-yiftachel.

Research Tracks in Urbanism: Dynamics, Planning and Design in Contemporary Urban Territories –
Allegri et al (eds)

Mobility, urban design and public space in intermediate cities

R. Hidalgo & F. Temtem
School of Architecture, Pontificia Universidad Católica de Chile, Chile

ABSTRACT: The new mobility infrastructures that face contemporary accessibility problems are potential elements of urban development and of improvement of public space in the city. This fact is verified to the extent that its design is integrated into the geography, history and particular dynamics of an urban environment and acquires particular relevance in the field of intermediate cities. Under this hypothesis, we developed a Research and Project Workshop on intermediate cities in Chile, whose results constitute the central body of this paper. Based on the territorial analysis and the formulation of projects for La Serena, Iquique, Alto Hospicio, Valdivia, Osorno, Puerto Montt and Puerto Varas, local solutions to global urban problems are discussed. The work is framed within the parameters of action research, emphasizing the exploratory nature of research and explicitly addressing the dimension of knowledge that arises from the project.

1 INTRODUCTION

The growing mobility demands require new urban and transport solutions, where infrastructures go beyond their mono-functional condition to become prominent components of complex urban projects. Railways, trams, promenades and cycle paths are designed today as unique parts of urban projects, capable of not only feeding the morphological structure and identity of each metropolis, but also to stimulate its economic growth and well-being (Andrés, 2008).

These projects receive special relevance in intermediate cities, comprising between 50,000 and one million inhabitants, which have experienced an important development during the last decades. As UNESCO states (1999), the intermediate cities develop remarkable physical and social mutations associated with the implementation of new mobility systems. A subject that deserves a particular analysis, as well as the formulation of new policies and design strategies. Even though studies about intermediate cities have contributed to organizing common settlement and mobility patterns, they have not been able to shed enough light on the local dynamics (Salazar et al. 2018).

Chile is not excluded from this phenomenon and presents a group of intermediate cities where local patterns clash with new habits, challenges and global urban conflicts. These are cities placed on diverse geographies, which reproduce Santiago's urban model without taking into account the local dynamics (Barton & Ramírez, 2017).

Facing this scenario, and framed within the Urban Project Master's degree of the Pontifical Catholic University of Chile's School of Architecture, we developed a Research and Project Workshop about intermediate cities in Chile, questioning: How may the implementation of new mobility systems stimulate specific development and urban design processes in those contexts? The main goal was to identify local issues capable of informing and generating urban projects to improve comprehensively the mobility conditions and public space quality of those cities.

We worked in the cities of Iquique, La Serena, Valdivia, Osorno, Puerto Montt and Puerto Varas. From their urban historical comprehension, current dynamics and future projects, as well as their morphology, geography, landscape and uses, key issues were identified that led to the creation of an urban project associated with the implementation of new mobility infrastructures.

Under these terms, the workshop led the development of a thesis-project, which explored the critical vision of the student, and at the same time, stimulated the development of design answers. Hence, our work is framed within the parameters of investigation-action, emphasizing the exploratory character of the research and addressing explicitly the dimension of knowledge that rises from the project. It is about an innovation related methodology, putting forward "operative truths", designs or objects with the potential of changing a city's reality.

2 GEOGRAPHICAL CONNECTIONS

2.1 *La Serena, tensions and opportunities between the canal and the river*

Bays, hillsides and rivers, among other geographical elements, have affected the form and development of

DOI: 10.1201/9781003220855-14

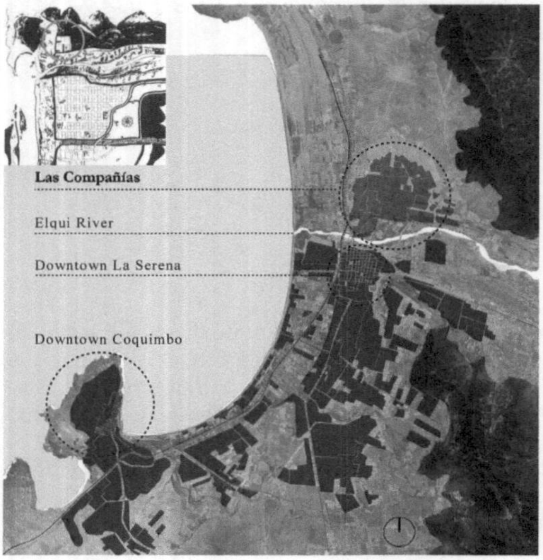

Figure 1. Coquimbo–La Serena conurbation and historical map of La Serena. Source: (Buglio, 2019).

our cities since their beginning. Such is the case for La Serena, founded by the Spanish in 1546, right by the Elqui river estuary, above the second marine terrace of the Coquimbo bay. A place from which, due to the rise of the mining industry and the arrival of the railway, the city grew and became a steady counterpart to its neighboring port Coquimbo, currently forming between the two a 450,000 inhabitant's metropolitan area (2017 census). Towards 1920, as a result of Chile's northern saltpeter crisis, La Serena experimented the arrival of immigrants that established themselves in an informal fashion in the historical center, giving way to the formulation of the Serena Plan between 1947 and 1952. Such a plan ordered and projected the city's growth, defining public infrastructure, avenues and public spaces in articulation with the slopes and watercourses of the territory. It also put forward the idea of recovering the historical center, ordering the relocation of the informal population in the Las Compañías sector on the north bank of the river, causing a socio spatial segregation, a phenomenon still present to this day.

Nowadays, Las Compañías concentrates 40% of La Serena's population, who are in great measure a vulnerable population, having to travel downtown on a daily basis to obtain jobs, services and public infrastructure. For this purpose, there are only three bridges above Elqui River and a deficient shared taxi and bus public transportation service, which added to the increase of private cars, creates long waiting lines towards the city's access, causing the rise in travel delays, traffic jams and environmental pollution. This metropolitan segregation phenomenon adds up with another less visible form of margination, inside Las Compañías. Just as the Elqui River hinders the connection between the city and the general conurbation, the Jaramillo canal, which runs at medium height along the northern bank of the river, determines socio-spatial segregation phenomena at a neighborhood level. Partially covered, the canal creates a break in the residential grid (Figure 2). It constitutes an abandoned cordon of closed walls and a focal point of crime and garbage landfill that divides neighborhoods and drives away the population from their local services.

Although La Serena initially got adapted to the geographical conditions of the territory, taking advantage of the natural watercourses and hillsides as natural protective elements, it has not integrated them into the latter development of the city, reinforcing them as a type of barrier and causing deep socio-spatial segregation problems. The workshop, through Paula Buglio's work, explored reconversion strategies of such barriers, by means of the formulation of mobility, public space and urban development projects at two levels.

At a metropolitan level, the extension of the currently projected Coquimbo-La Serena tram was suggested, crossing the historical center through Balmaceda Avenue and building an exclusive bridge for the tram, bicycles and pedestrians above the Elqui River, up to the foot of the hillside of Las Compañías. At this point, the development of an intermodal station was proposed to improve the accessibility conditions of the Las Compañías inhabitants, and to act as a service and public infrastructure sub-center for the community.

At a local level, a reconversion of the Jaramillo canal was explored, to act as a stitching axis of the broken residential grid of the hillside and as a new promenade reconnecting nearby neighborhoods, at walking distance from the proposed intermodal station. A regeneration of the blocks immediately close

Figure 2. Las Compañías, streets interrupted by the canal. Source: (Buglio, 2019).

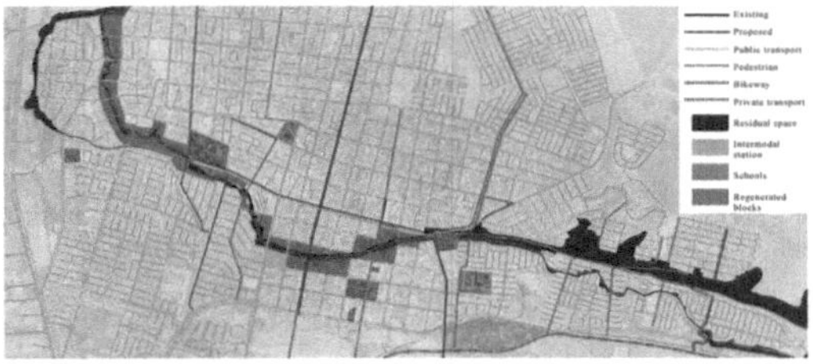

Figure 3. Jaramillo canal, residual spaces and crossings. Source: (Buglio, 2019).

Figure 4. Promenade and public infrastructure bases at the Jaramillo canal. Source: (Buglio, 2019).

to the canal was studied, setting out the recovery of their central voids as green spaces for collective use and the articulation of a compound building system, with commercial squares, public infrastructure and medium height residential blocks. Thus, the proposal explores, alongside the reconnection of the weaving and accessibility improvements, diversification of uses and residential densification propositions.

2.2 *Mobility and equity between Iquique and Alto Hospicio*

The socio-spatial segregation is observed in a more radical way between Iquique and Alto Hospicio, a conurbation of more than 320,000 inhabitants located next to the coastal cliff of the Atacama Desert, a steep slope that separates the coastal border from the pampas. Iquique was born in the first half of the 19th century as a Peruvian port enclave, occupying a small opening of the meagre coastal plain between the cliffs and the ocean. After the Pacific War (1879) and driven by the saltpeter and guano fields dispute among Bolivia, Perú and Chile, the city was annexed to the Chilean territory. From there, it grew as an important mining exportations commercial port, bringing in diverse merchants and workers from different countries, who stamped their culture into the historical center's architecture and character.

The invention of synthetic saltpeter during the First World War caused an unexpected setback on the city's development that lasted until 1975, with the creation of the *Zona Franca* of Iquique, a tax-free commercial enclave at a regional level. Iquique then became once more a national and international immigration focal point. Vulnerable residents coming from the inner parts of the pampas and other latitudes, flooded the city center and occupied the lands above the cliffs at the beginning of the 1980s, giving birth to Alto Hospicio, a settlement that got expanded over the desert plains through land occupation and makeshift social housings.

Alto Hospicio was born and developed under a poverty and abandonment stigma, with deep crime and drug addiction problems. It grew outside the urban laws and regulations, currently extending itself as a mono-functional residential outskirt of almost 125,000 inhabitants (2017 census).

While Iquique strengthened itself as a metropolitan public infrastructure and service center, and experienced vertical growth through residential towers along the coastal plain, Alto Hospicio established itself as a dormitory city for the most vulnerable, set apart from Iquique's urban attributes

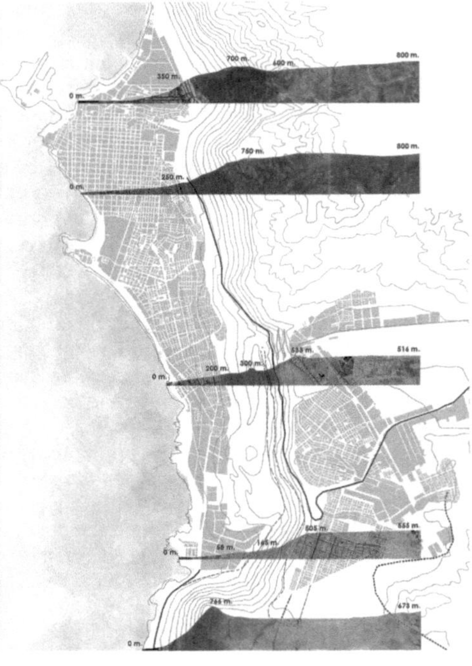

Figure 5. Iquique - Alto Hospicio conurbation and Coastal cliff segments. Source: (Bernal, 2019).

Figure 6. Park and cable car circuit between Iquique and Alto Hospicio. Source: (Bernal, 2019).

because of the steep slope of the cliff. Up to 2018, both cities were connected solely through the west access highway to Iquique, a winding path set into the cliff, leading into the historical center of the city. This path, constantly exposed to landslides, has to tolerate more than 160,000 daily trips from workers and students from Alto Hospicio, who travel in private cars or low-quality public transport services to get to Iquique. In 2019, a second road connection was opened into the south side of the conurbation, but only for private cars. The access for the inhabitants of Alto Hospicio to Iquique continues to be inadequate, especially through public transportation, restricting the possibilities of the most vulnerable people. Through the concessions system of the Ministry of Public Works, a proposal to improve the connectivity between both cities was made by means of building a cable car that would connect the northern side of Alto Hospicio with Iquique's city center in 2024.

Facing this scenario, inside the workshop, Maria Elisa Bernal's research suggested the construction of a second cable car into the south side of the conurbation. This proposal would establish a cable car and city park circuit, thus not only improving connectivity through public transportation between Iquique and Alto Hospicio, but also giving importance to the hillside as a scenic element and potential public space. The proposal consists of a cable car between the industrial area of Bajo Molle and the south area of Alto Hospicio, coupled by an intervention on the hillside, with public infrastructure bases and lookouts over the bay, which conclude with a park on the cliff's border, replicating above the coastal promenade of Iquique. The proposal opens up the discussion about the role of the desert hillside between both cities, exploring its integration to the conurbation's life, reversing its barrier condition through the construction of new connections, public spaces and infrastructure. In addition, it questions its possible usage as an urban growth space, as an alternative of urban densifying that the unlimited expansion of Alto Hospicio is facing into the Pampas.

3 AMPHIBIAN CITIES

We can observe in the south of Chile the urban expansion as a threat to the natural landscape in the wetlands of Valdivia, a city that currently holds about 155,000 inhabitants (2017 census). The problems of the residential outskirts – mono functionality, lack of services and restricted access – are joined by an unresolved dialog between the city and the river, and now between the city and the wetlands.

Valdivia is acknowledged as an inhabited area since pre-Columbian times, occupying a privileged location in the confluence of the Cau Cau, Cruces and Calle Calle rivers, that along with their wetlands form a water system that surrounds and runs through the city. It was founded by the Spanish in 1552, and from that point on, its population and financial development has been conditioned by different natural, political and economic events. We highlight the 1960 earthquake, since it determined the sinking of the ground and the flooding of vast land extensions and, among other factors, contributed to the decline of Valdivia as an industrial city. In the past 40 years, the city has experienced an important reconversion into the tertiary sector, standing out as a university development center.

Valdivia's growth shows, nonetheless, tendencies that put the natural environment at risk, and condition its future development. An increased use of private cars is observed (a 34% increase between 2002 and 2012 according to the Origin and Destiny survey), caused in part by the poor operation of

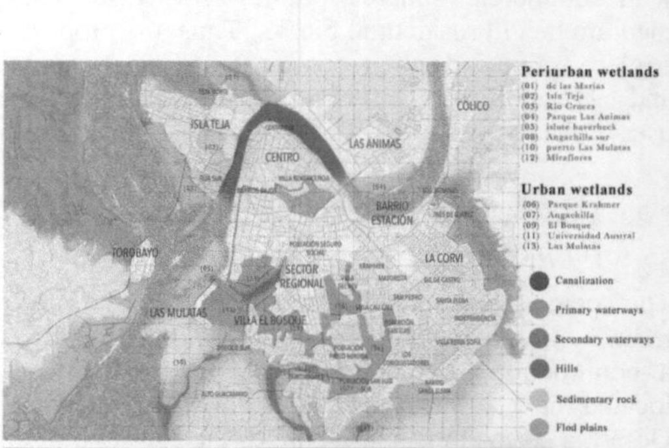

Figure 7. Location and images of the Valdivia wetlands. Source: (Alberti, 2019).

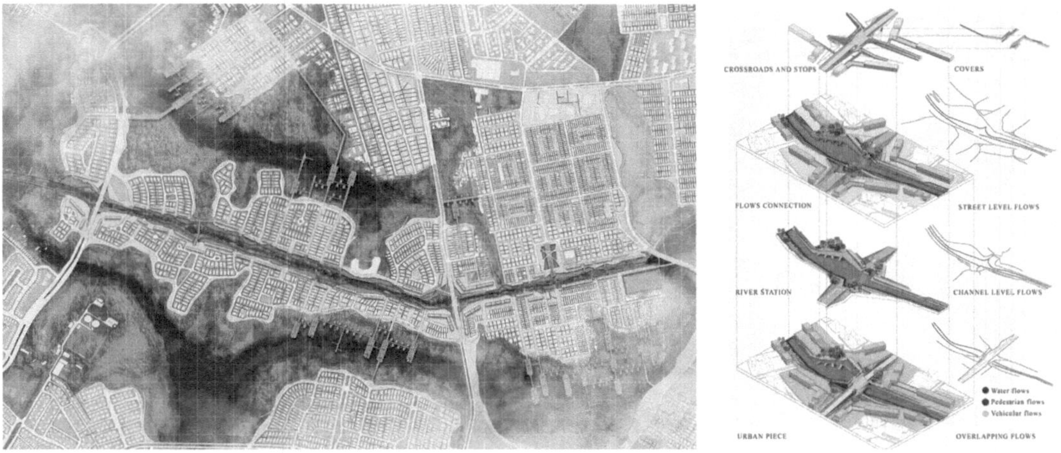

Figure 8. Canal plan and transfer station proposal. Source: (Alberti, 2019).

public transportation, with high traffic levels and increased travel time. Moreover, beyond the historical center, the urban expansion on the wetlands has generated a fractionated network of low density and high social and environmental inequity. The Valdivian territory, which was once formed by 65% water, has been urbanized by means of swamp and shore fillings, lowering the water proportion down to less than 45%. This alteration of the natural system has had secondary effects over the city, such as floods and exhaustion of the groundwater table.

Facing this scenario, the city's positioning as a scientific and university community has contributed to deliver innovative urban, environmental and mobility projects. In 2013, a SFT (Sustainable Fluvial Transport) was implemented, consisting of a taxi and bus solar powered network circulating around the Valdivia and Calle Calle rivers. This system does not only reduce costs and travel timing, but also contributes to develop the city's identity, oriented towards sustainability and the experience of the river. Additionally, the progressive urban occupation of the wetlands has motivated citizen organizations and the creation of scientific communities focused on its protection, restoration and conservation.

Along the lines of these initiatives, the work of Catalina Alberti in the workshop explored the formulation of integrated design strategies of urban growth and wetlands protection. Under the "irrigated cities" catchphrase, the integration of water and people flows were explored. Three objectives were set out: 1) reconnecting the city's wetland network; 2) improving public transportation; and 3) exploring sustainable alternatives to create new building territory.

To this end, the opening of an urban canal merging the courses of the Valdivia and Calle Calle rivers was proposed. It is about a new fluvial public transportation system that enhances the current SFT service, creating a circuit and "irrigating" the residential outskirts.

The canal's profile, 5.6 km long and 15 meters of average width, was designed alternating the natural borders with programmatic bases and crossings. Through its length, ten intermodal stations were arranged, to connect the water system with public and private transportation, pedestrian and bicycle systems. The stations were conceived simultaneously as service sub centers and public infrastructures of a new area of urban densification that explores the amphibian usage of the wetlands, based on the palafitte model. In this manner, the canal not only enables a new efficient and eco-friendly transportation means, but also transforms the positioning of the wetlands in the city, makes their presence evident and recovers the hydric quality of the territory.

4 LAYOUT, PAST AND RAILWAY NETWORKS

4.1 *Osorno, from industrial obsolescence to new fluvial facade*

Osorno's abandonment and growth can also be explained from the fluvial front, which was analyzed in the workshop as an unfinished case of the urban waterfront's evolution (Wren, 1983). Osorno was established in the 16th century by the shores of the Rahue River as an agricultural supply center for the Spanish approach on indigenous territory. The city was built and destroyed on several occasions due to the conflict between the Spanish and the Mapuche, and only by the second half of the 18th century it could establish itself as an urban center.

The arrival of German settlers drove the city's industrial development, giving way to the "development phase" of the fluvial front, which got shaped as

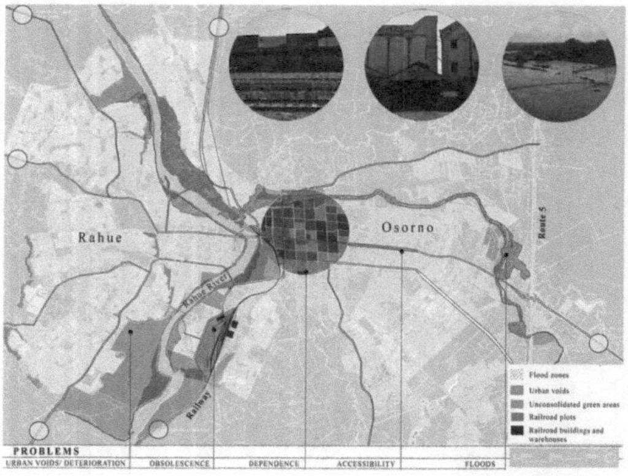

Figure 9. Relationship between Rahue and Osorno across the river. Source: (Bravo, 2019).

a commercial and transportation strategic space and generated important social and economic changes. Despite the arrival of the settlers, which were concentrated in the historical center, Osorno's population was mostly indigenous, constituted by workers that established themselves into the low zones and west bank of the river, giving birth to the Rahue sector. From that moment, the river became a socio-spatial barrier between the eastern richer riverbank and the western more floodable side where the working class lived.

The arrival of the railway at the end of the 19th century, parallel to the river and right by the historical center, consolidated the city's industrial character. Livestock, trade fairs, diary stores and windmills were established at the south side of the station, between the railway and the river. This led to the second development phase of the fluvial front, with an increase of the economic activity and a greater city growth. However, the new infrastructure determined a first distancing between the city and the river that lost its value as a transportation space compared to the railway and limited its access due to the presence of great industrial and railway infrastructures.

In the second half of the 20th century, the railway was replaced as a development agent by the Pan-American Highway construction and the boom of automobiles, which reoriented the city's growth into the west, alongside a de-industrialization process and economic outsourcing. This marked the beginning of the decline of the waterfront and the riverbank's industrial and railway installations obsolescence, which became empty urban spaces, characterized by the presence of massive abandoned infrastructures.

Nowadays, Osorno is an important service center at the regional level with more than 160,000 inhabitants (2017 census), with its operation altered by periodical river floods and its presence as a space of abandonment and city segregation. On the one hand, the river complicates the communication between Rahue (45% of the population) and Osorno's center, separating the most vulnerable neighborhoods from the commercial and service centers, linked by only three automotive bridges and two pedestrian footbridges. On the other hand, the same empty urban spaces and obsolete infrastructures of the riverbank hinder the access and the use of the natural space of the river. How can this dissociation be reversed and the socio-spatial relationship between the city and its waterfront reactivated? Camila Bravo's research in the workshop took on this question through the formulation of project strategies to put forward the final phase of development of the waterfronts: their rediscovery as a leading role in the cities' rehabilitation.

Interviews and field observations revealed the use of the river's empty urban spaces as a pedestrian mobility space. Multiple paths along the rail lines and abandoned industrial lots connect the residential with the commercial areas on a daily basis, albeit the risk of floods and deterioration of the place. From this place is where the waterfront's recovery was proposed, through the introduction of a new stitching with and along the river (Figure 11). A densification of the road and the area's public spaces was presented, through the conversion of obsolete segments of the railway into promenades, the opening of new intersecting streets and bridges, the construction of fluvial defenses, and the reuse of the river as an urban fluvial transportation space. Furthermore, a reconversion of specific obsolete industries into new programs and the densification of the area through the construction of new residential complexes and public infrastructure in remaining wastelands was presented. Thus, the proposal attempted to join efforts to spread from this point the comprehensive recovery of the river and its infrastructures as a central space of the city.

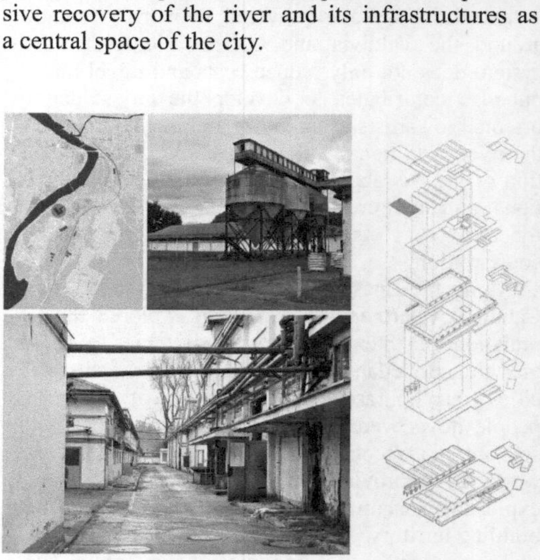

Figure 10. Obsolete industrial infrastructure: Nestlé plant. Source: (Bravo, 2019).

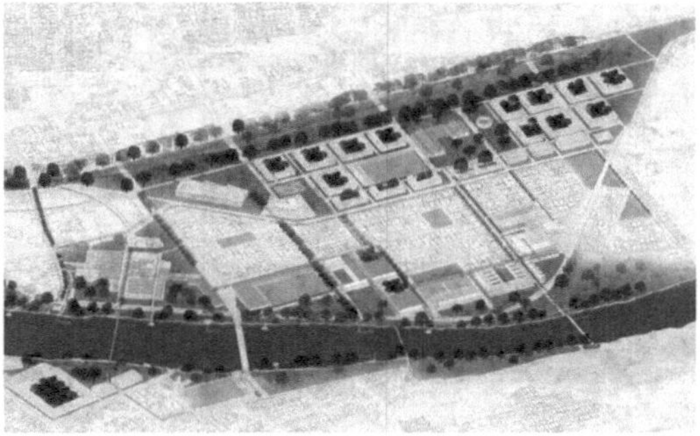

Figure 11. River front proposal. Source: (Bravo, 2019).

4.2 *Railway infrastructure in the urban sprawl between Puerto Montt and Puerto Varas*

The gradual retirement of railway services in the south of Chile, up until its final cancellation in 2010, produced obsolescence scenarios on the urban centers and the territory. The workshop addressed this subject by analyzing the abandoned railway segment between Puerto Montt and Puerto Varas, an emerging conurbation of about 350,000 inhabitants in the south of Chile. Both cities were founded during the mid-1800s, as a seaside port at the shores of the Reloncaví inlet and as a lacustrine port at the shores of the Llanquihue lake, respectively. Its growth was driven by the arrival of German settlers in the 1850s, who propelled the region's industrial development. At the beginning of the 20th century, the advent of the railway between Puerto Montt and Puerto Varas motivated the usage and settlement of the intermediate lands, broad topographic terraces of native forest, which were replaced by livestock, agricultural and forestry activity. The construction of the Pan-American Highway (route 5) during the 1960s, parallel to the railway, caused the abandonment of the train and its neighboring territories. The railway lost its meaning as a connecting axis and reinforced its condition as an urban barrier and decline space.

The town of Alerce, placed right by the railway, halfway between Puerto Montt and Puerto Varas, was particularly affected. The town emerged at the beginning of the 20th century associated with the exploitation of the larch forests of the area. In 1960, facing the demographic explosion of the region, Alerce was the target of a Municipal project that aimed at making it a "satellite city" of Puerto Montt. The project was abandoned in the 1970s, but Alerce continued to expand up to recent years by means of a random annexation of private housing estates and condominiums. Currently established as a "dorm city" of more than 45,000 inhabitants (2017 census), it comprises mostly of students and workers that travel 10 km on a daily basis to fulfill their duties in Puerto Montt or Puerto Varas. These two cities, on their part, have experienced a strong economic growth since the 1980s. Their development has led to the formation of a metropolitan area, which includes Alerce and other towns, and even though it does not become a continuous urban patch, it functions as one big city. Its inhabitants travel between the different centers on a daily basis using the highway and the V505 route intertwining with the railway. These travels are made mostly in private cars, due to the scarcity of interurban public transportation services. The situation has reached a critical point for the people of Alerce, who must travel up to 35 and 60 minutes back and forth to get to Puerto Montt or Puerto Varas, respectively.

Facing this situation, Denise Marco and Eduardo Arce, through the workshop, proposed the recovery of the railway layout by means of a new public transportation system for the conurbation: an interurban tram aimed at improving the accessibility conditions of the population and at the same time to become an urban development agent. The proposal consisted of an articulating axis, that overturns the railway's barrier condition through its stations and, along the V505 route, encourages the development of a new urban thickness, a "linear city" between Puerto Montt and Puerto Varas, with new services and public infrastructure. The tram's layout, shared with a bicycle lane, was presented as a new promenade and a transversely permeable public space, connected to the road network of the urban centers and the longitudinal road system of the suburban allotment.

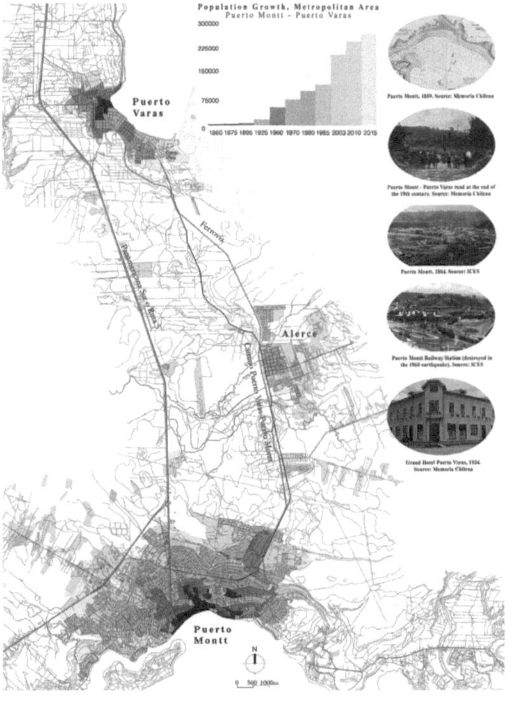

Figure 12. Conurbation growth stages and location of Alerce, next to the railway. Source: (Marco, 2019).

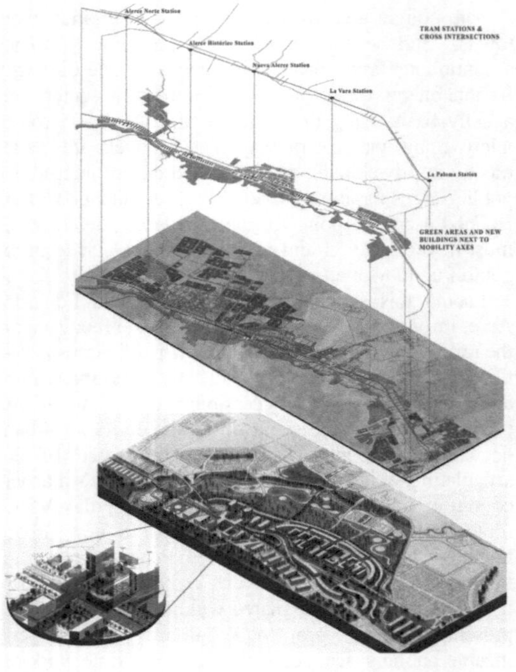

Figure 13. Linear city proposal in three scales. Source: (Arce, 2019).

5 CONCLUSIONS

The problems observed in each analyzed city, make evident the repetition of recurring urban problems. On the one hand, we notice an urban polarization problem, the coexistence of a rich and diverse city in contrast with a poor, distant, mono-functional and dependent city. While the first one gathers most of the goods and features by use of the working force of the second, the latter is not able to cover its inhabitant's minimal needs, forcing them to commute daily to the first. This polarization is based on the pre-existence of important geographical and infrastructural barriers that complicate the connection and mobility between these two realities, giving way to inequity and social exclusion scenarios. Jiron & Mansilla (2013) name these obstacles as "accessibility barriers" which raise the difficulty and travel time creating a "thickness" that must be crossed through to move "between" or "inside" determined urban contexts.

On the other hand, we verified a persistent urban expansion phenomenon that threatens the ecosystem and the natural landscape's integrity in the interior of the city as well as on the outside. This low-density growth is comprised of increasingly distant and disconnected residential fragments, which interrupt the natural systems and force the extension of artificial ones, creating unbalance and inefficiency on both realities. According to Bartorila (2009), this dynamic generates tensions produced by two opposing forces: the urban pressure on the land and the natural resistance to this pressure. However, this tension cannot be understood only as a space of conflict, but also as a potential exchange space.

The workshop faced these issues, exploring new connective tissue creation strategies, new linking pieces between the polarized areas and between the natural and the artificial. The geographical elements are reinterpreted as "urban ecotones" (Bartorila 2009), presenting artifacts that answer to the urban reality as well as the natural one, potential articulators and convergence elements between geography and city. These are compound urban projects, which propose the insertion of new mobility infrastructures associated with the creation of public space that revert the barrier condition of the identified obstacles and that encourage the creation of new urban enclaves and environmental balance. They are precise urban pieces, which by their extension and continuity impact the structure and the global functionality of middle-sized cities.

REFERENCES

Andrés López, G. 2008. ¿Las infraestructuras como proyecto de ciudad? algunas reflexiones sobre Burgos, ciudad intermedia de Castilla y León. *Revista del Instituto Universitario de Urbanística de la Universidad de Valladolid*, Nº 11.

Alberti, C. 2019. *Ciudades Irrigadas. Flujos de agua y movilidad en la ciudad de Valdivia*. Master's Thesis, PUC, Chile.

Arce, E. 2019. *De ciudad satélite a ciudad lineal. El caso de la ciudad bi-satélite de Alerce*. Master's Thesis, PUC, Chile.

Barton, J. & RAMÍREZ 2017. El futuro urbano de Chile está en el desarrollo sustentable de ciudades intermedias. http://www.plataformaurbana.cl/archive/2017/01/30/opinion-el-futuro-urbano-de-chile-esta-en-el-desarrollo-sustentable-de-ciudades-intermedias/

Bartorila, M. A. 2009. Articulaciones entre artefacto urbano y ecosistemas naturales. Doctoral Thesis, Universidad Politécnica de Cataluña, Spain.

Bernal, M. E. 2019. *Barreras topográficas. Movilidad y equidad entre Iquique y Alto Hospicio*. Master's Thesis in Urban Project, PUC, Chile.

Bravo, C. 2019. *El borde río urbano y la reconversión de su fachada fluvial*. Master's Thesis, PUC, Chile.

Buglio, P. 2019. *Fronteras urbano-geográficas. El caso de Las Compañías, Coquimbo*. Master's Thesis, PUC, Chile.

Jirón, P., & Mansila, P. 2013. Atravesando la espesura de la ciudad: vida cotidiana y barreras de accesibilidad de los habitantes de la periferia urbana de Santiago de Chile. *Geografía Norte Grande* (56), 53–74.

Marco, D. 2019. *Infraestructura ferroviaria en la ciudad difusa: la vuelta del tren entre Puerto Varas y Alerce*. Master's Thesis, PUC, Chile.

Salazar, G., Fonck, M. & Vergara, L. 2018. Ciudades intermedias: dinámicas de intermediación desde la noción de lugar. *Revista de geografía Norte Grande*, Nº 70, 109–130.

UNESCO 1999. *Ciudades Intermedias y urbanización mundial*. Lleida: Alcaldía de Lleida, UNESCO, Unión International de Arquitectos, Ministerio de Relaciones Exteriores de España.

Wrenn, D., Casazza, J. A. & Smart, J. E. 1983. *Urban Waterfront Development*. Washington: Urban Land Institute.

Research Tracks in Urbanism: Dynamics, Planning and Design in Contemporary Urban Territories – Allegri et al (eds)
 ISBN 978-1-032-10788-2

Formation, dynamics and urban configuration of the metropolis of Recife

S. Maciel & J. Crespo
CIAUD, Lisbon School of Architecture, Universidade de Lisboa, Portugal

ABSTRACT: The Metropolitan Region of Recife presents an urban development process with incomplete networks of metropolitan public infrastructure, fragmentation and territorial dispersion, with a low level of integration, resulting in increasing environmental and economic costs. In addition to such issues, other challenges persist, although they are resolved in developed countries, such as environmental sanitation or housing deficit. All of this contributes to a disorderly, discontinuous, unregulated and incomplete urban expansion. The speculated scientific hypothesis is that the metropolitan urbanization of Recife, both continuous anddispersed and fragmented. It has its own characteristics and was induced, in its formation, by changes in the dynamics of urban growth and by the construction and expansion of its infrastructure networks. This article addresses the complexity and characteristics of the formation, expansion and urban structuring of the metropolis of Recife.

1 INTRODUCTION

In Brazil, over one third of the population lives in metropolitan areas, approximately 98 million people, and it is where half of the country's GBP is generated[1]. The first metropolitan areas of Brazil were created by the federal government in the 1970s in a context of centralization of urban and regional planning. These were characterized by an urban growth with a continuous shift from the center to the peripheral areas. Institutional changes originated by the Federal Constitution of 1988 gave more power to municipalities and metropolitan management to the state, allowing for various institutional arrangements, often without the protagonism and leadership necessary to drive the metropolitan management. Despite the institutional competency, municipalities have a limited executive capacity to attend to the various challenges imposed, thus resulting in a significant institutional void.

Brazil's metropolitan areas showed a spatial organization based on a continuously expanding center and a peripheral area, resulting in a specific form of economic, social and political take on the territory. In the center would occur the main economic activities, governmental institutions, universities, urban infrastructure and high-income homes. In the peripheral areas, we can observe the dynamic process of urban expansion motivated by the industry sector, metropolitan logistics and services, which also demanded for new homes near such areas.

The territorial and institutional cut-out proposed in this work is the Recife Metropolitan Area – RMA, as considered similarly on other territorial planning models for metropolitan urban expansion. Such take on the subject can be further studied through the Doctorate in Urbanism of the Lisbon School of Architecture, and the knowledge taken from other studies on the Lisbon Metropolitan Area. These studies approach the formation of metropolitan Lisbon from unoccupied space as an element of great importance (Morgado, 2005), metropolitan infrastructures – the backbone of its expansion and territorial organization (Santos, 2018), partitioning as a base of transformation and its management model (Crespo and Cabral, 2010). Furthermost, the choice for such territory is motivated by the metropolitan take on Brazil's territory with the approval of the Status of Metropolis (Federal Law nr. 13.089/2015) which states the laws for the Integrated Urban Development Plan – IUDP and the formulation of planning instruments for Brazil's metropolitan areas.

Recife's Metropolitan Area (RMA) was instituted in 1973 and currently limited by 14 municipalities, with a total population of 3.98 million (according to the Brazilian Institute of Geography and Statistics – IBGE, 2018), which represents 41% of the Pernambuco state in Brazil. The metropolitan territory, the

1 The Brazilian Institute of Geography and Statistics - IBGE - estimates that Brazil has 208.5 million inhabitants at the end of 2018 and a population growth rate of 0.82% between 2017 and 2018. Source: https://agenciadenoticias.ibge.gov.br/agencia-sala-de-imprensa/2013-agencia-dnoticias/releases/22374ibge divulga-as-estimativas-de-populacao-dos-municipios-para 2018.

DOI: 10.1201/9781003220855-15

backbone of the area's urbanization, has one 117 kilometers of littoral area with a plane bathed by the Capibaribe and Beberibe rivers to the center, Jaboatão e Ipojuca rivers to the south, and the Igarassu river to the North. The hills, following the westside, present itself covered in Atlantic forests which hold the water reserves that supply the metropolitan city.

The evolutionary reading of the RMA is organized through a periodic board and cartographic support developed for the research. Cartographic reading is an important morphological record of the evolution and configuration of the RMA, and the foundation of the hypothesis formulated. The timeline proposed for the research is set considering the main periods of formation and occupation of the Recife area. The choice of two procedures used to indicate different territorial analysis perspectives is proposed, which are complementary: the study on the formation of the metropolis considering the defining elements of each period (infrastructures, planning, projects, legislation and economic factors), and the study of the current state of the territory with the proposed test of the metropolitan area. This will sustain the research's development, the analysis of current territorial occupation and urban expansion through several time periods and the interaction with the installment of metropolitan infrastructures.

The research in this work attempts to build a narrative of the RMA over time, between periods of significant territorial transformations, elaborated based on cartographic interpretation of the metropolitan configuration and three observation guidelines. This research intends to develop an interpretative matrix aiming to analyze the processes of fragmentation and dispersion, the level of complementarity and territorial integration of the test model, which will guide and set the development of the doctoral thesis "Urban Expansion in Metropolitan Areas: the configuration of the Recife's Metropolitan Area (Brazil)".

2 FORMATION AND STRUCTURING OF THE RECIFE METROPOLIS

The metropolitan area of Recife has, like other Brazilian areas, its institutional origin in the 1970s, but Recife's metropolitan formation is the result of a historical process of over four centuries, with evidence of the various phases, divided into five main periods: i) 17th-19th centuries– the ways of sugar harvest to the harbor; ii) 20^{th}century: 1900-1950 – The formation of infrastructures and industrial activities; iii) 20th century: 1951-1973 – The phenomenon of metropolization; iv) 20th century: 1973-1999 – acknowledgement of the metropolis; v) 21st century: 2000-2018 – the incomplete metropolis processing for complementarity. From the 21st century on, the urban expansion of the RMA intensifies when spatially disperse urban areas start to appear. The economic agents operate autonomously in the metropolitan area, articulating independently with the public sector, occupying the areas located further from the central core, although these areas were regulated by the municipal region plan.

The metropolitan phenomenon has been observed since the 1950s, although still without the dynamic scale of recent times – the relationship between the state capital and its neighboring municipalities. Currently the urban patch is spread throughout several municipalities, continuously expanding, irradiating linearly and heterogeneously in peripheral areas. With an eminently urban profile, the region presents a population and services with little expression. Its environmental collection is rich and diverse, despite the constant threats and risks that arise from the process of metropolization and economic exploration of the territory. The natural environment allows for differences in the metropolitan area: plains and hills. Such fact establishes the two parallel and predominant occupation factors: a littoral area, where predominantly urban and industrial areas operate, natural ecosystems (mango fields), sugar cane and coconut crops, and fragments of the Atlantic forest; and the other area further West from the RMA, where we find the polyculture, sugar cane fields, farms, and the remainder of the Atlantic forests, with different development and urban areas. According to such premises, two main ecosystems stand out from the coastal area and hill sections.

The landscapes that result from the sugar cane plantations surround some cities that make the metropolis of Recife. Cane fields are significant landscapes in the northern metropolitan municipalities of Itapissuma, Igarassu, Araçoiaba and Abreu e Lima; at the center: São Lourenço and Moreno; and in the south: Cabo and Ipojuca. Either sugar cane plantations or natural reserves define the picture of green and blue colors. Such conditions result in a set of continuous and intermediate areas, formed by plains, elevations and embankments, limited by rivers, damns and drainages, passing through roads and urban areas. Such facts allow for the understanding of its contents and its impact on the territory since the 17th century, as pictured in Figure 1: 1st period – 17th-19th centuries.

The RMA was formed from the city of Recife, and was the center that irradiated expansions to South, North, alongside the BR-101 and following old roads. Several factors contributed to expansion of the RMA territory, among these, geographical factors that defined the environmental areas, and plantations. Those that contributed the most for the expansion and deserve to be noted: the topography of the region defined by great plains and the proximity to water courses which favor territorial occupancy and new infrastructures, mainly new roadways. A limitation to urban growth were the great agricultural facilities and factories destined to harvest and processing sugar cane. Such limiting factor has been reduced over time, since the beginning of the 21st century we can observe a significant

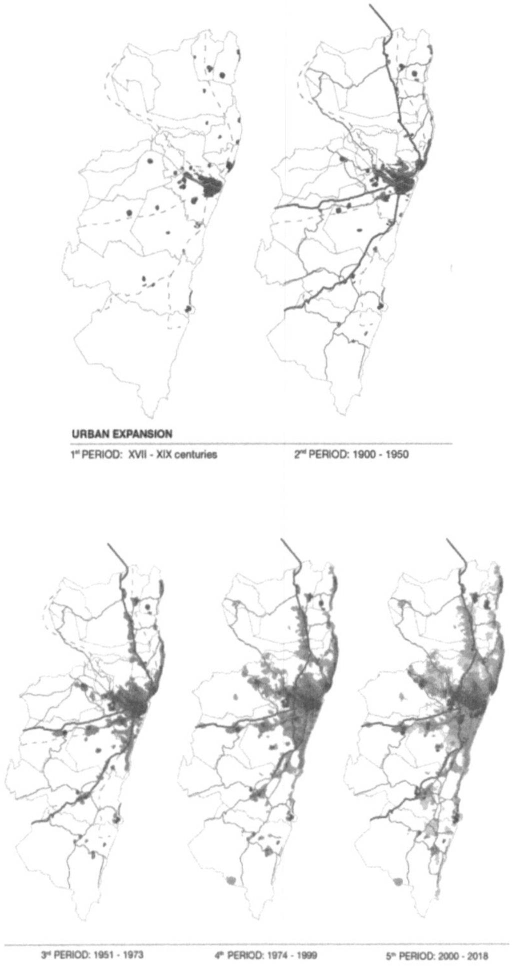

Figure 1. The process of urban expansion in the Recife metropolitan area. Source: Author (2019).

urban expansion replacing the crops. The municipalities of Itapissuma and Igarassu stand out in the northern metropolitan area, mainly the brewery pole. In the southern metropolitan area, the municipalities of Cabo and Ipojuca stand out, with the advancement of summer tourism and the development of the harbor and the industrial district of Suape, alongside other factors such as the investments made in housing programs.

With such basis, we can identify and map the transformations occurring in Recife's metropolitan territory. Such mapping completed with a systematic process of research and scientific observation is expected to answer the following questions: i) which are the specific characteristics for the fragmentation and territorial dispersion identified in the urban metropolitan expansion in Recife? ii) is there a link between the urban metropolitan patch and the expansion of infrastructures in Recife? Did these induce the current configuration of Recife's metropolitan area?

In this topic, the timeframe of this research is made of significant periods in the formation and construction of big infrastructures, and the metropolitan expansion, accompanied by the interpretation of such timeframe. The territorial structure of Recife at a metropolitan scale must be analyzed considering several space and time scales. The timeframe set for this research was established considering the time periods of occupation of Recife's metropolitan area, as detailed on the five maps below (Table 1), that depict the metropolitan expansion (Figure 1). The methodology proposed uses two distinct concepts that complete each other, to identify different perspectives in the analysis of the territory: the study of a specific time period (synchronous analysis) following the course of the tests proposed for the metropolitan area. This will support the analysis of the current territory occupancy and urban expansion through time periods (diachronic analysis) and the interface with the metropolitan territory's infrastructures. The proposed research is presented in a simplified manner through existing pictures, cartographical readings and basic bibliography, serving as a guide for future research.

Recife and Olinda were born in the 16th century as twin cities. The hills of Olinda, a Portuguese city, served as an observation point, the plains of Recife with the harbor where the rivers of Capibaribe and Beberibe meet. The neck of Olinda enabled the advances of Dutch occupation, expanding following the same guidelines of its Portuguese layout. In

Table 1. Cutout from the ongoing research, author's work, 2019.

PARTS	THE METROPOLITAN TERRITORY	PERIODS
The formation process of the metropolitan structure	The ways of harvesting to the harbor: the first villages and the expansion of Recife	XVII - XIX
	The formation of infrastructures and the first industrial activities: textile products and sugar factories	1900 - 1950
Singularities of the expansion and metropolitan configuration	The phenomenon of metropolization: the intensification of urban activity	1951 - 1979
	The acknowledgement of Metropolis: planning and conception	1974 - 1999
	The incomplete metropolis: infrastructures, services and technology	2000 - 2018

1549, the village of Camaragibe is formed, followed by Igarassu in 1564. In that moment, the paths that connect to the harbor of Recife for transporting sugar cane byproducts were established. The religious buildings are significant urban elements of such period. In 1654, the village of São Lourenço is born, whose economical foundation was the exploration of brazilwood and the plantation of sugar cane. The sugar industry helps develop the villages of Goiana and Igarassu, with the villages of Itapissuma and Itamaracá also advancing in that time.

From the second half of the 18th century, the municipality of Recife is developing, the Capibaribe river is crossed and the plains allow for the urban expansion to advance into the continent. In 1858, the railroad infrastructure was also established. The notes of Vauthier's trip through the province of Pernambuco in the second half of the 19th century emphasize the distinct levels of urbanization throughout the territory, and the typological diversities between the urban, suburban and rural areas. Occupation of the territory follows the neck through the lines of the insipient Portuguese layout (Reynaldo, 2017). Count Nassau, governor in Brazilian lands from 1637 to 1644, confirms the strategic location of the harbor and expands the urbanization to the island of Antônio Vaz, expanding the communication with the productive areas of South and West, through Afogados, crossing the river networks. The perspective for urban growth includes the construction of the first bridge (1944), connecting the harbor and the island of Antônio Vaz, thus expanding the urbanization from Recife's harbor.

In the first half of the 20th century, we can observe the influence of the network of roadways irradiating from Recife's harbor in the process of urban expansion of the region. In that time, two main urban infrastructures stand out: The project for remodeling and modernization of Recife's harbor, adopting a radial growth model from Alfredo Lisboa in 1909 and the project of Recife's sanitary infrastructure from Saturnino de Brito in 1910. Animal-Powered Transportation that functioned from 1840 to 1914 was replaced by streetcars from 1914 on.

The metropolitan city is then formed and the occupation of the waterfront to the south, firstly, in 1925, as a second home and supporting the construction industry, from Beira Mar avenue, that separates the ocean from the grounds to be urbanized. An increase of factories run by the state could then be observed with working class villages and big infrastructures in the state capital of Pernambuco, like the Federal University of Pernambuco in the 1940s and the international airport in the 1950s. Between 1950 and 1970 the state highway BR-101 was built. Notably the 1960s and 1970s are characterized by a big urban growth in Brazil, specifically in Recife from the coast to the West, to the North and to the South, along the Atlantic coast. Such fact resulted in a territory like that of an oil stain continuously expanding its limits in a tentacular form. The resulting urban patch is filled with voids and low occupation areas. Such areas were left as reserved space for when the value of land would rise to be put in the market. The exceeding value was capitalized by the means of public expenses in infrastructures and urban services necessary for the newly occupied areas (Lacerda and Zanquetti, 1999).

1. Olinda
2. XIX century, the networks of roadways and railways
3. Network of streetcars, 1914-1960
4. Port complex of Suape, 2000.

Figure 2. Reading of the metropolitan territory and timeline, 2019.

The metropolization is completed with the conurbation of the urban centers to the central city. In 1973, the metropolis of Recife is then acknowledged. In 1974 the Metropolitan Region of Recife is instituted, formed by 10 municipalities. In 1983, the construction of Recife's subway begins, with the West and South lines serving three metropolitan municipalities.

Currently, the RMA is comprised of 15 municipalities. To the South, the agricultural sugar cane harvest is partially substituted, and a renovation of soil use takes place. Industries are settled along the BR-101 to the South. The industrial port complex of Suape is institutionalized in 1978. The digital port sets the standard as one of the main technological innovation parks of Brazil. Cities are expanding, commerce and services, and shopping malls are set alongside the main runways. Beaches in the littoral South start to grow with beach houses and resorts, although the coast of Paulista, Olinda and Jaboatão already suffer from the advancement of the waterfront into the land.

The networks of interconnecting services and industries expand with the duplication of two important axis of mobility to the west (BR-232) and parallel to the North and South coast (BR-101) the agricultural areas of sugar cane are replaced by housing, industry and service complexes. Still, the main social indicators show a big inequality when it comes to paying rent and access to basic services. The deficits in housing and sanitary infrastructures remain. The population of RMA reaches 4.06 million. New investments in housing occur through the "Minha Casa Minha Vida" program, but are still insufficient. The urban infrastructure is incomplete, the mobility and sanitary infrastructures compose the biggest challenges the metropolis must overcome.

3 PLANS AND THE CONCEPTION OF THE RECIFE METROPOLIS

The fact that the RMA has a tradition in planning and a significant archive of studies and plans is relevant, with a great number of such studies elaborated by the state through the Pernambuco State Planning and Research Agency Condepe/Fidem. In the set of plans is the Plan for a Strategic Metropolis, elaborated in 2002, which serves as base for the current Plan for the Integrated Urban Development of the Metropolitan Area of Recife – PDUI. Such plan has the federal law Nr. 13.098/2015 as a guideline, the status of the Metropolis, which presents the concept of metropolis, metropolitan area and interstate governance[2], presenting one main challenge in the way of articulating the constitutional powers of federal offices. This status also stipulates the integrated urban development tools and manages the common interest public services, enveloping the cases where more than one office, municipality or state share the territorial power outside its individual region.

The Recife Metropolitan Area has a history of plans for territorial development dated since the 1950s. The study entitled "Development and Settlement of Industries Relevant for Pernambuco and the Northeast" from Padre Lebret (2001) pointed to the need for structuring a new regional port for the metropolitan development of the South and the balance of the Recife metropolis.

In the case of the Recife Metropolitan Area, we can observe new plans regarding the rational occupation of the urban space and strengthening and dynamization of the economy from its institutionalization (1973). The Plan for the Metropolitan Development – PDM/RMR (1983) presents the concept of attracting new opportunities for the development of metropolitan areas surrounding Recife, to the North, West and South. Some strategies were then winning strategies.

The chosen documents for the analysis correspond to the periods of 1973/1978 – Plan for the Integrated Development – PDI (1976); 1979/1989 – Plan for the Metropolitan Development – PDM/RMR (1983). The plan includes the concept of territorial organization and the proposal for a central North, South and West metropolis, the settlement of the Industrial Port

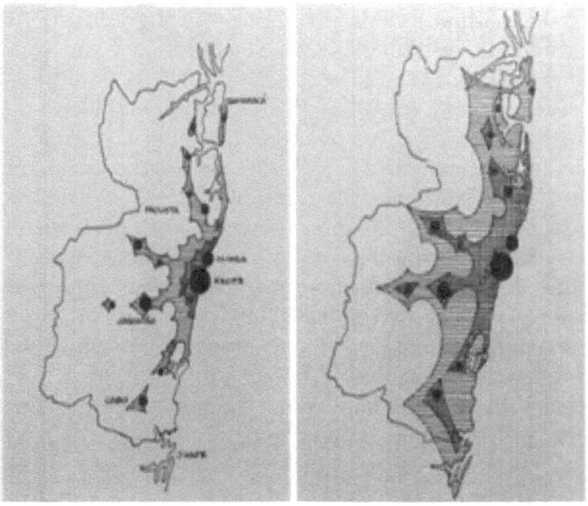

Figure 3. Plan for the Metropolitan Development PDM/RMR (1983). Source: Acervo Técnico 1974 - 2018 - Condepe/Fidem, Recife (2019).

Complex of Suape and the enlargement of the industrial districts. The Metropolitan Strategic Plan – PRDRM (1998) presents the concept of Sustainable Development and the Strategic Metropolis Plan (2002) which present the strategic plan: A Competitive Axis with Social Inclusion and the Axis of Habitability and Inclusion.

To analyze a territory is to dig deep in society itself, its potential for the relationship and conflict, contingency and permanent instability (Domingues, 2015). The rhetorical narrative admits an infinite plurality of language sets, setting a guideline as we can observe by reading the technical archive of the Condepe/Fidem agency. A converging strategy can be identified in the plans: rational occupation of the urban space, strengthening and dynamization of the economy, strengthening of the planning process, and the coordination of the metropolitan administration, to guarantee the mobility of people, goods and services, amongst others. The challenges are as follows: urban control, social inclusion and social security, sanitary network, commuting from home to work, the shared management of public services, and the involvement and participation of municipalities and funding.

4 CONCLUSIONS: THE INCOMPLETE METROPOLIS IN THE PROCESS FOR COMPLEMENTARITY

The current configuration of the Recife metropolis results from its territorial support and consequences

2 Inter-federative governance of public functions of common interest: sharing of responsibilities and actions between entities of the Federation in terms of organization, planning and execution of public functions of common interest, through the execution of an integrated and articulated system of planning, projects, structuring financial, implementation, operation and management.

from the time periods it has been through during its formation. This reflects how several social and economic actors have utilized the territory, its first infrastructures, the location of large urban equipment, ports, and the archive of urban planning resulting or not from such actions. But the metropolis, as other Brazilian examples, had its formation and urban expansion disorderly, discontinuous, with incomplete networks of public infrastructures and partially regulated, despite the effort from urban planning institutions when elaborating the plans for urban expansion and strategic territorial projects. Some of the proposed strategies were successful and presented results that contribute to the dynamic of the metropolis. Still, some challenges remain unsolved, like the basic sanitary infrastructure, housing and the lower quality of life in the more fragile areas of the metropolitan territory.

This way, the metropolitan area of Recife presents an important process of spatial fragmentation, with a low level of integration, mainly between newer urbanizations and existing urban centers (Reynaldo et al., 2013). Like so, the thesis aims at furthering the questions of contribution to the infrastructures in the metropolitan area and the expansion of the Recife metropolis, based on the analysis of the relationship between the evolution of the urban patch of Recife and enlargement of such infrastructures. This means to assess the specifics of the metropolitan urban expansion and investigate if the introduction of infrastructural support allows for the generation of opportunities for the settling of activities and new urban patches (Santos, 2018).

Thus, the question on the metropolitan urban expansion can be posed. Which levels of fragmentation and territorial dispersion can be identified in the process of urban expansion of the RMA? Which indicators and morphological elements should be selected when designing it? How should we represent and classify them? To analyze the urban structure we require the assessment of the elements with strategic values to the urban reality, which vary in time and in space. For that, we must consider the geographical scale, the timeframe, and in some situations, the singular context of each urban system given by its historical process and accumulation of past systems residues (Tourinho, 2011). The expression "urban space structure" is then adequate to identify the way the strategic elements of spatial organization are arranged and relate to one another, in each moment of the continuous structuring process.

To Morgado (2005), the acknowledgement of environmental values in the metropolis and the effective need of space organization contribute to a situation of complementarity between open spaces, urban patches and big infrastructures. The culture for urban planning has been constructed over time, although alternatives are necessary to motivate and involve actors (technical and representatives of society) in the proposal of an institutional environment and system of governance that is appropriate for the development of the metropolis. Specifically, a system with governance models and processes of territorial planning (place-based), in order to avoid the urban fragmentation of communities and the unregulated urbanization, which is potentiated by neoliberal state reforms and financing of space (Cabral, 2018).

REFERENCES

Cabral, J. 2018. Progress(es) in planning – theory and practise and the quest for activist modes. *PHI 2017 - Progress(es) – Theories and Practices, Proceedings of the 3rd International Multidisciplinary Congress*, CRC Press/ Balkema,October 4- 7.

CONDEPE/FIDEM – Agência Estadual de Planejamento e Pesquisas de Pernambuco. 2002. Plano Metrópole Estratégica. Recife.

CONDEPE/FIDEM – Acervo Técnico. 2019. Plano de Desenvolvimento Urbano Integrado - PDUI. Recife.

Crespo, J. & Cabral, J. 2010. The institutional dimension of governance in the Lisbon metropolitan area. *Análise Social*, 197: 639–662.

Domingues, Á. 2015. *Retóricas do Território. Território Casa Comum*. Catálogo de Exposição, Casa do Território. Famalicão: C.M.V.N.

Lacerda, N. & Zanchetti, S. 1999. Planejamento territorial na Região Metropolitana do Recife. *Cadernos de Estudos Sociais*, Recife: Editora Massangana, 15: 375–409.

Lebret, L. 2001. *Estudo sobre desenvolvimento e implantação de indústrias, interessando a Pernambuco e ao Nordeste*. 1ª ed. (1954), 2ª ed. (1976), 3ª ed. rev. Recife: CONDEPE.

Morgado, S. 2005. *Protagonismo de la ausencia: interpretácion urbanística de la formación metropolitana de Lisboa desde lo desocupado*. PhD Thesis in Urbanism, Escuela Técnica Superior de Arquitectura de Barcelona, Universidad Politecnica de Catalunya.

Reynaldo, A.; Lacerda, N. & Jucá, S. 2013. A Reconfiguração Espacial da Região Metropolitana do Recife: processos de fragmentação e integração. In R. Caracciolo; P. Elinbaum; B. H. Estarellas & M. Debat (Org.), *La metrópolis iberoamericana en sus propios términos*: 117–129. Barcelona: riURB_editores.

Reynaldo, A. 2017. *As catedrais continuam brancas: planos e projetos do século XX para o centro do Recife*. Recife: Cepe.

Santos, J. 2018. *Espaços de Mediação infraestrutural: interpretação e Projecto na Produção do Urbanismo no Territorio Metropolitano de Lisboa*. Lisboa: Edição publicada pela AML, Papel, Press Lda.

Tourinho, H. 2011) *Estrutura urbana de cidades médias amazônicas: análise considerando a articulação das escalas interurbana e intraurbana*. PhD Thesis, Programa de Pós-Graduação em Desenvolvimento Urbano, Universidade Federal de Pernambuco, Recife.

Research Tracks in Urbanism: Dynamics, Planning and Design in Contemporary Urban Territories – Allegri et al (eds)

Program for integral urbanization of favelas: Colinas D'Oeste – Osasco city – São Paulo State

V.M. Rubio
School of Architecture and Urbanism, Mackenzie Presbyterian University, Brazil

A.T.B. Alvim
PPGAU, School of Architecture and Urbanism, Mackenzie Presbyterian University, Brazil

ABSTRACT: This paper presents part of a doctoral study evaluating the Program for Integral Urbanization of Favelas, a housing policy instrument implemented between 2005 and 2016 in Osasco city, in the Metropolitan Region of São Paulo. This instrument is considered an important strategy to promote equity of land occupation, fulfil the social function of property and secure the right to the city for inhabitants (Lefebvre, 2001 *apud* Rubio, 2017b). We report the urbanization process of Colinas D'Oeste, the largest precarious settlement in Osasco and one of five favelas supported by the Growth Acceleration Program (PAC). The study entailed a qualitative evaluation of the public policy implemented using a methodology based on Dunn's theory (2007). The data collected and analyzed helped identify the challenges, advances, and shortcomings of the instrument, and put forward recommendations for improving the policy's scope and maintaining the interventions carried out in precarious settlements of Osasco.

1 INTRODUCTION[1]

In Brazil, favelas[2] are the product of the use of poor models and of the urbanization process, as well as a failure to implement effective public urban and housing policies (Rubio, 2017).

After the enactment of the Federal Constitution in 1988, public policies addressing the state of precarious settlements in cities gained greater attention, where actions for the urbanization of favelas became an integral part of planning agendas in many cities throughout Brazil.

In the 2000s, numerous cities stood out for their efforts reorganizing administrative structures and devising programs that provided integrated integral actions within favelas, underpinned by instruments of the City Statute. From 2007, the introduction of the Accelerated Growth Program (PAC) further promoted these actions.

The Program for Urbanization of Favelas is regarded as a key strategy, providing equality in land access through an instrument ensuring the "social function" of urban property and securing inhabitants' right to the city (Lefebvre, 2001 *apud* Rubio, 2017b).

In 2005, Osasco city authorities, in an unprecedented move since emancipation in 1962, implemented the Housing Policy and created the *Novo Bairro* (New Neighborhood) Program. The Osasco housing policy drew not only on the guidelines of the Federal Constitution and City Statute, but also the municipal urban development policy.

At the time of its implementation, a third of the city's population (200,000 inhabitants) was living within precarious settlements under poor living conditions on illegally occupied land. Over the space of 12 years, the city gained new roadways, public lighting, green spaces, along with health and education facilities, and a social welfare

1 This paper is the result of the doctoral thesis *Programa de Urbanização e os Desafios Para Integração da Favela à Cidade. Avanços e descompassos de uma política pública em Osasco – SP* (2005-2016), defended as part of the Graduate Program in Architecture and Urbanism, authored by Viviane Manzione Rubio, under the supervision of Angélica Benatti Alvim. This research was supported by the Mackenzie Presbyterian Institut and the Formula Santander Program to period of study at the University of Lisbon. The publication of the paper was supported by the CAPES Academic Excellence Program (PROEX).

2 Favelas, regarded as a largely Brazilian phenomenon, are a type of precarious settlement found in many cities throughout Brazil and in numerous countries worldwide. The present doctoral thesis centered on interventions carried out within five favelas in Osasco city.

DOI: 10.1201/9781003220855-16

service. The precarious settlements[3] were served by the subprograms of the *Bairro Novo* (New Neighborhood) Program: Integral Urbanization; Gradual Urbanization; Housing Provision; and Land Tenure Regularization, with actions targeted to address the specific situations and conditions of land occupation and irregularity.

Despite concerted efforts and investment by city authorities toward tackling the local precarious land occupation, since 2012 there has been a rise in the number of new settlements, an increase in density and expansion of urbanized favelas.

The present paper provides a summary of the evaluation of the Program for Integral Urbanization's implementation within the Colinas D'Oeste Favela, the largest precarious settlement in Osasco. The public policy was evaluated using a methodology based on Dunn's theory (2007), highlighting challenges, advances and shortcomings of its implementation, and providing key recommendations for improving the program's scope and maintaining interventions already in place.

2 PROGRAM FOR URBANIZATION OF FAVELAS: INSTRUMENT SECURING THE RIGHT TO THE CITY

The 1970s and 1980s were pivotal decades regarding the implementation of basic urban infrastructure and of instruments for land tenure regularization, as well as for housing provision within Brazil's favelas (Rubio, 2017).

According to Denaldi (2003), the most notable programs were: the PROFILURB and PROMORAR, marking a milestone in Federal Government involvement and a shift in approach to the issue; the PRODECOM run by the Government authorities of Minas Gerais; the PROJETO RIO within the Maré Favela in Rio de Janeiro; the PROFACE/PROSANEAR favelas program of the Sewage and Water Company (CEDAE) in Rio de Janeiro; the PROFAVELA program in Belo Horizonte; and the PREZEIS program in Recife.

Following the enactment of the Federal Constitution in 1988, actions for the provision of housing and implementation of basic sanitation infrastructure within the favelas were carried out with the support of the government. Between 1989 and 1992, a number of cities within the Metropolitan Region of São Paulo developed policies for the urbanization of favelas (Denaldi, 2003).

According to Rubio (2013a), the programs addressing the needs of favelas implemented between 1985 and 1990, with national support employing local labor in voluntary or paid community campaigns, implemented urban infrastructure, but only in a palliative manner.

The *Favela Bairro* (Neighborhood) Program, implemented in Rio de Janeiro during the period spanning from 1996 to 2000, was considered pioneering in actions promoting integral urbanization of favelas. The Program brought together a number of municipal government bodies and agents in projects whose interventions specifically targeted favelas with the population as their principal interlocutor (Rubio, 2013a).

The *Bairro Legal* Program, implemented in São Paulo between 2001 and 2004, was structured based on the Rio program. Both programs, although serving only a few favelas, were emblematic in that they led to restructuring of the government and received overseas funding.

The program for integral urbanization of favelas constitutes a key urban instrument for the regularization of city land occupation and became an important urban public policy in many cities, encompassing restructuring of municipal management to post-settlement monitoring.

According to Farah and Barboza (2001), between 2000 and 2007, there was a change of direction regarding implementation of public policies, when those developed in the states and cities promoted actions by the Federal Government.

After publication of the City Statute in 2001, the Federal Government created the Ministry of Cities (MCidades). In 2005, the National Social Housing System and Fund (SNHIS/FNHIS) were established, signalling the intention to invest in improving the living conditions of the low-income population, allowing a greater injection of funds.

The Federal actions culminated in 2007 with the inclusion of the urbanization of favelas as one of its central planks (Support for Improvement in Living Conditions of Precarious Settlements) under the Growth Acceleration Program (PAC) (Rubio, 2013a).

The program for integral urbanization of favelas is regarded as an important strategy to provide equity in land occupation, recognizing precarious settlements as an integral part of the contemporary city, allowing their physical and social integration.

The actions incorporated in this program included identifying the settlements, registering their populations and classifying the settlements, providing land tenure and urban regularization, and assigning official addresses, thereby promoting access to social programs and opportunities, via a process of full citizen participation. This process secured the residents' right

3 This study adopted the term Precarious Settlement, as formally defined by the Ministry of Cities, to denote an area of urban land physically and irregularly occupied in breach of prevailing land use legislation in the city. Typically, such settlements are situated in a spatially distinct manner having various levels of legal irregularity in terms of land ownership and consist of structurally substandard constructions, which pose risk to health and the environment. Precarious Settlements include: 1) favelas; 2) irregular dwellings housing low-income persons; 3) overcrowded houses, and 4) dilapidated housing estates (MCidades, 2007).

to the city, together with all the benefits and adversities associated with the Brazilian urban setting[4].

> "Although countless favelas have been urbanized since introduction of the PAC up until 2015, and despite the evolution of the legal framework that ensures and underpins the urbanization and regularization of these areas, there has been an imbalance between the policies, amount of resources and consolidation of interventions, as well as in the process of property tenure regularization of these areas in the cities" (Rubio, 2017b:132).

In Osasco city, part of the Metropolitan Region of São Paulo, the application of PAC resources provided urbanization and land tenure regularization, particularly in favelas. However, 12 years after these concerted efforts and heavy investment, an expansion of illegal settlements has taken place, with a breakdown of the interventions and reoccupation of the areas of environmental risk and protection in the urbanized favelas.

3 OSASCO AND URBANIZATION OF FAVELAS

Osasco, a city situated in the Western part of the Metropolitan Region of São Paulo (Figure 1), covers a land area of 64.95 km2 and has a total population of 666,740 inhabitants (IBGE, 2010). The city grew to become an important industrial hub of the Metropolitan Region of São Paulo in the 1950s. However, the city's urban infrastructure in 1962, the year in which it gained independence, remained substandard. This was mainly due to Osasco's reliance on São Paulo for investment in infrastructure and public services and goods since its inception in the late 19th century.

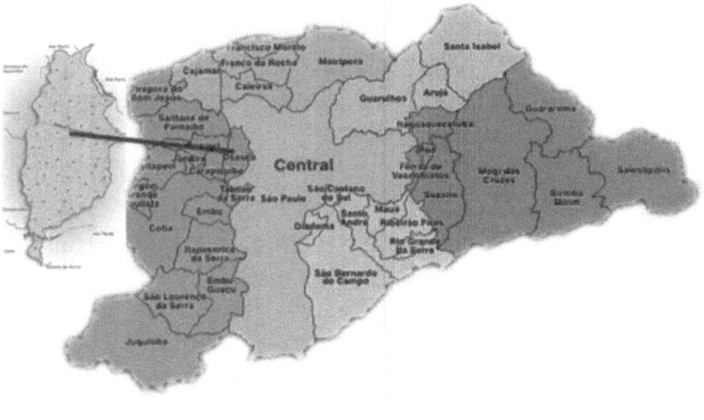

Figure 1. Location of Osasco city within Metropolitan Region of São Paulo. Source: RUBIO, 2017b.

Table 1. Osasco population growth (1934-2010).

Date	Population	Growth	Growth rate (%)
1934	12.091	-	-
1940	15.528	3.167	3,95
1950	41.326	26.068	10,48
1960	114.828	73.508	10,76
1970	283.073	168.245	9,44
1980	474.543	191.470	5,30
1991	568.225	93.682	1,81
2000	652.593	84.368	1,55
2010	666,740	14.147	0,23

Source: PLHIS. (SEHDU, 2012b)

Through a wave of migration and the steady growth driven by São Paulo city's economy, Osasco attracted a large population, which settled not only on formal land plots, but also on public land, giving rise to irregular housing developments. Between 1940 and 1960, Osasco's population increased around 10-fold (Table 1).

From the 1970s, the process of settling Osasco accelerated, promoted by the metropolitan dynamic fueled, by the demand for housing to accommodate workers in São Paulo city's industries and businesses.

The pattern of settlement of the neighborhoods differed, as did the level of urbanization, giving rise to developed areas alongside others with poorer infrastructure. This disparity was further exacerbated by the physical division by the railway line and the Tiete River dividing the land area, isolating part of the population from the central, more developed area of the city.

The public areas of the settlements, denoted "free areas", were occupied by the low-income population during the 1980s, consolidating the unbalance of urbanization and overly dense settlement of city land. This trend was stimulated by government authorities, which began granting change of use, legally redefining land for other purpose such as housing and commercial sale (Rubio, 2017b).

In 1970, Osasco had two favelas housing around 378 residents. By 1980, this number had swelled to 40 settlements housing some 18,397 residents. In 1985, there were a total of 94 settlements with approximately 35,000 dwellers, rising by 2005 to 170 settlements with 200,000 residents (SEHDU, 2012b). The maps depicted in Figure 2 show the timeline of occupation of the city and geographic location of the precarious settlements.

The major management challenge faced by Osasco city in 2005 was addressing the issue of

4 The concept of integrated urbanization, encompassing interventions for implementation of roadways, sanitation infrastructure, solutions for remedying geotechnical risks and for environmental recuperation, besides construction of housing units, can be found to varying degrees throughout the country.

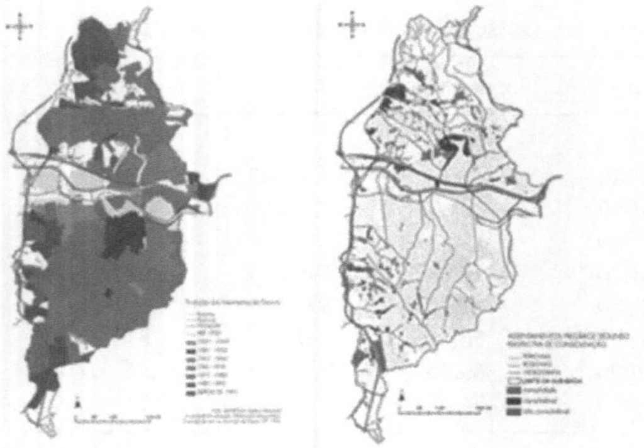

Figure 2. Evolution of subdivisions in Osasco compared to the location of precarious settlements. Source: PLHIS, (SEHDU, 2012b).

substandard housing, given that 200,000 of its inhabitants resided in 45,000 households within free areas, translating to 32.5% of the overall population seeking government assistance at various levels (Consórcio HAGAPLAN/GERIS, 2015).

According to data from the Osasco City Authorities, the majority of the city´s informal settlements were located on public land plots and non-regularized areas. These land areas possessed numerous restrictions and obstacles to regularization, both under city and state legislature, e.g. being situated within risk areas, on steep land, or along the banks of watercourses, inaccessible to public services and goods.

The land tenure status of the precarious settlements, by region, is displayed in Table 2, showing occupation of public land in both Northern and Southern areas of the city.

The implementation of Public Policies for improving social, physical, environmental and economic aspects took place on four action fronts: 1) institutional strengthening; 2) land tenure regularization and urban requalification; 3) housing provision, and 4) democratic management of the city.

The Osasco city authorities reorganized the administrative structure and, through Decree No. 9498 of 23 November 2005, articulated two planning sectors within the new structure of the Secretariat for Housing and Urban Development (SEHDU), to restore land access and equity in the city.

The SEHDU (2012b) set forth principles, guidelines, specific competencies and scales, with the aim of better catering for the population in an integral fashion, by addressing a range of issues at different levels, making optimal use of resources, while fostering action involving the precarious settlements in the city. The urban instruments and programs, based on the City Statute, such as the ZEIS (Zones of Special Social Interest), were employed to ensure structural urban interventions, social improvements and environmental enhancement of consolidated urban sectors.

In order to identify and classify the precarious settlements, the SEHDU (2009a) defined a set of characteristics to determine the level of irregularity and priority for assistance, in addition to the type of intervention undertaken, namely: total or partial lack of urban infrastructure, public services and equipment; occupation within land areas of risk, with steep gradient, or on banks of watercourses; settlements located on private land or property with tenure disputes; legally insecure ownership rights, and settlements in environmentally protected areas, requiring partial or total resettlement.

To meet the specific needs of each precarious settlement, the SEHDU implemented the *Bairro Novo* Program, comprising of four integrated complementary subprograms, namely: Gradual Urbanization of Favelas; Integral Urbanization of Favelas; Housing Provision; and Land Tenure Regularization.

The Program for Integral Urbanization of Favelas sought to provide urban land tenure regularization of favelas, whose area and complexity of occupation dictated actions prescribed for the implementation of urban infrastructure, including works for environmental protection, the provision of housing and other public services and social programs.

The goal of the Gradual Urbanization of Favelas Program was to implement urban infrastructure and equipment within settlements eligible for land tenure regularization.

The aim of the Housing Provision Program was to build housing units in areas equipped with infrastructure and services, providing resettlement of displaced families in areas receiving interventions, and to address the housing shortage in the city.

The objective of the Land Tenure Regularization Program was to promote recognition of security of title ownership by residents of precarious settlements, incorporating these into the formal fabric of the city.

Table 2. Land tenure status of settlements by region in Osasco.

Region	Public	%	Private	%	Mixed	%	Undefined	%	Total	(%)
North	48	47	14	13	25	24	17	16	104	49.5
South	44	42	33	31	24	23	5	4	106	50.5
Total	92	44	47	22	49	24	22	10	210	100.0

Source: PLHIS. (SEHDU, 2012b)

These subprograms sought to promote urban and social integration of the informal settlements into the formal land of the city at different scales and scopes, sometimes implemented concomitantly.

4 ASSESSMENT OF THE PROGRAM FOR INTEGRAL URBANIZATION OF FAVELAS: RECOMMENDATIONS FOR IMPROVEMENT

The methodology employed for analysis of the cases was based on Dunn's theory (2007), which states that evaluating public policies is a process of multidisciplinary inquiry to solve practical problems. According to Dunn (2007), concepts of social sciences and political philosophy can aid in devising a simplified evaluation process, thereby helping to determine the effectiveness of the policy being assessed.

The process is aimed at the creation, critical assessment and communication of relevant information for understanding and improving policies. The process should include not only purely objective analysis backed by scientific methods, but also incorporate empirical observations to inform public administration in planning actions in the city (Dunn, 2007).

According to Alvim and Castro (2010), there are a number of ways of defining the evaluation concept, but all seek to improve services and programs offered and applied by government authorities. This approach can enhance management performance of the State, underpinned by a set of instruments and processes, which includes monitoring these and their results.

Evaluation entails understanding all dimensions and implications of the policy assessed. Given that evaluating public policies can involve numerous modalities, it is vital to appreciate that this goes beyond comparing expected and observed outcomes, requiring an understanding of the course taken, as well as the elements and agents involved. Dunn (2007) added that the political process is made up of complex cycles, where each stage of the policies' cycle is linked to the next, both backwards and forwards, such that the process as a whole has no clearly defined beginning or end.

This cycle involves a multitude of interests, both individual and collective, in addition to numerous public and private bodies, modulated by information, problems, crises and disputes. The other cycle regards the policies *per se*, where new and old organizations draw on prior policies to make new ones, and the discontinuation of each policy implies the cessation of its associated programs and actions (Rubio, 2017b).

Dunn (2007) noted that, irrespective of the problem in hand, some questions can serve to guide the process, initially to ascertain its scope, then define the guidelines and instruments for solving the problem or predicting possible deviations, in addition to prescribing recommendations and establishing expected outcomes. The main objective of evaluating policies is to improve their formulation, where this is a complex task insofar as changes are gradual, disconnected and may exacerbate the problem.

Dunn (2007) presented five basic questions that can be posed in any evaluation: 1) What is the problem for which a solution is sought?; 2) What path should be followed to solve the problem?; 3) What are the results of choosing this path? 4); Which outcomes will help resolve the problem?; and 5) Which future outcomes can be expected, should the path be changed?

4.1 *Methodology*

The graph depicted in Figure 3 illustrates that, based on evaluation of the performance of the policy applied by observing the practical inferences and results achieved, the policy problems can be outlined. After elucidating the problem, this can be structured, revealing the difficulties and possible ways forward which, after following the recommendations prescribed, can be realigned and expected outcomes planned.

As described by Dunn (2007), the procedures in policy analysis have special names: monitoring, forecasting, evaluation, prescription/recommendation and problem structuring.

- Monitoring produces information on observed policy outcomes;
- Forecasting produces information on expected policy outcomes;
- Evaluation produces information about the value of expected and observed policy outcomes;
- Prescription (recommendation) creates information about preferred policies; and
- Problem structuring produces information about which problem to solve (Rubio, 2017b: 250).

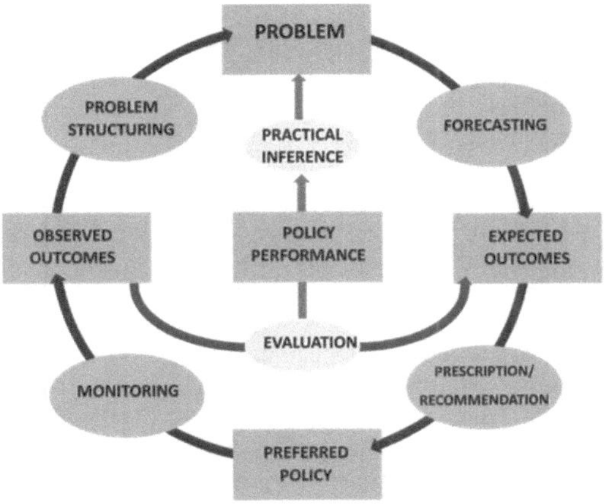

Figure 3. The process of public policy analysis. Source: Dunn, 2007, adapted by Rubio, 2017b.

From Dunn's theory (2007), the research has drawn methodology (policy problems and their structuring, preferred policies, observed outcomes and prescription/recommendations). Given the multidisciplinary nature of the policy problem evaluated, the data was structured under 4 headings: 1) land tenure regularization; 2) social inclusion;. 3) institutional organization; and 4) resources applied.

Two groups were defined for pooling the information collected according to cut-off dates: 1) occupation characteristics in 2005: physical environment, land tenure and social situation, and agents involved; and 2) actions taken: urban plan and project; interventions deployed and situation in 2016.

Four components were elected for more in-depth analysis: 1) the challenges faced implementing the policy, i.e. the problem; 2) the advances, actions and objectives achieved; 3) the shortcomings, such as setbacks and reversals seen in the urbanized areas; and 4) the perspectives, such as future recommendations beyond the study endpoint.

In the case of the integral urbanization of favelas in Osasco, mapping the situation that existed before implementation of the process (2005), i.e. the social, legal and physical situation, together with the prior actions performed, as well as the situation after implementation of the actions planned and executed in the projects (2016) involving building works and resources applied, proved vital.

Empirical observation plays a key role in analyzing the data compiled, since much of the information is not objective data that can be measured numerically, yet directly impacts outcomes. Therefore, the study author's empirical experience in the capacity of an agent was brought to bear as an analysis tool.

4.2 *Urbanization of Colinas D'Oeste*

The Colinas D'Oeste in the Jardim Bonança Neighborhood, the Northern part of Osasco, the city's largest precarious settlement, is classified as consolidable according to Local Plan for Social Housing (PLHIS) criteria (SEHDU, 2012b). In 2005, the area comprised around 300,000 m² of hilly terrain and large swathes of irregularly deforested land stripped of the vegetation supporting slopes and embankments (Figure 4).

This situation was exacerbated by the division of land plots carried out by residents without technical expertise to provide the foundations for building houses, creating areas of potential risk, including at sites that were hitherto safe. The "*non aedificandi*" areas bordering water sources and along the banks of water courses were also occupied.

The urban infrastructure was inadequate, where irregular narrow roadways conveyed both motor vehicle and pedestrian traffic. Many houses were situated on the edges and slopes of watercourses in an unsafe and hazardous manner. The local water and sewer systems were irregular, resulting in wasted water and contamination of existing watercourses.

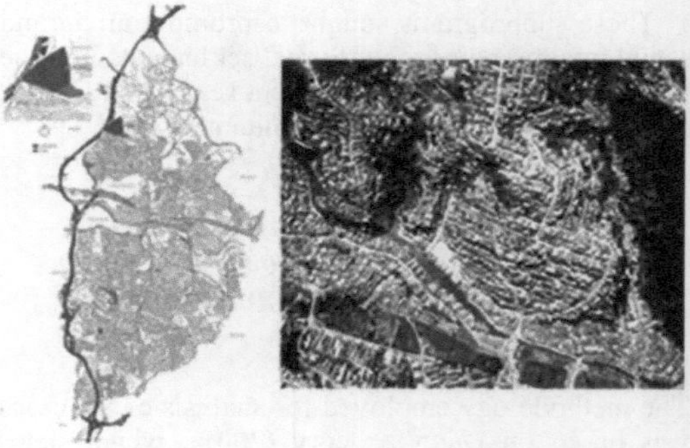

Figure 4. Location and aerial view of Colinas D'Oeste Favela. Source: Rubio, 2017b.

A total of 2,641 families, or 8,698 inhabitants, resided in the favela, 49.70% of whom were female. In 21.96% of cases, the head of the household was female. Most families had a monthly income of less than one minimum wage, while 37.23% of individuals aged 14-24 years had never been to school.

4.3 *Application of the Methodology: Challenges, Advances, Setbacks and Recommendations*

A summary of the evaluation methodology, as applied to the Program for Integral Urbanization of the Colinas D'Oeste Favela, is provided below.

4.3.1 *Challenges*

The physical situation of the Colinas D'Oeste favela was found to be as complex as the land tenure irregularities encountered. Given the complexity of occupation, allied with need to maintain the vast majority of dwellings, the challenges faced were considered. The primary goal concerning land tenure status was to formalize ownership titles to the land occupied by the favela. This step allowed buildings to be made code compliant, with subsequent consolidation of the dwellings and their formal incorporation as part of city land. Building rates were then levied, providing a source of funds for investment in public services, including maintenance of the urbanized areas (Figure 5).

4.3.2 *Advances*

The main advance was addressing the poor physical and social situations encountered at the Colinas favela. All urban instruments available in the municipal and federal spheres were applied, allowing investment of resources for restructuring the system of roadways, surfacing existing roads and installing water, sewer and storm water drainage systems, while also funding the construction of further housing units and public equipment. The actions carried out under the program paved the way for future formalization and consolidation of the favela as a neighborhood of the city, where

Favela: Colinas D'Oeste	
Dimensions of the area	312.279.02 m^2
Number of inhabitants	8.698
Number of households	2.641
Number of planned housing units	402
Number of housing units built by 2014:	230
Number of housing units built by 2016:	172
Project date:	2006/2007
Date of beginning of Civil Works	2008
Date of completion of first stage of civil works:	2012 (civil works in progress to complete the number of HUs and infrastructure in 02 sectors)
Monitoring of program	Equipe Secretaria de Habitação e Desenvolvimento Urbano (Técnica e Social)
Urban Project Author:	Architects Hector Vigliecca & Associados SC LTDA
Infrastructure Project Author:	Geométrica Engenharia de Projetos LTDA
Construction company responsible for the Works:	CONSTRUCAP CCPS Engenharia e Comercio S.A.
Project and Civil Works Management:	HAGAPLAN Engenharia e Serviços LTDA
Land regularization Management:	GERIS Engenharia e Serviços LTDA
Social Technical Work Management:	Consórcio GERIS-HAGAPLAN
Financial resources:	
Basic project:	R$ 840.122.64
Executive project:	R$ 1.603.109.52
Total Investment Cost (including Social Work and Land Regularization)	R$ 80.199.603.04
Cost of civil works:	R$ 67.961.077.95
Federal Government resources:	Resource source: Programa de Aceleração do Crescimento – PAC (Modalidade Urbanização de Assentamentos Precários) – R$ 50.428.866.29
Municipal Government resources:	Resource source: General Budget of the Municipality and the Municipal Fund of SIH – R$ 29.770.736.75

Figure 5. Technical data on Urbanization of Colinas D'Oeste Favela. Source: Rubio, 2017b.

urban demarcation permitted implementation of the plan and effective land tenure regularization (Figures 6 and 7).

4.3.3 *Shortcomings*

The goals originally set for tackling the physical and land tenure aspects were not fully met, since the interventions deployed failed to cover the full area of the settlement. This was partly owing to its extensive size

	FAVELA	COLINAS D'OESTE	
OCCUPATION CHARACTERISTICS	**Favela Location**	North Zone of the Municipality of Osasco, Jardim Bonança neighborhood, bordering the Osasco Landfill to the North, and part of a private area also in Osasco; to the west, Chico Mendes Avenue, and part of the border with the city of Barueri, to the east, part of the blocks of the Loteamento Portal D'Oeste and Portais favela, and to the south, the CTEEP (Electricity Transmission Company Paulista) land strip.	
	Planning Unit	ZEIS 2 & 3	
	Intervention Area (m2)	**312,279.02 m2 (*)**	
	Settlement Type	Favela	
	Consolidation Level	Consolidable	
	Level of complexity	Complex Urbanization	
	Beneficiary Population / Socioeconomic Profile	8,698 inhabitants, 2641 families, of which 43.7% live on an income of 1 to 2 minimum wages and only 12.8% with an income of 3 minimum wages. 4.42% of the families are supported by social income transfer programs (Bolsa Familia and Bolsa Escola) and 60.97% are not included in any type of social program.	
	Situation encountered	Part of the area is occupied by houses of good construction quality in consolidable lots. Occupation of areas of risk and environmental protection	
AGENTS INVOLVED AND ACTIONS TAKEN	**Sector responsible**	Secretariat for Housing and Urban Development - SEHDU	
	Source program	**Bairro Novo Program** Urbanização Integral de Favelas (Intergral Urbanization of Favelas) - Subprogram Project 2006 /2007	
	Deployment Period	1st stage of Civil Works: 2008 / 2012 1st stage of Civil Works: 2012 – 2016 - in progress	
	Fronts of action	**Social**	Social monitoring
		Physical environment	Projects, Civil Works and Technical Assistance
		Land	Land tenure regularization plan. Delivery of 148 titles out of 281 generated.
		Environment	Landscape treatment of the urbanized area and restoration of preservation areas
		Institutional	Articulation between municipal sectors for the integration of actions. Articulation between the spheres of government to obtain resources. New legislation to regulate interventions was created.
	Basic Legislation / Instruments	Master Plan of Osasco. Demarcation of a zone of special social interest (ZEIS). Urban Demarcation	
	Type of intervention	Reconfiguration of urban design with the implementation of a road system with thoroughfares for pedestrians and vehicles. Implementation of basic sanitation infrastructure, rainwater drainage and slope containment. Stream channeling. Construction of housing units and public educational and cultural facilities	
	Total investment value	R$ 80.199.603,04	
	Origin of resources	Federal government – PAC. Municipal government - FMPUH	
	Post-occupation / deployment situation	Urbanization not completed, with unpaved alleys, incomplete infrastructure and undeployed equipment; Reoccupied areas, including environmental protection areas; Land regularization not completed with plan in preparation; Remaining areas of urbanization without treatment; Environmental recovery project with incomplete implementation; 148 Tenure guarantee certificates delivered to residents.	

Figure 6. Chart outlining actions, part of methodology applied. Source: Rubio, 2017b.

Figure 7. Interventions carried out under the Program for Integral Urbanization of Favelas. Source: Rubio, 2017b.

and associated physical and land tenure complexities, hampering control by the Osasco city authorities of reoccupation of risk areas and environmentally protected areas that were cleared during the urbanization process.

Although the urban plan envisaged consolidation of the whole favela, with the clearing of houses situated in risk areas and designation of part of the area for environmental recuperation, with clearing of environmental protection areas, the actions were limited in the land tract that had existing major roads and denser housing areas.

The incompatibility of the timeframes for devising the project, raising funds, tendering and execution of building works, with those required to ensure public participation, also hindered the full embracing of the interventions by the population, leading to misuse of the housing units built and area urbanized shortly after inauguration.

The reoccupation of risk and environmental protection areas, which took place from 2012 onwards, compounded the risk posed by haphazard building along watercourses and on unstable land, with partial return to the poor levels seen in 2005.

Figure 8. Occupation of areas of risk and environmental protection in 2016. Source: Rubio, 2017b.

4.3.4 *Recommendations*

The results of this evaluation of the re-urbanization process point to the need for immediate resumption of the land tenure regularization plan, given that urban demarcation had been completed. This move will yield a blueprint of the land parcels to be restructured and consolidated, providing the basis for continuity of the project and for control of land occupation by the Osasco city authorities.

Concomitantly, specific legislation must be provided to prevent the expansion of the occupied area, based on investigation of the prevailing conditions, both physical and social. The urbanization process should be resumed, with the development of a new project including a timeline for delivery of the pending building works.

Contracts with project managers should incorporate follow-up of meetings discussing the population's demands, with greater attention paid to these in a bid to avoid future changes that can lead to further building of poor housing and deterioration of the land.

Developers should, together with the project designer, oversee the projects to identify potential issues that may impact final costs of the services or the timeline for implementation. In addition, the developer should carry out more careful planning of the execution of the services to prevent delays and overspending and unforeseen stoppages. In addition, provisions should be made for implementing the Technical Assistance Program to monitor betterments and refurbishments which householders plan to make in their own homes.

The setting up of the Social and Urban Guidance Centre (POUSO) in Colinas will serve both to oversee refurbishments and adaptations made by dwellers and contribute toward the upkeep of public spaces and of the urbanization implemented to date.

5 CONCLUSION

The city of Osasco has gained prominence in Brazil since 2005 for its implementation of the housing policy, introducing the City Stature (2001) and innovative practices. Between 2005 and 2016, the Secretariat for Housing and Urban Development has adapted its structure to play an integral role in the informal and formal city, expanding and training the technical and administrative teams, implementing program management, thereby allowing investment of resources from the Federal Government.

The *Bairro Novo* (New Neighborhood) Program and its subprograms, as part of Osasco´s housing and urban development policy, sought to address the specific issues of each type of precarious settlement in the city, deploying components and actions aimed at urban and land tenure regularization.

The Program for Integral Urbanization of Favelas consists of a complex series of concerted actions aimed at achieving the main goal of incorporating precarious settlements to become an integral part of the formal city.

The appraisal of the implementation of the Program for Integral Urbanization of Favelas in Osasco, between 2005 and 2016, has led to several recommendations, which can serve to inform policy in general:

- Ensure that modernizing and expansion of the administrative structure fosters intra and inter-institutional articulation in the implementation of the programs;
- Define priorities in the assistance delivered, along with the limits of the intervention;
- Carry out in-depth vetting of the capacity of companies tendered for hire;
- Improve the timing and methods of public participation to ensure the urbanization process is better embraced by the population;
- Provide continuity to the process of land tenure regularization so as to guarantee planning of further urbanization actions;
- Set up a steering committee for the interventions within precarious settlements, according to their specificities, and create instruments for defining the remit of its agents to produce an integrated model of involvement;
- Implement the ATHIS program to serve families during the post-urbanization period;
- Implement the POUSO instrument for controlling expansion and density of the occupations; and
- Update the PLHIS (SEHDU, 2012b) data on a regular basis (currently based on data from 2010), with the inclusion of proposals outlining effective actions for each type of settlement.

Taken together, the results showed that the urbanization actions implemented under the programs secured the right to the city for favela dwellers, not only implementing urban infrastructure, public equipment and services, but also including them in the discussion process about their place of abode, empowering them to engage in community participation and organization, and thus in citizenship practices.

There is still much to be done to achieve full integration of the precarious settlements into the formal city of Osasco, but policy evaluation can help elucidate potential ways forward in furthering this goal.

REFERENCES

Alvim, A. T. B. e Castro, L. R. 2010. *Avaliação de Políticas Urbanas: Contexto e Perspectivas*. São Paulo: Mackenzie Presbyterian University and Romano Guerra.

Brasil. 2016. PAC – *3º Balanço 2015-2018*. Brasília: Ministério do Planejamento, Desenvolvimento e Gestão.

Consórcio HAGAPLAN/GERIS. 2015. *Proposta Técnica de prestação de Serviços Técnicos Especializados de Gerenciamento e Assessoria na Execução dos Planos de Regularização Fundiária, Implantação dos Planos de Trabalho Social, Acompanhamento, Elaboração de Projetos e Fiscalização de Obras no Âmbito dos Programas e Empreendimentos da Secretaria de Habitação e Desenvolvimento Urbano, de acordo com o objeto do RDC Presencial nº 004/2016*. Osasco: CONSORCIO HAGA-PLAN/GERIS.

Denaldi, R. 2003. *Políticas de Urbanização de Favelas: evolução e impasses*. Tese (Doutorado em Arquitetura e Urbanismo). São Paulo: Universidade de São Paulo.

Dunn, W. N. 2007. *Public Policy Analysis. An Introduction*. 4th edition. New Jersey: Pearson Education Ltd.

Farah, M. F. S.; Barboza, H. B. (ed.) 2001. *Novas Experiências de Gestão Pública e Cidadania*. Rio de Janeiro: FGV.

IBGE. Instituto Brasileiro de Geografia e Estatística. 2010. *CENSO*.

Lefebvre, H. 2001. *O Direito à Cidade*. Tradução Rubens Eduardo Frias. São Paulo: Centauro, versão do original em francês, 1969

Ministério Das Cidades (Mcidades). 2007. *Assentamentos Precários no Brasil Urbano*. Brasília: Secretaria Nacional de Habitação/MCidades/Centro de Estudos da Metrópole - CEBRAP. Available: https://direito.mppr.mp.br/arquivos/File/relatorio-assentamentos-precatorios-no-brasil-urbano.pdf. Accessed: March 08th, 2020.

Rubio, V.M. 2013a. Projeto de Urbanização de Assentamentos Precários no Rio de Janeiro e São Paulo: Um Instrumento de Construção de Cidade. *CADERNOS DE PÓS GRADUAÇÃO EM ARQUITETURA E URBANISMO (MACKENZIE. ONLINE)*, v. 13, p. 65–102, 2013. Available: http://dspace.mackenzie.br/handle/10899/25901. Accessed: March 08th, 2020

Rubio, V.M. 2017b. *PROGRAMA DE URBANIZAÇÃO E OS DESAFIOS PARA INTEGRAÇÃO DA FAVELA À CIDADE. Avanços e descompassos de uma política pública em Osasco – SP (2005-2016)*. Tese (Doutorado em Arquitetura e Urbanismo) São Paulo: Universidade Presbiteriana Mackenzie.

Secretaria Municipal de Habitação e Desenvolvimento Urbano (SEHDU) 2009a. *SEHDU 2005-2008. Realizações Perspectivas*. Osasco: PMO/SEHDU.

Secretaria Municipal de Habitação e Desenvolvimento Urbano (SEHDU) 2012b. *Plano Local de Habitação de Interesse Social (PLHIS)*. Osasco: SEHDU/PMO.

Spinazolla, P. C. 2008. *Impactos da regularização Fundiária no espaço Urbano*. Dissertação (Mestrado em Habitat). São Paulo: Universidade de São Paulo, FAUUSP.

Research Tracks in Urbanism: Dynamics, Planning and Design in Contemporary Urban Territories – Allegri et al (eds)
© 2021 Copyright the Author(s), ISBN 978-1-032-10788-2

The urban palimpsest in three renewed neighborhoods of Santiago, Chile

X. Arizaga
Facultad de Arquitectura Diseño y Estudios Urbanos, Pontificia Universidad Católica de Chile, Chile

ABSTRACT: The proposal analyses three sectors of Santiago de Chile, representative of three types of urban renewal that are also three forms of space production. In these neighborhoods, the treatment granted to the permanences and irreducibilities of the city is different in each urban renewal policy. The palimpsest of the three neighborhoods is reconstructed in plans that show the different periods of building to understand the composition of their different urban fabrics. The experience today is expressed in interviews with the inhabitants. The proposal presented seeks to verify whether the reading of the inhabitants coincides with the overlapping of buildings from different eras in the neighborhoods studied, and understand what this reveals about the forms of renovation of the city.

1 INTRODUCTION

Urban renewal policies account for different ways of treating the urban space. In these, the permanences – what remains unchanged – and irreducibilities of the city – elements that cannot be changed – acquire different values when the built complex is intervened upon. The way of treating the interstices and the permanences of the past evidences results in different disciplinary visions of the urban space. At the same time, the incidence of permanences and irreducibilities plays an important role in the experience of the metropolitan neighborhoods, where the inscription of the urban renovation in the traditional urban fabric results in the mixture of different temporalities and materialities that make up the current city. As Rossi points out: "the difference between past and future from the point of view of the theory of knowledge consists precisely in the fact that the past is lived in part now, and that, from the point of view of urban science, can be the meaning that we have to give to permanences; they are a past that we still experience" (Rossi, 1999: 99).

Two extreme approaches are found at the origin of this analysis: the type "urban acupuncture" (Solá Morales, 2008) and the "*tabula rasa*" mode. In the first, there is a will to preserve the permanences, and on the contrary, promoted by the modern movement, a rejection of the existing city. Three sectors of the center of Santiago that correspond to three types of urban renewal policies that occur in three different historical moments (Arizaga, 2019) were analyzed. A first moment, marked by hygienism, is when the desire to structure and beautify the city is materialized through Vicuña Mackena's "Santiago Transformation Plan" (1872). A second moment is marked by the model that takes its origins in the International Congress of Modern Architecture (CIAM), and corresponds to the period of the Urban Improvement Corporation (CORMU; 1965-1969) and its policy of "urban remodeling" marked by the Welfare State, social logics and housing policies. A third moment, that begins with the 1985 earthquake – and is expressed today in its greatest amplitude –modifies the urban scale and density, conserving the network of urban blocks and subverting it, at the same time, in its third dimension: height.

The results of a previous investigation in which the conducting of interviews allowed to verify the importance that the inhabitants give to the mixture between the new and the old, which takes on special significance in their speech and their preferences, was the starting point. The research presented seeks to verify whether these preferences are effectively anchored in the density of the permanences that were conserved when each of these neighborhoods was renewed.

Based on the plans, building permits and field validation, the objective is to understand the composition of the urban fabric of different neighborhoods of Santiago that are experienced today. The representation of the different building layers seeks to recompose the thickness of the urban blocks. The reading of the urban palimpsest, as "stratification of layers inscribed in time but (re)updated by a subject" (Bähler & Frölisher, 2012) allows us to decompose the landscape of the city, and, the interviews allows us to understand how this is re-signified by the inhabitants. Through this analysis, the aim is to understand the values that underlie in the

DOI: 10.1201/9781003220855-17

preservation of continuities, and the breaks imposed by different renewal policies adopted in the consolidated center of the city of Santiago de Chile.

Layer overlapping plans were constructed to show the existence of buildings of different periods that compose the current fabric. These plans were built based on the planchets of the municipal cadaster (1910; 1939; 1960 and current) and were corrected or complemented with the observation in the field and the review of the building permits obtained in the Municipal Department.

The results of this exercise are counterintuitive. The neighborhood where, by rule, it is believed that the permanences were considered is Lastarria-Bellas Artes; however, when analyzed in detail, the plan reveals that in reality, all buildings built between 1930-1940 are preserved. On the other side, the San Borja neighborhood, which is in the imaginary of Santiago the perfect representative of the "*tabula rasa*", shows in detail a modern project that preserves representative pieces of the city that preceded it. Finally, the Santa Isabel neighborhood, which in the city of Santiago has the most intervened urban fabric in the context of the third urban renewal policy, is the one that conserves the largest number of buildings from 1910.

2 URBAN RENEWAL AS A HERITAGE CONSERVANCY PROCEDURE

Urban renewal has different origins: it is a tool for the requalification of the city that, since historical times, has served to solve social problems of hygiene and/or housing shortage, imbalances in metropolitan growth, obsolescence problems, and also political purposes; having urban beautification as an externality. Likewise, although the origins of the renovation are not exclusive, their objectives have aimed at different goals: economic reactivation, preservation of historical heritage, convergence of diverse uses or commitment to the coexistence of different socioeconomic categories. In all cases, urban renewal consists of an intervention in the existing city. In this sense, it is also a way of relating to the built heritage, of composing the future with the past, of re-signifying the urban fabric and its buildings. Through the preservation or denial of the preexistences, urban renewal chooses to preserve or renew valuable elements of the built fabric. In this way, it allows to re-create the conditions to guarantee the spatial relationship between the different urban elements.

There is an approach with a will to preserve the permanences, an urban renewal with a conservancy procedure; and, on the contrary, promoted by the modern movement, an approach where the preexisting city is rejected through the "*tabula rasa*". Rossi points out: "(. . .) the problem of the permanences has two aspects; on the one hand, permanent elements can be considered pathological elements; on the other, as propellants. Either we use these facts to try to understand the city in its totality or we end up being bound by a series of facts that we will not be able to later relate to an urban system" (Rossi, 1999: 101). Therefore, the challenge is to find a measure where the interventions carried out are open to change, and, at the same time, retain their intrinsic qualities. Lynch proposes that, while preserving has its virtues, change is also a necessary ingredient: "An environment that cannot be changed invites its own destruction. We prefer a world that can be progressively modified against the backdrop of valuable remains, a world, in short, in which one can leave a personal mark alongside the marks of history" (Lynch, 1975: 46).

This inclusion of the temporal dimension in the inhabited space refers to the concept of authenticity that Sukin (2010) signalizes. It allows us to bring past and present closer together, and therefore makes these neighborhoods rich and attractive: "You can feel the hum of another kind of time. The Greeks called it *kairos*: a sense of the past that intrudes on and challenges the present. It is different from *cronos*, our usual sense of time as a simple arc of endless progress between yesterday and tomorrow" (Sukin, 2010: 101).

The working hypothesis is that the preferences revealed by the inhabitants are based on their understanding of time, which manifests itself in space and its visual richness, what Cullen called "intricacy": "This quality (. . .) seems to have stopped the passage of time, with its block constructions, its individualized facades, its intrinsic banality, its light surfaces and its disconnected planes. But it is a quality that has the virtue of absorbing sight" (Cullen, 1981: 65).

This proposal sought to verify whether the reading of the inhabitants coincides with the imbrication of buildings from different periods in the neighborhoods under study. In addition, it aimed to understand what this reveals about the different forms of urban renovation, and about the preferences of individuals.

3 THREE NEIBORGHOODS IN INHABITANTS' DISCOURSES

The neighborhoods studied are located in the northeast of the central commune of the capital of Chile, and correspond to sectors inhabited by middle classes, with a level of education that is significantly higher than that of the commune and the country. Due to their location, they have good accessibility to and from the different destinations in the capital; all three have an underground station and good infrastructure, commerce, services and urban amenities. Also, they are adjacent to the Providencia commune, which gives them access to other quality urban facilities and influences the preferences of visit and choice as a place of residence for young people from the richer neighborhoods in the eastern part of the city.

Semi-structured interviews were conducted with three inhabitants per neighborhood in order to verify in their speech the assignment to certain spaces in the neighborhoods. At the end of the interviews, when it was possible, a tour to the places they described was done. At least two interviews from each neighborhood obtained optimal results and it was an exercise of interest both for the interviewer and for the interviewees, who discovered aspects of their adherence to the neighborhood that they had not verbalized before.

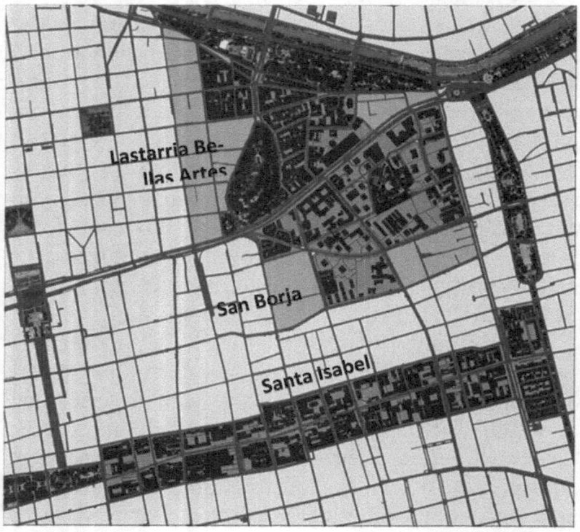

Figure 1. Three neighborhoods: Lastarria Bellas Artes, San Borja and Santa Isabel. (Source: Author).

3.1 *Lastarria Bellas Artes*

The Lastarria-Bellas Artes neighborhood corresponds to the north-east vertex of the central district of Santiago, and its plot preserves memory of the canal layout and the semi-agricultural character that differentiated it from the institutional and oligarchic center marked by the checkerboard street plan. Between 1870 and 1874, thanks to the efforts of the mayor Vicuña Mackena, the Santa Lucía hill, situated in the middle, was remodeled, becoming a park and promenade for the incipient bourgeoisie. This improvement affected the appreciation of the neighboring properties that then surrounded the hill. Between 1885 and 1918, the waters of the Mapocho River were channeled (Castillo, 2014), and the land gained allowed the construction of the *Parque Forestal* in 1901 and the creation of the Museum-Academy of Fine Arts in 1910 – both proposed in the Plan of Transformation of Santiago developed by Vicuña Mackenna in 1875. This gave an important boost to the neighborhood; on the edge of the park elegant chalets appeared, which were later replaced by buildings that maintain their standard and distinction to date. Between 1920 and 1930, a phase of renovation of the sector can be observed as some significant buildings remain. Then, a phase of re-change occurred in the decade of 1930-1940 and in the following decade, 1940-50. Of these decades a significant number of buildings are conserved and configure the aesthetics of the neighborhood: continuous facade and the same height, following the building standards defined by K. Brunner in the first Regulatory Plan of Santiago (1939). Later, an investment phase corresponding to the 1970s can be observed, matching with the inauguration of the underground. A last phase of renovation started at the beginning of the 21st century, with higher buildings that responded to the impulse given by the latest urban renewal policy.

The inhabitants of this neighborhood that were interviewed, Elisa, Jessica and Felipe, expressly stated that they live in a place that is "architecturally beautiful" and "harmonious", in which there is also a "general concern for beauty". The three participants identified the value of living in a place "where people know they live in a neighborhood" (Elisa), where people greet each other, greet the people who serve in the premises and maintain good relations with the neighbors. They also valued the cultural offer, diversity and "originality" (Elisa). Elisa specified the difference between new places that "appear to be something they are not" (on Lastarria street) and "real" places which are not "pretending to be something"; the ones they visit and are the true "neighborhood experience" (Felipe). Regarding these same new places on Lastarria Street, Jessica pointed out that they are designed for tourism. In this sense, one of the fears of the three interviewees is this "hyperventilation" (Felipe) of tourism, which makes prices more expensive and displaces the residents of the neighborhood. As a consequence, they believe that in the long run they will no longer be able to pay the rent. In addition, they pointed out the loss of stores and neighborhood uses such as the shoe repair shop or the greengrocer that they have seen disappear in recent years (Jessica). They also manifested a true concern for heritage, which they value for their "old" aesthetics and for the appearance of buildings that are shocking for their size and the massive office buildings, with a flow of people that did not exist before (Jessica).

3.2 *San Borja*

The San Borja neighborhood occupies what was previously the site of the San Francisco de Borja Hospital, of which only the chapel remains. In the neighborhood were also the Lyceum No. 5 for Girls and the Juan Antonio Ríos Market, whose buildings were acquired in 1976 by the University of Chile for the Faculty of

Architecture and Urbanism. The San Borja Remodeling project, promoted by the Urban Improvement Corporation (CORMU), included the construction of 21 towers and 12 lower-rise buildings between 1969 and 1976; and had inside it an important green area: Parque San Borja. This initiative based its actions on the "*tabula rasa*" approach and the insertion of the housing complex as a city in the city, through the consolidation of a mega-block. However, in the Torres San Borja sector, pre-existing neoclassical buildings from the beginning of the century are preserved, which combined with the passage of time and the towers themselves ended up constituting a homogeneous patina assembly. The complex defined was inserted into the city, offering a network of circulations that irrigate the neighborhood through walkways and commercial slabs that make up a public space very typical of functionalist architecture.

San Borja interviewees had disparate views of the neighborhood. In general, all highlighted the centrality and offer of services; but their adherence to the neighborhood and perception of a characteristic environment is less strong than in Lastarria Bellas Artes. Daniella and Gerardo were the ones who showed the greatest appreciation of "neighborhood life", the connection with "different types of neighborhood" (Gerardo), and the possibilities of walking through the neighborhood. They highlighted in particular the crossing through the San Borja Park. Daniella, who lived in the neighborhood when she was younger and frequently visits her mother who lives in Tower 12, recalls with nostalgia in particular the labyrinths and passageways that hide the towers. Gerardo, who has lived in the neighborhood for a year, especially values "people greet each other", that there is a "neighborhood life". Gerardo highlighted in particular that the neighborhood "has a personality" that he attributed to a "very strong concept of community" and believed that this has to do with "the time when the towers were built, and perhaps in other places it was lost but here it remains". In this same sense, he showed concern about a certain "standardization of public spaces" that he identified in acting from the State that would not be sensitive to the real interests and needs of people. Daniella and Paola emphasized the deterioration of the towers and they would like them to be conserved because they "represent something" (Paola). To recreate themselves, the three interviewees indicated that they visit the Lastarria neighborhood and the Parque Forestal; thus pointing out a lack of spaces that respond to the taste of the inhabitants of the towers.

Figure 2. Barrio Lastarria, where the old buildings meet the new buildings. (Source: Author).

Figure 3. Barrio San Borja: towers. (Source: Author).

Figure 4. Santa Isabel: the change of scale. (Source: Author).

3.3 *Santa Isabel*

The Santa Isabel neighborhood corresponds to the surroundings of the street of the same name and is representative of the latest urban renewal policy: characterized by the liberation of building heights. It concentrates a greater number of individual operations, which highlights the complex as a unit differentiated from the rest of the city, both in its morphology and density of taller buildings. It was also pointed out as a characteristic neighborhood of this urban renewal policy in the discourse of the inhabitants.

Santa Isabel Street was opened by joining sections of streets in the east-west direction through different blocks, to form a new alternative of vehicular flow. The opening of the street took 50 years and its linearity replaced a much more intricate fabric, giving space for the renovation that would come later. This sector was selected in the feasibility study of the Urban Renovation Subsidy as a priority, which prompted the construction of a significant number of residential buildings that replaced car workshops, parking lots and other uses such as cardboard deposits that had displaced the residential use. Of the modest buildings of that time, some vestiges prevail: *cités*, warehouses and large houses.

The inhabitants interviewed identified this modest heritage, and in their speech, express their concern about its disappearance. The three interviewees live in new apartments in the neighborhood' towers. Rosario and Javiera showed discomfort and concern about the high population and building densities that the neighborhood has, and, make "the city more difficult to inhabit". This translates into annoying noises and saturation of the services that are insufficient; for example, Rosario pointed out that going to the supermarket means "queuing for half an hour". In general, they pointed out that in their leisure and recreation times they prefer other neighborhoods such as República, Yungay, Lastarria and Bellavista because in Santa Isabel "there is not much to walk around" (Rosario). In particular, Javiera and Rosario commented in the interviews that they use the neighborhood stores that they prefer a lot over the new ones because "they are not abusers with the prices, they are more extensive and diverse". Rosario highlighted that there are still passages where "poverty" is hidden and that they are "stigmatized". Rosario and Javiera regretted that they are condemned to disappear. Natalia was afraid that all the "houses" that give account of a "neighborhood life" will be demolished; and, believed that the authorities should preserve them for a cultural use.

In this way, the residents of Santa Isabel clearly showed the difficulties and impacts of densification in height. It is highlighted that they do not choose the green areas built by real estate developers, taking refuge in other more consolidated spaces such as the Plaza San Isidro or the Parque Bustamante. They showed a concern for heritage, the loss of ways of life and spaces that allowed a "more dignified dwelling" (Javiera). They were distressed by the fact that Santa Isabel is a "bedroom neighborhood" where the inhabitant has to make a real effort to "occupy the space they inhabit" (Rosario).

4 RE-CONSTRUCTING THE PALIMPSEST

Through an exercise of overlapping layers, the aim was to unveil the palimpsest that emerges in the inhabitants' discourse about these three neighborhoods. According to the Dictionary of the Spanich Royal Academy, the palimpsest is an "ancient manuscript that preserves traces of a previous artificially erased writing". In this sense, it sought to bring to light the existence of traces of previous buildings and irreducibilities present in the experience of these neighborhoods.

Plans that show the existence of buildings built at different times and that are preserved in the current fabric were drawn. These plans were built on the basis of the planchets of the municipal cadaster (1910; 1939; 1960 and current) and were corrected or complemented with fieldwork observation and the review of the building permits obtained in the Municipal Public Works Department (Figure 5: Lastarria-Bellas Artes; Figure 6: San Borja, and Figure 7: Santa Isabel).

The permanences, as well as the building heights, make up the urban fabric, accounting for different building temporalities: their expression in the present accounts for the sum of previous interventions.

The results of this exercise are counterintuitive. The neighborhood where it is believed that the permanences were considered is Lastarria-Bellas Artes. Notwithstanding, in reality, most of the buildings preserved were built between 1930-1940 as shown in Figure 5, and correspond to the re-conversion of the Mapocho River and the implementation of the *Parque Forestal* proposed in the Vicuña Mackena Plan. These buildings give character and distinction to the neighborhood, they correspond to a homogeneous height and to the interpretation of the rules contained in the regulatory plan devised by K. Brunner in 1939. In contrast, very few buildings are preserved prior to the "Transformation Plan" of Vicuña Mackena: the Church of the Vera Cruz and its parish house, some constructions, houses very intervened in the same block that possibly conserve vestiges of adobe walls from the early twentieth century. Despite these changes in the built complex, this neighborhood has been able to integrate new buildings both from the 1960s and 1970s as well as those built at the beginning of the 21st century. Both the intricate fabric and the respectful intervention manage

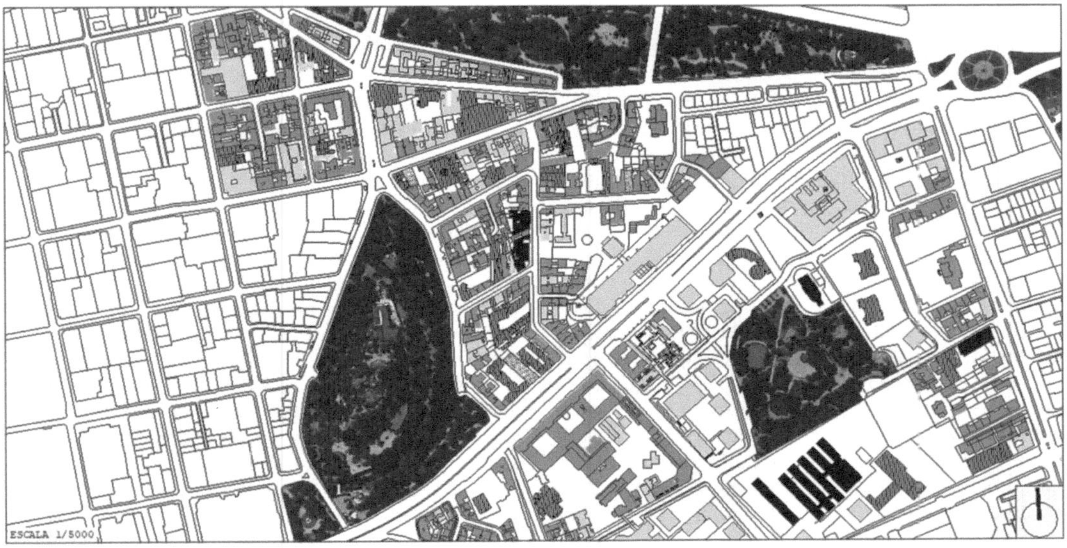

Figure 5. Plan of the permanences of Lastarria-Bellas Artes.

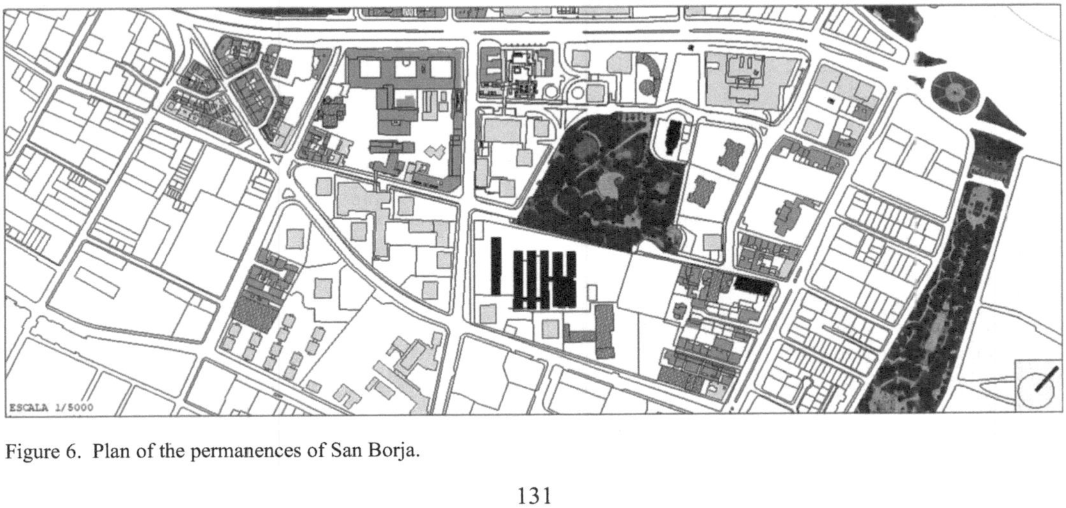

Figure 6. Plan of the permanences of San Borja.

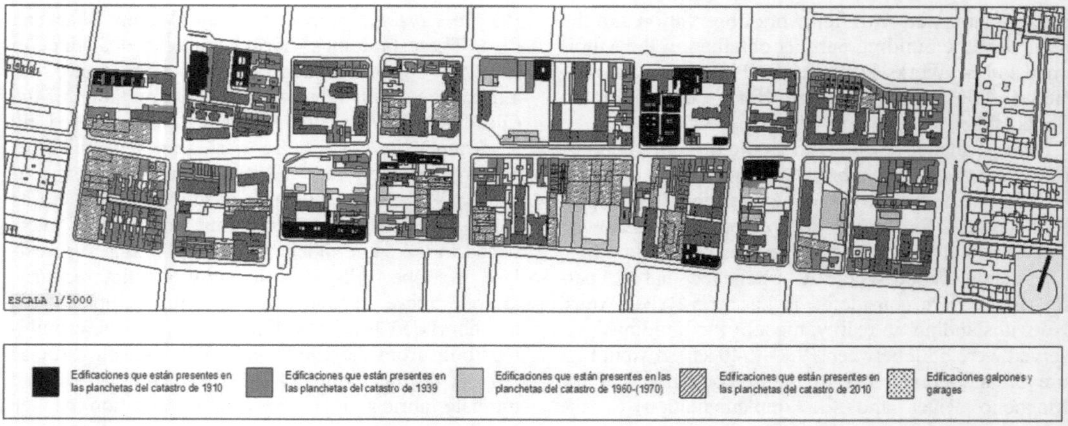

Figure 7. Plan of the permanences of Santa Isabel. (Source: Author)

to create an urban space that the interviefig8_3wed inhabitants identified as "harmonious" (Figure 2).

At the same time, in the San Borja neighborhood, which is the largest representative of the "*tabula rasa*" proposed by the modern movement in the local imaginary, we find that there are representative pieces of the city that preceded this urban design. The church of the San Borja Hospital, the Juan Antonio Ríos market, the cloister of the Sagrado Corazón congregation between towers 23 and 24 are pieces preserved from the early twentieth century. Likewise, buildings built in the first half of the twentieth century, of four to six floors, were integrated from the beginning in the treatment of the mega-block.

On the contrary, the Santa Isabel neighborhood, which, in the context of the city of Santiago, is one of the most intervened upon in the framework of the third urban renewal policy, is the neighborhood that preserves the largest number of buildings from 1910. As it is shown in Figure 7, the pieces conserved are smaller and modest: private houses, *cités* and passageways that do not constitute a monumental complex or a significant critical mass as would be the case of the Juan Antonio Ríos Market from the San Borja neighborhood. This condition of modest, deteriorated houses, simple in their style and construction, is precisely what motivated the application of the urban renewal subsidy in this neighborhood as a priority case. These buildings, without distinctive elements that stand out from others due to their aesthetics, were then invisible behind the deterioration and functional obsolescence of the neighborhood. Today, their situation remains the same, oppressed between towers they host a smaller trade of warehouses and workshops that nevertheless feed neighborhood life. The inhabitants, through their daily actions and their discourse, express their concern for this disappearing heritage and for the fewer person resources that it hosts.

In fact, in the first neighborhood, Lastarria Bellas-Artes, the traces of the irrigation canals of the first occupation of this triangle between two arms of the Mapocho river, and, preserved in the trace of the streets, provide the intricacy pointed out by G. Cullen. These invisible traces, irreducibilities, more than the buildings allow inhabitants a harmonious understanding of the past.

In the San Borja neighborhood, the dimensions of the mega-block, heir of the hospital and the grounds of the religious congregation which preceded it, as well as the permanence of monumental constructions, allow the insertion of the towers of the Modern Movement in a context at their scale. Thus, the reading of the past remains invisible to the inhabitants.

Finally, in the case of Santa Isabel neighborhood, the piercing of the street erased an intricate fabric leaving room for real estate developers to transform a tight urban fabric into lots that can accommodate towers. This change of scale left the permanences in a category of vestige which does not go unnoticed by the inhabitants.

5 CONCLUSIONS

The interviews made possible to illustrate the differences between neighborhoods, which materialize in the analysis of the plans and take a special meaning in the discourse of the inhabitants. Lastarria-Bellas Artes is the neighborhood where, for the inhabitants, this mixture between the new and the old appears most strongly. In San Borja, it was suggested that, although the interviewees did not express the aesthetic values of the towers and the complex, they are interested in their conservation and concerned about their deterioration. In Santa Isabel, the feeling of an insufficient response to residents expectations of inhabiting an urban space is notorious. The towers combined with the existence of deteriorated and precarious mansions and *cités* threaten "what remains" of a neighborhood life and a city scale that inhabitants seek precisely in other sectors of Santiago.

The understanding of the valuations that underlie the conservation or denial of urban preexistences shows that urban renewal is also a way of making heritage, of composing the future with the past, of re-signifying the urban fabric and its buildings. The inhabitants are very aware of this edge of public policies and express it in their concerns about the future of their neighborhoods. Therefore, urban renewal finds itself in a new perspective where it is not about conserving at "all costs" nor is it about demolishing on a large scale. Its new challenges would reside in the need to think "the new" anchored "in the old" not in a perspective of "conservation" but in the perspective of creating a "new context" (Bourdin, 2009: 151) that forces us to think the interrelationships with the past and the long term. It is therefore a question of finding a measure between "preserve" and "renew".

For this reason, a respectful urban renewal should protect the differences, those valuable but sometimes subtle distinctive elements of the built fabric that allow to re-create those conditions that guarantee the spatial relationship between the different built elements. And it is perhaps this aspect that is most strongly manifested in the reconstruction of the palimpsest in the proposed plans: for this harmony to emerge, it is necessary to preserve the subtle relationship between the elements, enhance the urban layout, take care of the scale between the new buildings and existing constructions, to accommodate the modest and put in scene the monumental. When this relationship in scale and/or height is not respected, as in the case of Santa Isabel, the daily understanding of the neighborhoods makes invisible those pieces of the past that would be called upon to give the renovated neighborhood the "authenticity" that ensures its attractiveness as a residential area and its valuation within the real estate market. As Corboz (2001 [1983]) would say, understanding these traces of the past "gives the opportunity for a more intelligent intervention".

REFERENCES

Arizaga, X. 2019. Propuesta de caracterizacioón de la renovacioón urbana en Chile. El caso de la comuna de Santiago centro. EURE Vol 45, N° 134, 169–212.

Bähler, U., Frölisher, P. 2012. Figurations de la Ville Palimpseste. Alemania: Gunther Narr Verlag Ed.

Bourdin, A. 2009. Du Bon Usage De La Ville. Paris, Francia, Descartes Et Cie.

Castillo, S. 2014. El río Mapocho y sus riberas. Espacio público e intervención urbana en Santiago de Chile (1885-1918). Santiago, Chile: Ediciones U. Alberto Hurtado.

Corboz A. 2001, Le Territoire comme palimpseste et autres essais, Besançon, Francia, Editions de l'Imprimeur.

Cullen, G. 1981. El paisaje urbano: tratado de esteótica urbanióstica. Barcelona: Edit. Blume.

Lynch, K. 1975 [1972]. ¿De Qué Tiempo Es Este Lugar? Para Una Nueva Definición Del Ambiente. Barcelona, España: Editorial Gustavo Gili.

Rossi, Al. 1999. La Arquitectura de la ciudad. Barcelona, España, Edición, GG.

Solá-Morales I Rubió, M. 2008. De cosas urbanas. Barcelona, España: Editorial Gustavo Gili.

Sukin, S. 2010. Naked city: the death and life of authentic urban places. Oxford, USA: Oxford University Press.

Vicuña Mackena, B. 1872. La Transformación De Santiago. La transformación de Santiago: notas e indicaciones respetuosamente sometidas a la Ilustre Municipalidad, al Supremo Gobierno y al Congreso Nacional. Santiago, Chile: Imprenta de la Librería del Mercurio De Orestes L. Tornero.

Brazil

Research Tracks in Urbanism: Dynamics, Planning and Design in Contemporary Urban Territories – Allegri et al (eds)
 ISBN 978-1-032-10788-2

Social housing, real estate and environmental protection in Pará, Brazil

A.P. Furo & J.J. Lima
PPGAU UFPA, Postgraduate in Architecture and Urbanism, Universidade Federal do Pará, Brazil

ABSTRACT: In the United Nations 2030 Agenda, the right to housing is one of the greatest needs for promoting sustainable development. Using the concept of the right to the city and Brazilian legislation to reconcile housing/real estate development with protecting the environment and guaranteeing access to urban goods, this paper aims to analyse the implementation of the Quinta dos Paricás Housing Project, under the Minha Casa, Minha Vida Program (known as PMCMV), located in the Maracacuera neighbourhood of the Brazilian municipality of Belém, Pará. This project facilitated a series of normative changes – most notably in the city Master Plan – for the area in which the project was implemented, and enabled major changes to the urban fabric of the region, providing conditions – previously unavailable – for real estate development, thereby contributing to the ineffectiveness of urban policy, and moving the poorest communities away from the city to which they have rights.

1 INTRODUCTION

Since the enactment of the 1988 Brazilian Constitution, the Brazilian State has assumed the responsibility and commitment of promoting a wide range of rights, all aimed at combating social inequalities - in all its forms - and promoting full citizenship for all, through guaranteeing human dignity, the social function of property and environmental protection.

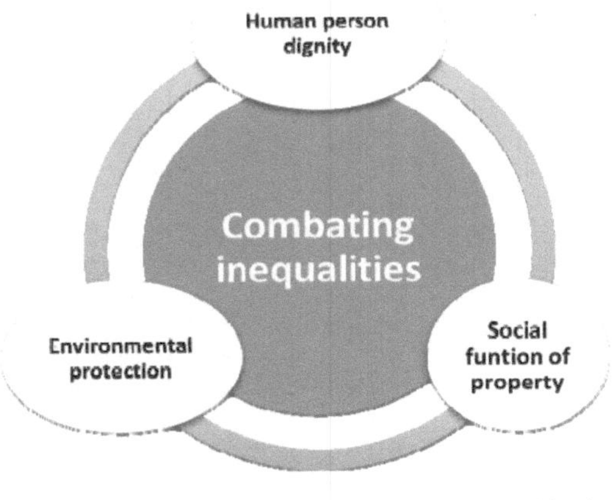

Figure 1. Diagram depicting the Constitutional Objective to Combat Inequalities. Source: Produced by the authors (2018).

Based on this premise, the constitution determines the role of each social agent to balance and reconcile their interests so as to contribute to a social project for the country, which is viable for all Thus, it is the responsibility of the State to create and implement public policies, to adapt the interests and objectives of civil society, and to regulate the activities of the country's productive forces towards these public policy guidelines. These objectives and commitments need to be constantly reaffirmed, at least in theory and in legal norms (Brazil 1988).

This imposition also exists, by extension and in theory, within the context of housing development and guaranteeing the right to the city. However, in-depth analysis of some of the projects arising from this policy, paying particular attention to the Quinta dos Paricás Social Housing Development (hereafter referred to as QPSHD) built in the Northern state of Pará, Brazil, it is clear that the agents involved were unable to guarantee the inclusion of all spheres of civil society among those considered by these objectives and minimum guarantees of citizenship at any point in the project.

2 THE ORIGINS OF RESIDENCIAL QUINTA DOS PARICÁS

The Maracacuera neighbourhood, the location of the QPSHD, is situated within the limits of what was conventionally called the "Colônia Pinheiro", in the

DOI: 10.1201/9781003220855-18

Second League of Land of the Municipality of Belém. The "Colônia Pinheiro", on July 6, 1895, was converted to "Vila do Pinheiro", but its interdistrict boundaries were only demarcated on October 31, 1938. Demarcation of the area began with Val-de-Cans and Mosqueiro, covering the subdistrict of Outeiro. The name change to "Icoaraci", a word of native Tupi origin meaning "sun of the river", occurred by Decree No. 4,505 of December 30, 1943.

Currently part of the Administrative District of Icoaraci, the Maracacuera neighbourhood was created through the promulgation of Municipal Law 7.802 of 1996, which demarcated the areas comprising the 71 neighbourhoods of the municipality (Belém 1996). Historically, the area that makes up its boundaries was part of two large farms bordering the Pinheiro railway branch line: with the lands of the Pinheiro Farm towards the north/northwest, and the lands of Maracacuera Farm towards the south/southeast– with the latter providing the name of the main access road to the neighbourhood, as well as the neighbourhood itself.

One of the areas, the largest of the unincorporated areas in the region, belonged to EIDAI do Brasil Madeiras S/A[I] (Pará 2019a). This company had used the area to house its infrastructure for processing and trading hardwood extracted from the Amazon region and, at the same time, also used a considerable part of the area for replanting Amazonian hardwood species. This production model persisted, with reasonable success, until the mid-2000s, when the company fell into financial difficulties, leading to bankruptcy. Its assets were basically made up of large amounts of land and equipment for the wood extraction, processing and export industry, and included the land along Maracacuera Road.

Part of the land in question, a total area of 935,052.18m², was sold to the construction company Status Construções, after a purchase proposal of R-$4,404,000.00 was submitted to the bankruptcy administrator (Pará 2018). A few months after acquiring the land, the construction company announced the launch of a new real estate project called "Bougainville", which occupied the area purchased from the bankrupt estate of EIDAI and another non-adjacent plot of land facing the Augusto Montenegro Highway covering 44,125.00m². The project, after being suspended for a period, was relaunched in 2017/2018 with a new format, this time with an urbanized neighbourhood that included two horizontal subdivisions with free access and two horizontal gated communities, with the advance sale of 2,949 lots, made up of 2,849 residential lots and 100 commercial lots - later changed to 3250 residential lots.

Figure 2. Map of the Icoaraci Industrial District area and surroundings, including the ZEIA V area - Orla do Rio Maguari. Source: (Google Earth Pro 2019).

I EIDAI do Brasil Madeiras S/A was one of the Brazilian companies of the multinational wood processing and manufacturing conglomerate EIDAI Co., Ltd, headquartered in Japan, and which together with Maracacuera Florestal S/A extracted and cut green lumber for processing by the company's Japanese headquarters for the production of floors, custom-built cabinets and composite wood materials.

The remaining area, still owned by EIDAI, was the subject of a judicial dispute with the State Government of Pará, which had sought to expropriate the region in order to implement a social housing project justified by the "need to build decent housing for low-income families". After notification of the urgent need to expropriate the area – under the Minha Casa, Minha Vida Program (My Home, My Life Program) - the State Government of Pará requested the provisional legal imposition for possession[II] of the EIDAI area, which the court granted and conceded the filing of a new registration, thereby effecting the subdivision of the area and the transference of exclusive land use rights to the public administration. This move ensured the minimum conditions necessary for beginning the implementation process of the project (Pará 2019e).

3 THE RESIDENTIAL PROJECT(S) AT RESIDENCIAL QUINTA DOS PARICÁS

Initially designed to receive approximately 7000 housing units across its entire area, the QPSHD has, over time, undergone a number of adaptations. Given the complexity of the project and host of factors that needed to be considered for its execution, as well as the needs of the Municipality of Belém - governed by the same political party as the state government, with the electoral promise of enabling partnerships to be established more easily between the two levels of government - the State of Pará decided to divide the implementation of the residential plan into two different projects aimed at completely different population groups.

The first development, considered the first stage of the QPSHD, was built on land subdivided from the total area of the 1,235,319.00 m² expropriated by the State of Pará (Pará 2019b). The subdivided land, located on the left side of the Maracacuera Road, comprising an area of approximately 167,101m², was selected to receive housing units intended for the beneficiaries of the housing program Viver Belém - Minha Casa Minha Vida (Live Belém – My Home, My Life) created by Ordinary Municipal Law 9.014 (Belém 2013) - promoted by the Municipality of Belém to assist families that were either homeless, or received no housing benefit and whose total income did not exceed R$1,800.00[III]. The units were expected to be built in accordance with the provisions of Ordinance No. 168/2013 of the Ministry of Cities (Brazil 2013), taking into account the characteristics and the production model previously consolidated for social housing implemented by the PMCMV, along the lines of other projects in the State, either under construction or already delivered.

Responsibility for construction of the project was assigned to SPE Paricás Construção Civil, a special-purpose entity, formed on 10/22/2013 as a consortium between the companies Laje Construções, Atan Engenharia, Engefix Construções and Mape Engenharia. Initially, the project to be built comprised a residential unit, with characteristics as described in the registration of the subdivision, as follows: 1,360 parking spaces in total, with each housing lot comprising 6-12 blocks/buildings, each having four floors and four housing units per floor, with each unit corresponding to the ideal share of 0.015426% of the overall land area, measuring 43.80m² of private area, 6.73m² of common area and 50.53m² of total built area, containing living/dining room, kitchen/laundry area, two bedrooms and bathroom for shared use. The estimated cost for executing the residential works, as per contractual agreement with SPE Paricás Building Works, was R $ 166,494,857.98, as stated in the project registration document (Pará 2019c).

The second project, designed to be built on the remaining area of the land, covering approximately 1,068,218.00 m², occupied the right side of the Maracacuera Road. However, the target group for this stage of the project was completely different to that intended in the first phase. Initially considered as giving priority to families of state civil servants with a monthly income of up to R$ 5,500.00 – with the possibility at a later date, of being extended to include other population groups - the second stage was designed to be a larger, more complex project, comprising a complete planned neighbourhood along the lines of projects that were found in the real estate market in the metropolitan region of Belém.

The intention of the second stage of the development to conceive a residential project that would mirror the private enterprises being developed and commercialized by the real estate market becomes evident to the point that, when the state government announced the residential project, it submitted a design that included more than one housing type and urban equipment uncommon in government housing projects, such as footpaths for walking, commercial areas, an amphitheatre and even a marina. Innovations were also introduced which, at the time, were just beginning to be implemented in the housing projects of major national construction companies - such as MRV and Cyrela/Living - such as systems for collecting and reusing rainwater and for solar energy.

To date, only the work in the area corresponding to the first stage has been completed and is currently

II The institution of the provisional imposition of possession, provided for in Art. 15 of the Expropriation for Public Utility Law (Decree-Law 3.365 of 1941), grants the guaranteed possession of a given area of land in favour of the expropriating power, when an urgent need to obtain it has been alleged, provided that the expropriator pays the value of the asset to the court within the hypotheses provided for in Art. 15, § 1 of the respective decree-law.

III At the time of writing, the value of 1 Brazilian Real was equivalent to U$0.19.

in the process of progressive occupation by the recipients of the housing program Viver Belém - Minha Casa Minha Vida by the Municipality of Belém. The project was handed over to the community with two schools, two crèches, a health centre, eleven community centres, a multi-sports court, eight playgrounds, an outdoor gym and its own individual water supply and sewage treatment system, which was transferred to the State of Pará Sanitation Company (Pará 2019d).

The keys to the first apartments were handed over on March 21, 2019, the high point of which was the symbolic delivery of the project, which took place on June 13, 2019, attended by the President of the Republic, Jair Bolsonaro, the State Governor Helder Barbalho - both elected in 2018 - and the mayor, Zenaldo Coutinho. It should be mentioned that the mayor was the only head of government present on this occasion who had participated in all stages of the project, from its conception through to delivery to the beneficiaries.

The location chosen by the government for implementing the QPSHD called upon the need to discuss the urban-environmental regulation applicable to the area occupied by the project, particularly in view of its previous use of the area purchased by Status Empreendimentos and later expropriated by the State of Pará.

3.1 *The impact of the urban-environmental regulation at Residencial Quinta dos Paricás and insets*

The area that had been occupied by EIDAI Madeiras do Brasil S.A., and amongst various economic and other uses, was used for planting Paricá seedlings - in addition to hardwood species from the Amazon region – and for handling and processing this species. Because of this characteristic and since it preserved a considerable portion of native vegetation within it and borders a tributary of the Maguari River, the 2008 Master Plan for the Municipality of Belém began to include the non-anthropized area of the EIDAI land in a zoning system prescribed as a Special Area of Environmental Interest (known as ZEIAs) that encompassed a substantial area of the Maracacuera neighbourhood, extending from the boundary of Sector II of the Special Economic Promotion Zone 1 (ZEPE1) of the Icoaraci Industrial District to the boundaries of the Tenoné neighbourhood.

The ZEIAs, as part of zoning ordinances, can be defined as "free or occupied spaces, on public or private land, whose elements of the natural environment contribute to the public good, since they are important for maintaining the socio-environmental balance of the Municipality". Moreover, according to the provision of Art. 110, IX of the municipal Master Plan, and as outlined in the guidelines, they should also include, amongst other aspects, the need to preserve "riparian forest areas, the margins of watercourses and unoccupied areas of vegetation". Thus, analysis of the conditions implemented by the Ordinary Municipal Law 8.655 (Belém 2008), together with Art. 3, XX of Law 12.651 (Brazil 2012), which sets forth the conditions for use of urban green areas, makes clear that the occupation of these areas for the purposes of housing should only

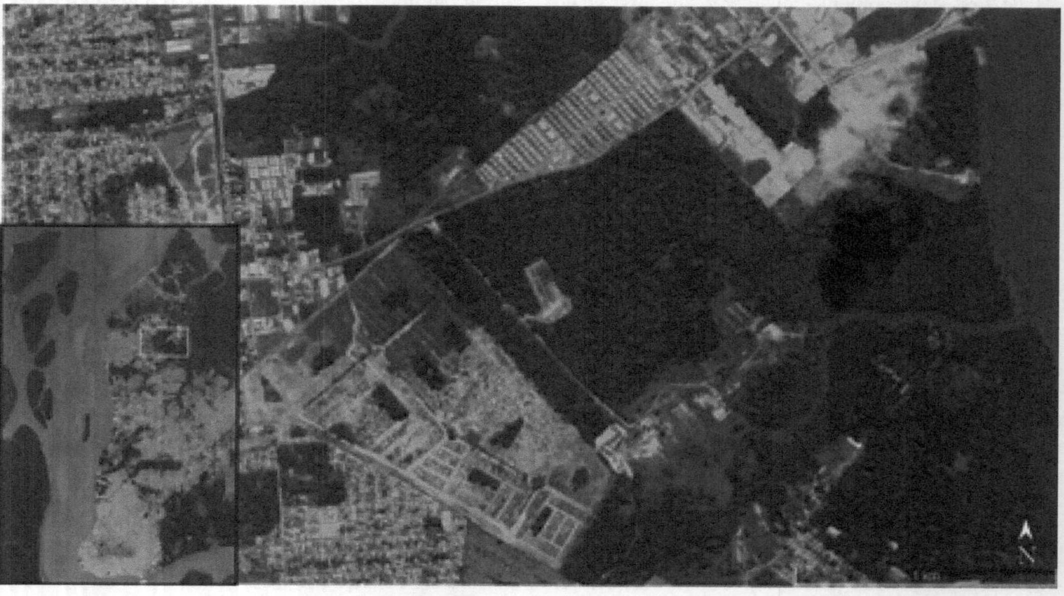

Figure 3. Satellite image of the land belonging to EIDAI do Brasil Madeiras S/A, subdivided, partially deforested and with the housing units of the first phase of Residencial Quinta dos Paricás built. Source: (Google Earth Pro 2019).

occur for promoting land regularization of the already consolidated settlements, in areas occupied for urban use, thereby highlighting the need to promote the preservation of the banks of streams, rivers and the natural environment as a whole (Brazil 2012).

These rulings could have posed serious obstacles to implementing the QPSHD in the area expropriated by the State of Pará, as well as to implementing the real estate development in the area purchased from Status Construções, since the two were within the polygonal area of the respective ZEIA, and therefore, would be subject to legal impediments for any housing/real estate development within the area of environmental preservation. The solution found to appease these needs, including those of the municipal administration, for the urbanization and housing/real estate development in the region circumscribed for ZEIA V - Orla do Maguari, was to bring about a change in the uses provided for in the respective zone. Such a change was made possible by the city council voting through Ordinary Law No. 8.883/2011. The respective instrument containing four articles, proposed specific changes in ZEIA V, called Orla do Maguari (or Maguari riverfront) which would, amongst other things, add to the uses of the area, thereby allowing:

> Commerce and services, and single-family or multi-family housing units within the "polygon" area between the limits of Igarapé Taboquinha, which flows into the Piraíba River (tributary of the Maguari River), the left bank of the Maguari River, Igarapé Maguari (a branch of the Maguari River) and the Augusto Montenegro Highway, on the municipal boundary between the Tenoné and Coqueiro neighbourhoods, observing the environmental preservation limit of 10% of the green area (Belém 2011).

Although the hypothesis of occupying permanent protected areas, it is not new in the Brazilian legal system as this hypothesis has already been envisaged, albeit precariously, by the CONAMA (National Council for the Environment) Resolution 396/2006. This possibility was guaranteed to the cities municipalities, and whenever necessary, replacing it, for the tates in article 54, paragraph 1 and paragraph 3 of the law that instituted the PMCMV, which was in force until the year 2017, when a specific law was created to deal with Urban Land Regularization (Reurb) with the enactment of Law 13.465/2017 (Azevedo & Oliveira 2014, Brazil 2017a,b). The states and municipalities took advantage of this possibility in order to regularize and urbanize the occupations that had existed until that point, which until then had been provided for in the law. However, many others used this understanding with the clear intention of forging development on new fronts through the creation of new PMCMV enterprises in sections of permanent protection area, by justifying that it would significantly "improve the environmental conditions with regard to the situation of irregular occupations" (Shimbo 2010, Maricato 2015).

The aster Plan of the Municipality of Belém seeks to balance these objectives by introducing special zones with specific purposes within the territory. The aim of the Special Zones of Social Interest (ZEIS) was to promote the regularization of land tenure and the occupation of new areas, guaranteeing that the necessary infrastructure would be implemented to promote the right to the city. The Special Zones of Environmental Interest (ZEIA) were aimed at the environmental maintenance of non-urbanized areas and at recovering degraded areas and those with no human occupation for housing purposes. The municipal legislator is obliged by federal law - particularly the City Statute - to choose the delimitation of these areas in order to identify the potential of each and ensure that they converge within a development plan of the municipality, as laid out in the 1988 Constitution and the City Statute. As a result of this objective, the Master Plans were created in order to have a ten-year "expiry date", allowing, at the end of this period, the planning and development of the city to be updated in order to reflect the real needs of the population.

It is apparent that the Residencial Quinta dos Paricás did not consider this chain of principles when designing and implementing the project. Choices of an environmental nature, for example, demonstrate that the intentions to promote a sustainable development model were opposed from the outset of the project (Lefebvre 2001, United Nations 2017). It was decided to expropriate an area previously intended for the management and preservation of native species in the Amazon region and, subsequently, to completely remove the vegetation of the region for a building development project comprising 170 buildings and 2,720 units. This did not even comply with the normative provision for an environmental reserve of 10% of the original green area, established by the previously mentioned Municipal Law 8.883/2011, which changed the uses of the ZEIA V - Orla do Maguari (Belém 2011).

Observing the project "in loco", there is no indication that it has managed to promote the preservation of the remaining pockets of vegetation in the interior of the municipality, nor preservation of the riparian forest areas, the banks of the watercourses or the remaining unoccupied patches of vegetation in the Maracacuera region. Amendments made to the municipal Master Plan three years after it was approved, reveal an evident imbalance between the intentions to promote social housing and environmental protection. Altering the proposals for ZEIA V - Orla do Maguari, promoted by the Municipality of Belém in 2011, completely breaks with the necessary stringency of the Master Plan in order for it to be applied. Thus, this demonstrates the preference shown by the public authorities to allow the agents of housing construction and real-estate developers to access cheap land over the commitment to preserve the natural landscape and environmental balance made by the same public authority when zoning it as a ZEIA.

Although a discourse of sustainability was widely used by the advertising campaigns for the project and by the official discourse of the political representatives when the project was delivered to the first residents, the initial objectives of demarcating ZEIA V in the Belém Master Plan were completely ignored with the implementation of the Residencial Quinta dos Paricás project, as presented in the image below.

By changing the current legislation outside the time-window stipulated for such changes, lawmakers have completely distorted the necessary perspective of legal security, the continuity of urban policy, the guarantee of sustainable intergenerational development and the right to the city for those who live in municipalities that promote this type of change. In the case in question, in order for the municipal legislator to grant permission to the public entity to remove vegetation - and subsequently, due to the size of the area circumscribed for the ZEIA V, to engage private entities wishing to be involved – so as to implement equipment for commerce and services and single-family or multi-family housing with the need to guarantee the preservation of just 10% of the original green area, it was necessary to draft an additional normative mechanism for the Master Plan. It transformed the special areas of environmental interest into veritable “land reserves” which may be exploited by whoever is interested and possesses sufficient funds and power to do so. This breaks with the constitutional objectives and imposes the agenda of a few agents on the urban environment that defines the direction of cities, thus distorting the participative, democratic and plural nature defined by the Master Plans.

Furthermore, it should be considered that the addition to the municipal Master Plan in 2011 was made both formally and legally - since it is the responsibility of the legislative branch to inspect and amend municipal laws that present problems of material applicability. Hence, in addition to allowing the removal of the native vegetation, the alteration to the Master Plan also allowed the region to be released for the construction of social housing and for the real-estate market directed towards housing, commerce and services. It is therefore questionable whether there was any proportionality in the decision to build more than 7000 housing units, of which 2720 units were in the first stage alone, within a new area hitherto completely devoid of accessibility, basic infrastructure and any guarantee of urban mobility (Carvalho 2015).

4 THE PROJECT AND REALITY OF QUINTA DOS PARICÁS HOUSING DEVELOPMENT

The QPSHD project, in addition to the number of housing units and the thousands of square meters that make up the project, its effects have expanded beyond the limits of land expropriated by the State of Pará and it has been an initial milestone in the urban occupation of the Maracacuera neighbourhood. The project was built in an area that lacked many of the requirements that are currently part of Annex III of Ordinance No. 269/2017 of the Ministry of Cities, and that required the construction of a series of new urban equipment, but did not allow this new area of city, designed to support almost 11,000 residents, to become effectively integrated

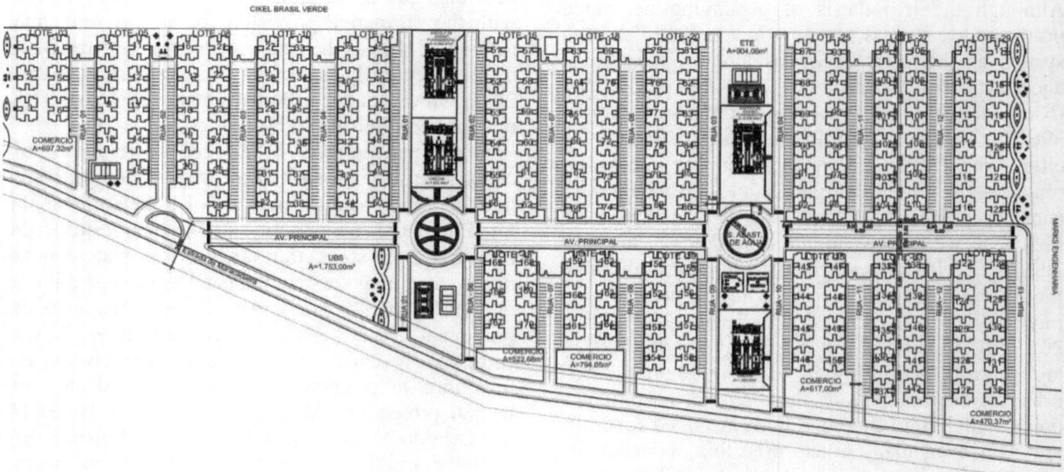

Figure 4. Layout of urban design of first stage of Residencial Quinta dos Paricás. Source: (SPE Paricás Construções 2018).

within the urban fabric of the city. The infrastructure was implemented only in the area of the housing project, and no changes were made to the access roads to the project, nor was there any effective improvement to the access options in the region, especially by public transport (Ferreira et al 2019).

Up until July 2019, there had been no provision of public transport to serve more than 1800 residents who had already received the keys to their units at the QPSHD. This situation improved somewhat over the second half of the year, but fell far short of what was necessary in both quantity and quality, according to the allegations of residents in their continual interviews with the local newspapers. Currently, motorcycle taxis still predominate as the mode of access to the development.

Furthermore, no move was made to resurface the main access road of the project, the Maracacuera Road, in order to provide adequate access for vehicles that could occupy more than 1,000 rotating parking spaces. The project was delivered with no bike lanes or preferential lanes for cyclists, despite the fact that for many of the residents, this form of transport was the most accessible. The project did not include areas that allowed activities of a commercial or service nature, which is gradually causing residents to occupy the public and community areas of the project, and use them to establish small businesses, or when this is not the case, they have changed the use of one of the rooms in the unit they received to accommodate this new use (Maricato 2001, Ventura Neto 2012).

The guaranteed access to water and sewage treatment was conducted by implementing two stations, one aimed at capturing and treating water obtained directly from the water table below the residential unit in a well 223 meters deep, and the other dedicated to treating the sewage collected from the residential area. Both stations were located at the central point of the project delivered in the first stage, as described in the authorization to collect water provided by the State Secretariat for the Environment and Sustainability (SEMAS). There is no indication that, in the event of a failure in the system implemented, there is any other means of guaranteeing the necessary water security in a project of this size.

5 FINAL CONSIDERATIONS

The revision of the region's zoning rules created a unique situation because, although the area is zoned as a ZEIA, the municipality permitted the implementation of equipment for commerce and services and housing. Moreover, even though the region has the largest social housing project in the municipality, it nonetheless is not zoned as a ZEIS and allows projects to be implemented that are aimed at higher income brackets, with larger layout, larger lots and a much lower housing density per m^2 than in the QPSHD, thereby fostering inequality in access to urbanized land in the region. The effects of flexible urban regulation in favour of the QPSHD - according to a public statement by the municipal mayor, in a televised speech while delivering housing units in the first phase of the residential project has already allowed private sector developments that lay within the then ZEIA V - Orla do Maguari area, such as the horizontal condominium Bougainville Belém by Construtora Status, to be approved, registered and sold freely by the companies without any consideration for improving the surroundings of these projects (Shimbo 2010). The changes that Quinta dos Paricás has enabled in the region have been so wide sweeping

Figure 5. Satellite images, taken in 2013 and 2019, of the area where the housing units of the first phase of Residencial Quinta dos Paricás were built. Source: (Google Earth, 2019).

that they were able to foster the creation of a new neighbourhood, Antônio Lemos (Belém, 2019). This encompasses the areas of the then ZEIA V - Orla do Maguari, symbolically recognizing the change of use assigned to the region of the former Maracacuera and Águas Negras, and completely disregards the uses originally defined and approved by civil society in the municipal Master Plan in 2008.

Thus, in summary, we may conclude that the actions taken by the group of public agents involved in the QPSHD, amongst them, the Municipality of Belém and the State Government of Pará, denote a concomitant choice that favours the valorisation of land and the promotion of an environmental imbalance and excludes the beneficiaries of such projects from the city, to which they have right. The design and implementation of the QPSHD expresses a very clear relationship between public authorities and private agents in defining the norms of housing policy, making social housing projects work as tools capable of promoting changes in the existing urban-environmental, which enables the speculation cycle of urban land to continue, to the detriment of the right to housing and the right to the city in its broadest sense. The implementation of the QPSHD created a completely irregular legislative loophole and deprived the regulatory instruments provided for in the Municipal Master Plan of legal certainty.

It can be observed that, in reality, the implementation of these large social housing projects, such as the QPSHD, has enabled the perpetuity of extreme poverty and the exclusion of the underprivileged from the city's production. Although beneficiaries may have their right to housing satisfied when receiving an MCMV unit, the manner in which the residential project was conceived has been responsible for maintaining these people on the margins of the political arena that defines the direction of the city (Lefebvre 1991). The public authorities deny the population the right to a balanced environment by surrendering conservation areas provided for in the Master Plan without proposing a viable alternative for replacing them. Thus, it is exploited to subvert the policy of promoting social housing in favour of the real estate market, to make the production of this market viable, violating the normative regulations and the guiding principles of urban policy.

REFERENCES

Azevedo, R. & Oliveira, V. 2014. Reflexos do novo Código Florestal nas Áreas de Preservação Permanente – APPs – urbanas. *Revista Desenvolvimento e Meio Ambiente*. 29: 71–91.

Belém (Municipality). 1996. Ordinary Law N.º 7.802.

Belém (Municipality). 2008. Ordinary Law N.º 8.655.

Belém (Municipality). 2011. Ordinary Law N.º 8.883.

Belém (Municipality). 2013. Ordinary Law N.º 9.014.

Belém (Municipality). 2019. Ordinary Law N.º 9.453.

Brazil. 1941. Decree-Law No. 3.365.

Brazil. 1988. Federative Republic of Brazil's Constitution.

Brazil. 2006. Resolution CONAMA No. 369.

Brazil. 2009. Law No. 11.977.

Brazil. 2012. Law No. 12.651.

Brazil. 2013. Ordinance MCIDADES No. 168.

Brazil. 2017a. Law No. 13.465.

Brazil. 2017b. Ordinance MCIDADES No. 269.

Carvalho, C. et al. 2015. *Minha Casa... e a Cidade? Avaliação do Programa Minha Casa Minha Vida em seis estados brasileiros*. Rio de Janeiro: Letra Capital.

Ferreira, G. et al. 2019. Política habitacional no Brasil: uma análise das coalizões de defesa do Sistema Nacional de Habitação de Interesse Social versus o Programa Minha Casa, Minha Vida. *Revista Brasileira de Gestão Urbana*. 11: 54–70.

Lefebvre, H. 1991. O Direito a Cidade. *São Paulo*: Ed. Moraes.

Maricato, E. 2001. *Brasil, cidades: alternativas para a crise urbana*. Petrópolis: Vozes.;

Maricato, E. 2015. *Para entender a crise urbana*. São Paulo: Expressão Popular.

Pará (State). 2012. Decree No. 417.

Pará (State). 2018. Pará State Court. Case No 00030518420088140201.

Pará (State). 2019a. 1st Property Registry Office. Property Registry No 53400.

Pará (State). 2019b. 2nd Property Registry Office. Property Registry No 20579KY.

Pará (State). 2019c. 2nd Property Registry Office. Property Registry No 20580KY.

Pará (State). 2019d. 2nd Property Registry Office. Property Registry No 35494.

Pará (State). 2019e. Pará State Attorney General. Intern Suit No 201100010011.

Shimbo, L. 2010. *Habitação social, habitação de mercado: a confluência entre Estado, empresas construtoras e capital financeiro*. São Carlos: Universidade de São Paulo.

SPE Paricás Construções. Descriptive Memorial and Technical Specifications. 2018

United Nations. 1948. Universal Declaration of Human Rights.

United Nations. 1976. Vancouver Declaration on Human Settlements.

United Nations. 1992. Istanbul Declaration of Human Settlements.

United Nations. 2017. New Urban Agenda.

Ventura Neto, R. 2012. *Circuito Imobiliário e a Cidade Coalizões Urbanas e Dinâmicas de Acumulação do Capital no Espaço Intraurbano de Belém*. Belém: Universidade Federal do Pará.

Research Tracks in Urbanism: Dynamics, Planning and Design in Contemporary Urban Territories – Alegri et al (eds)
 ISBN 978-1-032-10788-2

Project strategies for territory development in degraded coastlines. The case of Caleta Tumbes, Talcahuano, Chile

A.C.A. Hernández
College of Architecture and Urbanism, Universidade Presbiteriana Mackenzie, Brazil

E. Salinas, M. Nazar & E. Matuschka
School of Architecture, Universidad del Desarrollo, Chile

ABSTRACT: The case of Tumbes is analyzed in the context of one of the over 70 coves in the Region del Bío-Bío considered "fragile territories"; this forms a territory system with unique characteristics, besides a high geographical, cultural and microeconomic potential. Through the use of project strategy methodology, approaches to current problems are determined, but above all, these are understood as opportunities for future projection. The hypothesis suggests that urban infrastructures, increasingly complex, encourage territory restructuring based on urban scenarios that meet the needs of users in complex, dynamic and degraded territories. In this context, the work done can be designed, applying the methodology in different "fragile territories" that make up the identity of the regional border, building long-term strategies, operating from the perspective of a territory system.

1 INTRODUCTION

This study presents the Methodology of project strategies applied in 2019 to the coastal town of Tumbes in the Región del Bío-Bío, Chile, through the holding of the "International Workshop 2019", in which a group of students and professors, along with government professionals, connect with real contexts concerning territory, culture and society. This workshop was organized by the Universidad del Desarrollo, through its School of Architecture Sur-Sur in Concepción, Chile, along with the Laboratory of Project Strategies (LABSTRATEGY) of the Universidade Presbiteriana Mackenzie in Sao Paulo, Brazil. This project was supported by the MINVU/SERVIU, and by both Architecture and Urbanism Schools.

In the current global and regional urban hierarchies, there is a wide diversity of territories that have become increasingly excluded from larger global processes that feed the world economy (Borja and Castell, 1999). In Latin America, these territories are cited in the document "2018 Revision of Urbanization Prospects" (UN, 2018), where rural areas and non-urban towns tend to lose population given their geographical, social, political, demographic, and income characteristics; this situation contrasts with that of urban areas, where there is a clear growth in population in projections for 2050.

In this context, Caleta Tumbes is analyzed among the over 70 coves in the area and the 240 coves in the country, as a territory of great geographic and micro-economic potential. The interest for this location is explained by its value within the territory made up of several coves with unique characteristics, which are on the verge of socio-demographic extinction due to recent major economic, territorial, and social changes.

In the face of these changes, this article aims to propose strategies that promote the development of these degraded coastal areas, identifying possible situations that can create and incorporate strategies as an urban area in the context of current urban design guidelines. Therefore, this article seeks to provide strategic comprehension in terms of how to conduct this kind of intervention in contrast with other methods that tend to identify architectural projects. Thus, this process aims: (1) to delineate opportunities for these vast territories; and (2) to show possible urban interventions, by determining strategic elements through the implementation of tactical urbanism.

Therefore, it is envisaged that the projection of this study, by applying the aforementioned methodology to different "fragile territories" that are part of the regional coastal identity, can result in a greater understanding of the potential of these areas as a long-term strategy, working from a local towards a territorial system.

2 CASE STUDY: TUMBES

Caleta Tumbes is located in the central-south area of Chile, in the commune of Talcahuano, Región del

DOI: 10.1201/9781003220855-19

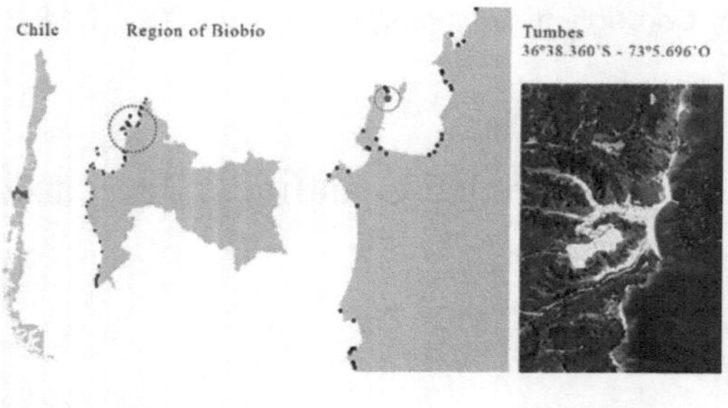

Figure 1. Maps showing location of Tumbes within the country and the cove system of the Biobío Region. Source: prepared by the authors.

Bío-Bío. It lies on the western border of the peninsula of Tumbes, facing Isla Quiriquina in the bay of Concepción, 12 kilometers from Talcahuano and 20 kilometers from Concepción, the capital city of the Region.

The town has a population of 1880 people and consists of over 600 houses (INE, 2019). Its main economic activity is related to exploiting resources from the sea seaweed harvesting, small-scale fishing, boat repair, and only recently, the increasing tourism and commercial activities, mainly associated with gastronomy involving sea produce.

This cove is located between two gorges called Maryland and San Juan, the latter of which provides access to the cove; this is the only paved two-way access. Its geographical characteristics, along with the proximity of the houses to the coastline (a common characteristic of most coves), has made the cove vulnerable to tsunamis. The last tsunami of February 27th, 2010 hit most of the coastline of the region. It was because of this natural disaster that the cove suffered a dramatic restructuring, as the tsunami destroyed all the buildings - homes and stores - near the coastline. Given its vulnerability, this also resulted in the abandoning of three minor settlements in the north located between the hill and the coast, part of Tumbes. Also, the relocation of a group of families from San Juan promoted a new urban structure, different from the one that had been built organically by the population, and which used to live just a few steps away from the sea.

3 METHODOLOGY OF PROJECT STRATEGIES

The methodological process drew on Hernández' doctoral thesis "Estratégias Projetuais no Território do Porto de Santos"[1] (Hernández, 2012), whose research prompted the formation of the research group Estratégias Projetuais em Territórios Urbanos/Degradados e Portuários (LABSTRATEGY), and has since 2013 led to 27 workshops in different geographical areas in Latin America and Africa, using a specific methodology for degraded urban areas.

The methodological process holds that the vision of systemic [urban] planning is based on the principles of the production of territory control systems from the 1960s, in contrast with the processes of strategic planning that define the urban planning approach of the 1980s.

As suggested by Borja (1988), the exhaustion of the classic territory planning and ambiguity of a major project, which was supposedly isolated, forced the idea of planning according to the nature of the interventions that correspond to new spaces.

The present study is based conceptually on the views presented by Brian McLaughlin in his book "Urban and Regional Planning: A systems Approach" (1969) and by José Fernández Güell in "Planificación Estratégica de Ciudades" (2006). Thus, the use and implementation of actions as strategic inducers allow the formulation of contemporary systems that operate in contexts whose components are interconnected with a set of common purposes determined by the individuals, companies and institutions that carry out their activities in these new urban systems; these require different resources, infrastructure and services that promote innovative spaces for the territories.

It is by means of this transformative process that both the language and conceptual structure should be determined and presented through strategies for future prospects; these will determine the possibilities of applying more tactical urbanism.

According to Fernández Güell (2006) a strategy is defined as a group of principles, activities and means that form a plan of action - following a particular itinerary-, to achieve a desired city model for a particular situation. Strategies are connected through the planning of diverse systems that define the fusion of the urban territory and the coastline: the interactive flows of pedestrians, light vehicles moving correspondingly in lanes, automobiles, bicycles, shipping lines, and small vessels, creating an open scenario that operates as a functional structure.

Planning is a systematic process that aims to offer a better future for a city, according to the identification of threats and opportunities, the optimization of its strengths and the minimalizing of its weaknesses (Andersen, 1984). Fernández (2006) states that strategic planning of cities is a systematic, creative and democratic process that builds up the basis for a long-term integrated action that defines a model for future development; this model formulates strategies and directives that aim to reach the appropriate models, and creates a permanent system of decision-making and participation of the local agents throughout the whole process.

1 "Project Strategies in the Territory of Porto de Santos".

Planning is aimed towards completing actions that make possible different proposals, determine the resources required, and carry out transformations. There is a marked difference between urban planning and traditional planning, where the first aims at the process and not the final product, while the second seeks separation between design and execution of proposals (Fernández, 2006). Strategic planning is the result of integration of the different local territory administration perspectives; according to Hudak (2006), the difference between these two approaches is that whereas traditional planning is characterized by the product as a target, structured in relation to a territory or area (this is directly linked to the norms oriented by the urban offer, subject to administrative limits; this development occurs in a late and diffuse manner), strategic planning is characterized by the planning of the whole process, in which actions are integrated and coordinated, working indicatively on the issues in response to urban demands. The latter covers territory beyond the administrative bounds, acting participatively, in an early and focused manner.

Conceptually, the use of design strategies aims to obtain benefits from the territory transformation; the relationship between action and local reality is defined as: (1) a definition in the long-term global vision among different sectors, (2) identification of trends and of opportunities, (3) promotion and coordination among public institutions, (4) community activities aimed at engaging further actions, and (5) the strengthening of the urban and social network of the territory.

a. *Project strategy: Conceptualization*

According to Gausa (2009), a strategy that allows the transformation and connection of the maritime border can be projected through a matrix, converted in a new virtual "urban network", which is also clear, complex, and open to simultaneous processes of redefinitions and programmatic reevaluations; thus, he defines:

(1) Coastline, an area that includes equipped facilities in the form of living spaces.
(2) Productive coastline, which includes a high density of identity elements, productive areas, and cultural and educational facilities.
(3) Urban coastline connectivity, which includes living, commercial, and multiple use areas. The integration of these spaces leads to the remodeling of the internal space in coexistence with the local network and the renewal of the different territorial fronts, consolidating the urban network and considering the structural elements as opportunity reactivators that result in enjoyable public places.

Based on the development of these aspects, it is necessary to point out the need to configure a process that defines: (1) the development of future scenarios, (2) the design and development of a desired strategic vision for the coastline, and (3) the identification of the critical issues that the territory must address to achieve a strategic vision.

In relation to the development of the work process, the actions applied to the coastline are:

1. The urban setting should be understood strategically as a long-term vision that exceeds an average 10-year time period,
2. The extension of the thematic scope of the setting, focused on an interdisciplinary basis; and,
3. The appropriateness of the local program in which the context must be developed by the joint action of both government and social agents.

The need to propose improvements within the urban limits is evident; this is based on a territory outline that considers the compositions of buildings, infrastructure and landscape for a coordinated development (Gausa, 2009). It is thus necessary to apply actions that promote urban restructuring - operations of territorial reassessment - in order to activate different contexts for the emergence of spaces with new identities. Activities integrate the collective spaces and the landscape, building a port city through collective activities and mixed operations that take place in the intersection of a range of phenomena.

The strategy is defined by the possibilities of a territory; these elements transform the landscape in the long-term and can be split into two conceptual groups:

1. Impulse: to reinvent the degraded space of the port, as it integrates elements that are characterized when applied as:
    1. Strategic booster: to promote strategic vectors of development and growth opportunities, providing new urban formulations, growth and flow inducers, density networks, articulated landscapes, and connectivity infrastructure;
    2. Coordinated infrastructure: new devices that act as inducers, providing efficiency and interaction among spaces, flows, and activities that operate between the old and the new infrastructure;
    3. Space relations: to articulate, at different scales, the possibility to integrate open spaces, operative, entertainment, recreational, sport, and experimental landscapes, providing new, continuous green areas through "landscaping";
    4. Structural impulse: recycling and restructuring of the connectivity elements, promoting the possibility of transforming into intermodal associating devices to the topography, the diverse platforms, and the coastline;
    5. Induced strategies: induction and promotion of operations that define strategic nodes,

   acting as qualitative reactivator elements, aimed at developing global strategies; and
   6. Three-dimensional mechanisms: in which the mechanisms must be applied in specific areas, by section, articulating smart mechanisms that are able to provide scenarios, landscapes, infrastructure, new areas, and topographies, to allow new territorial zoning.

2. Enabling: focused on creating the possibility of increasing actions by using elements that confer a new value to the degraded areas:
   1. New housing: in which the scanning of a vertical density is linked to the creation of strategic definitions that necessarily aim at the formulation of mixed activities, focused on the triad life-production-leisure;
   2. Economic assets: encourage the development of attractive centers for urban activities that act on the revaluation of existing places, such as heritage, landscape, education, gastronomy and social areas, through economic growth;
   3. Public programs: development of programs capable of developing public actions aimed at urban change, through the creation of innovation networks, and at the same time promoting the interconnection between new and existing spaces, through local and global innovations; and,
   4. Urban self-esteem: promotes the dynamism of the port space, allowing movements and paths from the creation of dynamic and imaginative environments, empowering new identities based on strategies that benefit collective actions.

Strategic definitions help develop guidelines to describe sets of elements that comprise the fabric of the city and the coastline as potentially unifying zones of transformation.

3.1 *Applicability*
Design strategies are based on the hypothesis raised conceptually and identified as a premise for the development of future scenarios that are considered on the basis of current circumstances. This article elucidates the possibility of redeveloping the cove space, according to the different relations promoted by the areas making up the present-day structure, which is characterized by a conformation of different infrastructures. This proposal considers different aspects, outlined previously, that define strategic compositions and solutions for the degraded urban fabric, established through conceptual tactical definitions, and which are mentioned throughout the present study. Strategic transformations are thus reinforced by the applicability of inducing and defining concepts, which are characterized by spaces through the following actions:

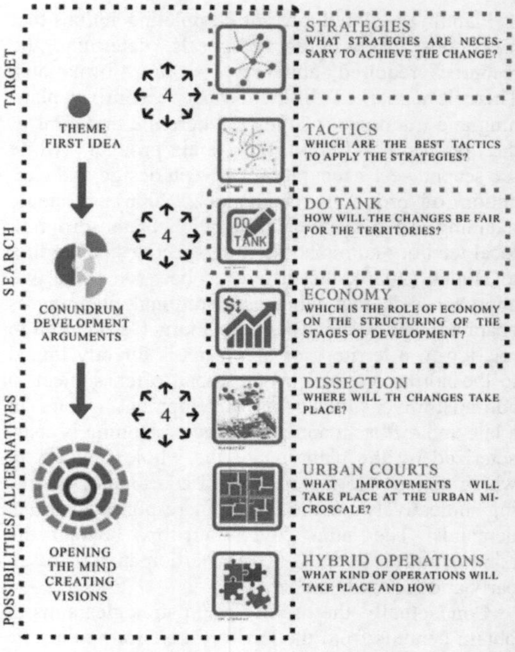

Figure 2. Diagram of project strategies. Source: LABSTRATEGY.

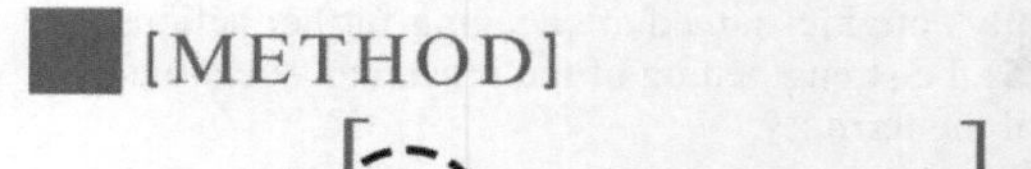

Figure 3. Diagram of project strategies. Source: LABSTRATEGY.

(1) Re-urbanization: the creation of a new urban border, which allows structures to remain open in themselves, promoting connections with networks that penetrate the urban fabric of a port center.
(2) Reconnection: which promotes the functionality of the limits of a port and the waters, through natural spaces that integrate both the urban fabric and the waters;
(3) Flows: the implementation of strategies that tactically promote the orientation of roads and other sectors, for the construction of new landscapes between the city and the port.
(4) Discontinuities: which promotes continuous spatiality in the horizontal plans, which occurs due to the absence of connectivity at different levels;
(5) Axes in Nodes: the development of urban networks articulated for their use in the creation of continuous urbanity, which promotes the connectivity between the city and the port;
(6) Urban recycling: by promoting strategies for the re-orienting of port development, thus improving the concentration of activities that promote new activities in urban spaces;
(7) Continuity: recognition of potentialities that originate from the spatial differences, thus promoting continuous connectivity between the territory and the induced insertion of green areas.

3.2 *Traditional model vs. workshop model*

In order to conduct this study, the "intensive workshop" model was selected; this essentially aims at, in a short period of time and with intense dedication, immersing in a real problem, whose development might lead to applicable results.

The previous works of the Sprechmann workshop have been taken as a reference; carried out between the 2002 and 2003 in the Vertical Workshop series in the University Ritter dos Reis, and which brought together Brazilian, Uruguayan, and Argentinean students; this workshop proposed a contemporaneous process in the teaching of architecture (Sprechmann, 2003).

The use of this model allows the establishing of creative practices, new models of experience, and the emergence of new systemic competencies for the visualization of components in a design process, moving away from the traditional model of a curricular workshop (Table 1). Drawing on this model, the aim was to answer the following research question: "How can this model be used to project in an emerging crisis situation, as opposed to the traditional project model?

Thus, the Project Strategies Model was applied to Caleta Tumbes by means of a workshop called "Triggering experience", in a four-day program of 8 - 10 hours per day, in which a group of 28 students, 7 teachers and 3 teaching assistants participated. The work was divided into 4 groups, each tackling the exercise according to 4 different theme areas:

- Economic - Social
- Economic - Population
- Sustainability - Endemic
- Urban - Disasters

The running of this exercise is explained in detail below:

Table 1. Comparative process between the traditional teaching model and the use of workshops as an immersion instrument. Source: prepared by the authors.

Comparison between the teaching method: traditional vs. workshop-based

Proposal keys	Traditional teaching	Territoriality to which the actions of the process respond	Workshop-based teaching
Process and development	No specific methodology	Geographical aspects/ degraded urban zones	Use of management methodology for processes (projectual strategy methodology)
Teaching space	Workshop	Weather and climate	Fieldwork as workshop
Projectual Space	Fictitious	Real social determinant in relation to their territory	Specific determinant
Interlocutors	Professor/lecturer	Various agents involved	Multidisciplinary group
Temporary period of development	Semester or bi-annual	Immediate response	Brief and intensive from 4 to 7 days
Studentships	Formation in terms of pre-requisites-separations by levels	Diverse areas of know-ledge and educations	Multiple levels
Teaching-learning relations	Academic relations profes-sor – student	Territorial practice by means of theatrical confrontations	Non-hierarchical -horizontal relations – various interlocutors (guidelines, stakeholders, guests, others)
Approaches	Architectural and Urbanism	Multiple knowledge areas	Multiple-scale territorial scenarios understood as landscape
Results	Projectual hypothesis	Protocols	Applied scenations

Day 1 - Probing the territory: field trip and survey of indicators

During this field trip, the following activities were conducted: identification of stakeholders of the territory, dialogue with the local community, identification of coastal and urban conditions, formal and informal commercial and housing relations, research on the action of the fishing community and of the network of women in the harvesting of algae, experience of consumption by the territory, and data collection through sketches, interviews, and photographs.

Day 2 - Doubt and doing as instruments of thought

Workshop activities were held within the university facilities; these activities included: the definition of concepts based on the Project Strategies model, mapping of aspects of the locale, thematic seminar and discussion of goals, structuring of working groups and initial demands of the SERVIU. Preliminary review of ideas and collection of macro visions to understand the territory. Structuring of indicators for the territory by theme, definition of zones of territorial appreciation, organization of tactical actions, workshop activities, and group debates with teachers and actors of SERVIU.

Day 3 - We do, think and execute

This day included an initial approach to the guiding questions, a review of the applied tactical actions of the territory, a debate on the meanings of applicability of the strategies, a thematic dialogue with the teachers, collection of visions of urbanity and reflections of experiences. This was an intensive day of reflections among the groups of students, conceptual revisions and decision-making, based on their knowledge; it concluded with a presentation of the activity results given to the teachers. The presentation encompassed a review of the final hypothesis, the results of the workshop with teachers and SERVIU representatives, application of strategies, and of technical concepts and formulation of scenarios for the community of Tumbes.

Day 4 - Tuesday great party

Printing of the results on presentation sheets and final display to government representatives, teachers and students.

4 RESULTS

The results of the workshop are described below, summarizing the development of each group according to theme, from devising the hypotheses to presenting a strategic proposal.

b. *Economic - Social*

This group built up the analysis from the indicators obtained mainly from a census of the present uses of the coastline, the uses of planned territory (Communal Regulatory Plan), commercial facilities, occupation, age range, and educational background of the population, along with geographic indicators and escape routes from the cove.

The main concern of this group was to promote the restructuring of the urban fabric using characteristic elements to foster the revaluation of elements in degraded areas. For this, they proposed the hypothesis that "the articulation of pre-existing nodes along with the recycling of disused land, generates a system that promotes the circular economy and social identity of fishing in Tumbes".

The main strategies and proposals are related to suggest a system of common spaces, recycling some deteriorated spaces, building a pedestrian promenade that connects the entire circuit, and adding transverse routes that lead to viewpoints in the hills.

c. *Economic - Population*

The main indicators of this group were related to the population (age range, sex, income) and the planned land uses, geographical conditions of the cove, location characteristics, and types of housing occupation.

The proposal of this group was to bolster the infrastructure, spatial relations and recycling, along with the restructuring of connectivity. Also, they sought to promote the economic centers, programs for public areas and urban self-esteem, seen as strategies for collective benefit.

Their hypothesis was based on the fact that "reducing the residency/work gap - achieved by relocating the houses at the top of the hill - will reconnect the fishermen with their trade, boosting local production."

The main strategies and proposals were: the intervention for the coastline, creating a defensive wall with steps (contemplation and meeting areas), a coastal pedestrian route, a funicular railway as a hill-coast connector, and work spaces for drying algae on the slopes of the hills, as a functional connector between housing and fishing.

d. *Sustainability - Endemic*

The indicators of this group were related to the roads around the cove, urban voids, and landscape elements and components.

It was agreed to promote the strategic vectors of development and opportunities, the spatial relations and the structuring and recycling of connectivities. It was also decided to promote centers of activities and flows as a collective benefit.

The hypothesis stated that "through urban reorganization and sustainable mobility, the value of the landscape and identity are incorporated, recovering the local maritime heritage."

As the proposed solution, they presented a series of measures for the management and control of waste and residues, as well as other elements related to health and wellbeing: green areas and infrastructure, a funicular railroad connecting public spaces and other facilities in hills with the coast, a coastal walk, cultural elements and bike paths.

e. *Urban - Disasters*

The main indicators for this group were roads, land use, building heights, evacuation routes, streams, and risk areas. Based on these indicators, they promoted coordinated infrastructure, spatial relations, and strategic and structural drives, while enhancing public programs and urban self-esteem.

The hypothesis stated that "the pressure to occupy the coastline, together with the geographical characteristics of the site, generate conflict and disorganization among the various agents of the place, resulting in increased vulnerability of the coast to the risks of natural disasters and exploitation of natural resources."

The result of the activity was the creation of a set of strategies that combine proposals for disaster prevention, tourism, resilience, and urbanism, such as the creation of escape routes and safe spaces - which also serve as squares and viewpoints -, defensive breakwaters that double as promenades, a funicular railroad, and connecting routes for pedestrians and vehicles.

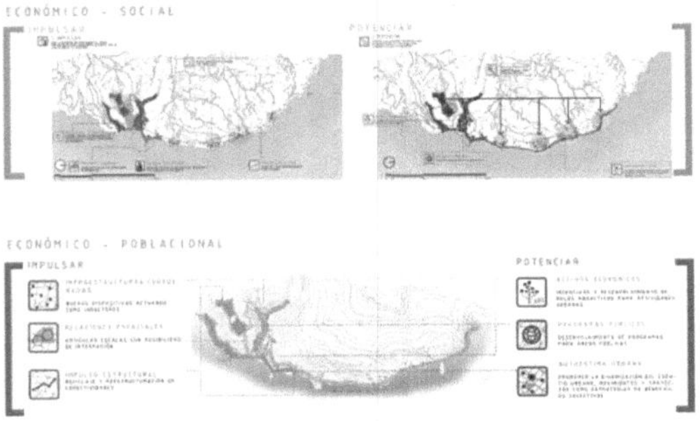

Figure 4. Indicators for each working group. Source: prepared by the authors.

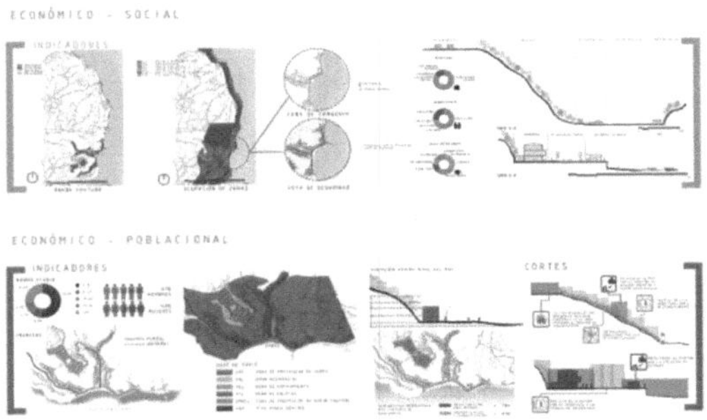

Figure 5. Elements to Promote/Potential of each work group. Source: prepared by the authors.

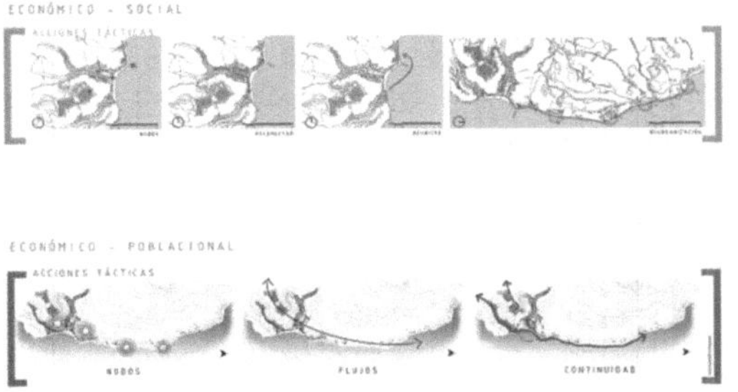

Figure 6. Hypothesis by work group. Source: prepared by the authors.

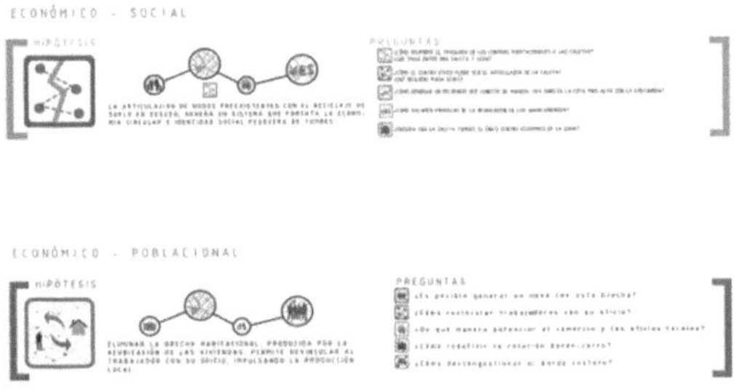

Figure 7. Tactical actions for each work group. Source: prepared by the authors.

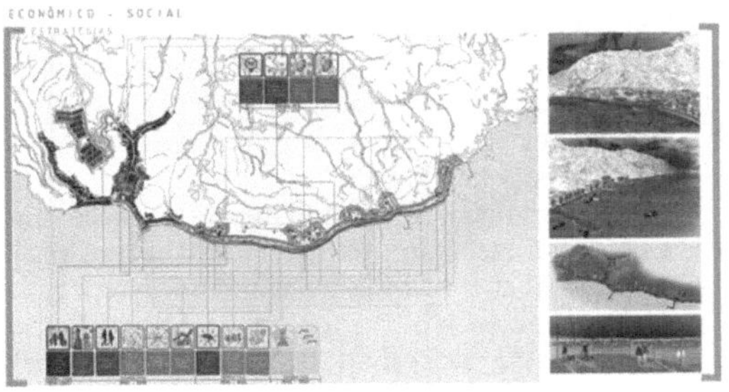

Figure 8. Strategies and scenarios of Economic/Social (top) and Economic/Population (bottom) groups. Source: prepared by the authors.

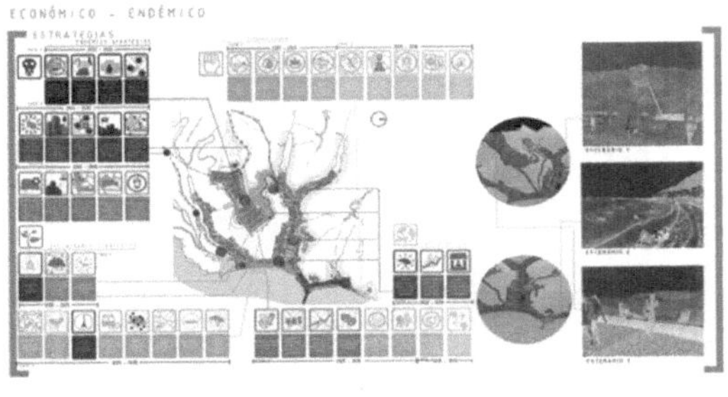

Figure 9. Group strategies and scenarios Sustainability/ Endemic (top) and Urban/Disasters (bottom). Source: prepared by the authors.

5 CONCLUSIONS

The process of knowledge production throughout the field workshop promoted a work structure based on four working groups, divided by specific topics with the common condition of resilience as a guiding element in the process of transformation of Caleta Tumbes.

The use of the methodology proposed in this territory allowed a specific approach to the existing economic, social, and cultural aspects by incorporating them into the proposal: making contact with the Algueras Cooperative (women dedicated to the collection and drying of seaweed) and with the network of local fishermen, along with creating the possibility of understanding the tourism aspects and the promotion of a significant gastronomic economy for the region, quickly built a base scenario on which to construct the proposal.

Getting to know the inhabitants, their spaces, activities and needs make it possible to promote and articulate the territory - through the work of the teams -, which will be organized in relation to specific views in the search for the formulation of scenarios based on the application of economic strategies, governance of sustainability and urbanity.

The use of indicators and data led to a detailed description and understanding of the territory, allowing us to promote or enhance local specific qualities and characteristics, which helped to formulate the hypothesis that aims to improve the area in the medium and long-terms. It is worth noting at this point the need to gather more data for other smaller sites, since much of the information is only available for medium or large cities, overlooking territories with smaller populations.

The development of the hypotheses allowed the researchers to structure the elements from local resilience, so that it has the contribution by specific actions based on each of the topics. At this point, the similarity of the workshop format with the real working conditions stands out, due to the collaboration of members of state organizations (SERVIU/MINVU), which allowed continuous reflection during the whole workshop, contributing with visions, regulations and knowledge from the professional work in the development of strategies.

This exercise is regarded as a starting point. It is set to continue with further workshops in other coastal locations, to build a body of knowledge around this territorial system, providing the communities with strategies and also proposing new forms of intervention and scenarios for decision-making by local governments.

REFERENCES

Andersen, A. (1984). Guide to Public Sector Strategic Planning. Chicago: Arthur Andersen & Co.

Borja, J.; Castells, M. (1999) La gestión de las ciudades en la era de la información: políticas urbanas de la globalización. Madrid: Taurus.

Fernández, J. (2006) Planificación Estratégica de Ciudades: nuevos instrumentos y procesos. Editora Reverté. Barcelona.

Gausa, M. (2009). Multi-Barcelona Hiper-Catalunya. Estrategias para una nueva geo-urbanidad. Actar. Barcelona.

Hernández, C. (2012). Estratégias Projetuais no Território do Porto de Santos. 279 f. Tese (Doutorado) – Faculdade de Arquitetura e Urbanismo – Universidade Presbiteriana Mackenzie, São Paulo.

Hudak, J.; Sorkin, L.; Ferris, B.(2006). Strategies for cities and counties: a strategic planning guide, 1984. In: Fernández, J. Planificación Estratégica de Ciudades – nuevos instrumentos y procesos. Editora Reverté. Barcelona, p. 58.

INE (Instituto Nacional de Estadísticas de Chile). (2019). Ciudades, pueblos, aldeas y caseríos 2019.

Sprechmann, T. (2003). O Ensino da Arquitetura [en la enseñanza de la arquitectura] - Taller Sprechmann. Ritter dos Reis. Porto Alegre.

UN (United Nations) Department of Economic and Social Affairs. (2018). 2018 Revision of World Urbanization Prospects. Disponible online en https://www.un.org/development/desa/publications/2018-revision-of-world-urbanization-prospects.html.

Research Tracks in Urbanism: Dynamics, Planning and Design in Contemporary Urban Territories – Allegri et al (eds)

The centre of Campinas awaiting planning: Discourses and practices in municipal urban policies

C. Santos Chagas & M.C. Silva Schicchi
Postgraduate Programme in Architecture and Urbanism, Pontifical Catholic University of Campinas, Brazil

ABSTRACT: This article analyses the programs and projects formulated for the Campinas city centre by successive municipal administrations. The analysis of discourses, proposals and actions since the 1990s to the present day demonstrates that there have been no effective and long-term urban policies aimed at improving the built environment, securing the permanence of existing residents and the appropriation of space in the centre by all citizens.

1 INTRODUCTION

Urban centres are unique areas of the urban fabric of contemporary cities, witnesses to urban history itself. They are spaces endowed with a symbolic value that goes beyond their physical materiality and vast real estate, and bring together the identity and collective memory of the whole of society (Villaça 2001; Carrión 2009).

Whereas at a certain point in time public and private investments concentrated on new urban expansion fronts and the formation of new centralities, particularly in big cities, in the 1990s the idea of intervening in urban centres became part of the discourses and actions by municipal authorities of several Brazilian cities.

With this context in mind, the aim of the present article was to carry out a critical analysis of the ideas, discourses and actions on which urban policies for the centre of Campinas were based from the 1990s onwards, over successive municipal administrations. The article also seeks to highlight the nature of these policies, whose aim, above all, was the improvement of the urban image and reappropriation of the centre by local elites concerned with making it attractive to private investment.

The analysis of discourses by municipal management, private actors and local media reveals that for the "regeneration" of this space to take place, certain everyday uses and practices had to be eliminated. These discourses were mostly unconcerned with solutions to social problems and the forms of appropriation of the area indiscriminately by all citizens.

Our methodology was to carry out a historical analysis of the plans, programmes and projects formulated for the centre of Campinas, followed by a comparison among them. For the analysis of the urban policies, two important aspects were considered: the policies advocated – the sphere of discourses, objectives and justifications – and management practices – the modes of intervention and implementation (Alvim, Castro, and Zioni 2010).

The timeframe analysed ranged from the first proposal devised for the centre of Campinas in the 1990s, which was the Programa de Revitalização da Área Central do Município de Campinas (Urban Renewal Programme for the Central Area of the Municipality of Campinas) [1993], to the Projeto de Requalificação da Avenida Francisco Glicério (Redevelopment Project of Avenida Francisco Glicério) [2015]. Due to space constraints, this article will not address all the proposals (plans, programs and projects) formulated, discussed or implemented– only those that fit the aims outlined above.

As well as discussing the ideas, principles and discourses that have driven the interventions in the centre of Campinas, the intention of this article is also to enable a general reflection on the many problems that also occurred in other Brazilian cities from the 1990s onwards, especially because it will reveal why projects and plans systematically conceived over almost thirty years have not resulted in permanent urban policies for the urban centres.

2 URBAN CENTRES AS PROTAGONISTS OF URBAN DEVELOPMENT

Particularly since the 1990s, urban centres have resumed their central role in the municipal urban

DOI: 10.1201/9781003220855-20

policies of several Brazilian cities. These policies, termed "renewal", "redevelopment" or "regeneration" policies, have generally been concerned with improving the urban image, preserving the built heritage, recovering cultural and tourist activities and, above all, seeking strategies for attracting private investment.

The implementation of these policies in urban centres in Brazil from the 1990s onwards coincided with important changes in the political and economic contexts. The first change was the autonomy granted to municipalities to devise and implement urban policies by the new Federal Constitution of 1988, which now had mechanisms to drive democratic management and social justice. The second was the advance of neoliberal policies at the international level, which directly influenced the design of these policies through the coordination of private investments in strategic regions and large urban projects.

In the early 1990s, the urban centres of large Brazilian capitals such as Rio de Janeiro (RJ), São Paulo (SP) and Salvador (BA) became the target of these specific projects with the implementation of programs called Rio-Cidade (Rio-City), Programa de Requalificação Urbana e Funcional do Centro de São Paulo (Programme of Urban and Functional Redevelopment of the Centre of São Paulo) and Programa de Recuperação do Centro Histórico de Salvador (Regeneration Programme of the Historical Centre of Salvador), respectively.

It should be noted that the 1970s and 1980s were marked by great growth in Brazilian cities, a process that resulted in the consolidation of new centralities or sub-centres, which in many cases occurred concomitantly to the abandonment of public and private investments and the reduction or abolishing of certain uses and activities in urban centres by higher-income classes. From that moment on, urban centres started to become stigmatised as "deteriorated" "decadent" or "degraded" areas, which generally refers to the appropriation of an area for less productive or informal uses, and to a concentration of social groups with lower purchasing power (Villaça 2001; Rojas 2004).

When discussing urban renewal processes in Brazil, Carlos (2007) points out that in many cases, in order to improve the urban image, and especially to favour private investments, these policies contributed to deepening socio-spatial segregation as they transformed and directed the uses and functions of sites at the expense of the expulsion of a certain population considered "unbefitting" or "decadent":

> Therefore, if on one hand this transformation is a formal one and refers to the possibility of restructuring urban spatial forms, on the other, the process is fundamentally social – a process of exclusion and segregation – and in this case, we can observe new types of appropriation of space by society.
>
> (Carlos 2007: 90)

The centre of Campinas (Figure 1) is part of this context, having been the object of several proposals, actions and attempts at implementing large urban projects by all the municipal administrations from the 1990s to the present day.

Although each municipal administration drew up its own proposals for the city centre, whether part of their urban policies for the city or otherwise, the constant element in all of them was an ideological discourse capable of transforming public actions into private interests, with the ultimate aim of reappropriation of the area by the hegemonic classes, in line with the interests of private business.

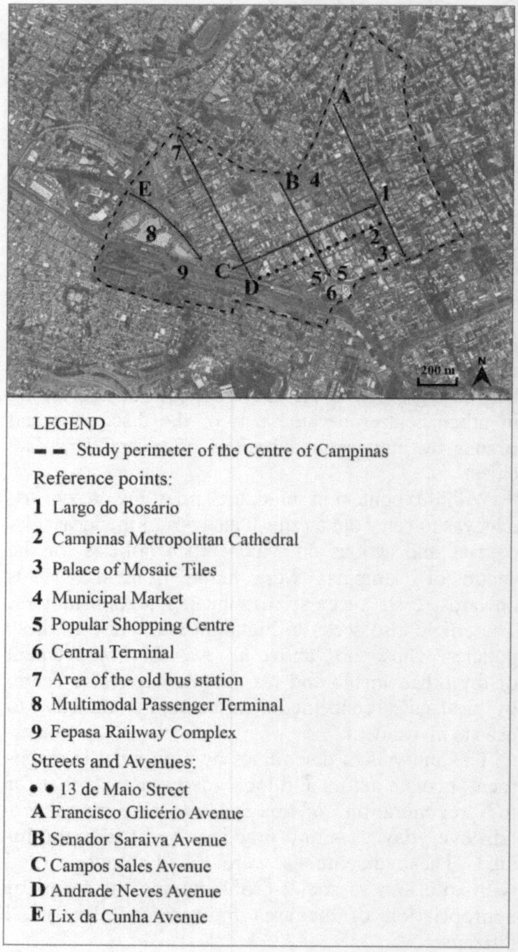

Figure 1. Aerial shot of the centre of Campinas showing some of the main avenues and important reference points. Source: Authors' own, based on an aerial photo from Google Earth, 2018.

3 PROPOSALS FOR THE CENTRE OF CAMPINAS FROM THE 1990S ONWARDS: DISCOURSES AND PRACTICES

3.1 *Delimit, grant, revitalise*

Launched in 1993 by the Planning and Coordination Department (Seplan), the Programa de Revitalização Urbana da Área Central do Município de Campinas (Programme for Urban Renewal of the Central Area of the Municipality of Campinas) had as one of its main objectives to reverse the deterioration process that characterized the city centre so that it would recover its symbolic power as the "heart of the city", and to restore its quality of life so that it would become an aesthetically pleasing area (Prefeitura Municipal de Campinas 1993a).

As stated in the Programme's Technical Notebook, three factors were responsible for the centre's deterioration: 1) During the 1950s and 1960s there was a high concentration of shops and services in the area, without there being secondary centres in the city; 2) In the 1970s, commercial activities and residential uses by the higher income classes relocated to other areas of the city; 3) In the 1980s, the implementation of public transport terminals contributed to the concentration of informal street vendors in the centre of Campinas (Prefeitura Municipal de Campinas 1993a).

In addition, the so-called "social crisis" of the 1980s and 1990s further aggravated the deterioration of the centre of Campinas, resulting in a

> concentration of a significant number of unemployed people, abandoned minors, beggars and prostitutes, which placed serious constraints on the valorisation and/or regeneration of the uses assigned to buildings in the centre.
>
> (Prefeitura Municipal de Campinas 1993a: 03)

Against this backdrop, the Programa de Revitalização Urbana da Área Central do Município de Campinas presented the project "Shopping de Rua" (Street Shopping Centre), to be implemented in Rua 13 de Maio, as one of its main strategies for renewing the area. It is important to point out that, at the time (and until the present day), Rua 13 de Maio was considered Campinas' main traditional shopping street. This intervention would give the street it translucent roofing supported by a metallic structure along its whole length, according to a project elaborated by the architect and urbanist Ruy Ohtake (Prefeitura Municipal de Campinas 1993b).

In actual fact, the project would have made it possible to privatise the street for the exclusive use of shopkeepers and consumers, providing a controlled space restricted to certain users, and demarcating an exclusive perimeter with the roofing. The project was to be implemented with resources from the owners and traders of Rua 13 de Maio, who were to form an association or cooperative for this purpose. The role of the city authorities would only be to grant the special administrative and land use concession of Rua 13 de Maio to the association through Law No. 7.580 of August 16, 1993. Although the Law was approved by the municipal legislature, most owners and shopkeepers failed to join in part due to the high cost of the investment and to negotiation difficulties. Thus, the "Shopping de Rua" failed to be implemented.

At the time, local newspaper Correio Popular conveyed Mayor José Roberto Magalhães Teixeira's discontent regarding the fact that the shopkeepers, according to him, had ignored the implementation of the "Shopping de Rua", and he was quoted as saying that this apathy contributed "to transform Rua 13 de Maio into a kind of Persian market, fought over by street vendors, beggars and prostitutes", and proved right the consumers who chose to do their shopping in the shopping centres recently built in the city (Verzignasse 1995).

3.2 *Displace, invest, and rebuild image*

The Programa Campinas Toda Nova (All New Campinas Programme) was another attempt, launched in 1997, to regenerate the Campinas centre under the municipal administration of Francisco Amaral (1997-2000). In the same way as the previous programme, the city authorities would act as a partner in the Programme and would be responsible only for formulating an intervention plan. A private finance initiative would be responsible for its implementation by setting up the Associação Campinas Toda Nova (All New Campinas Association), whose model was the Associação Viva o Centro (Association Long Live the Centre), created in 1991 for regenerating the centre of São Paulo.

At the time, the initiative to form an Association received support from bodies representing shopkeepers, the real estate sector and civil construction. Among its proposals was the idea of revisiting previous projects – such as "Shopping de Rua" – and studying ways to make them viable. One of the Association's strategies was to use urban marketing actions to ensure projects left the drawing board (Costa 1997b).

The preference for urban marketing actions and public-private partnerships is characteristic of what is known as strategic planning, whose model of "success" is the case of the city of Barcelona – which underwent a major urban reformulation in order to host the 1992 Olympic Games – and which was directly related to the neoliberal policies that followed in several European and Latin American countries. According to Vainer (2002: 87), this type of planning requires the protagonists of actions and decisions concerning a territory to be the same as the protagonists of market exploits, that is, it requires "the public-private partnership to ensure that the signs and interests of the market are sufficiently present and represented in the planning and decision-making process".

In accordance with the interests of the Association, two articles published in the local press at the time presented the ideas and discourses behind the proposals and stated that "more than [offering] a shopping spree, the bodies want a vibrant city centre that will foster the city's development and be the postcard of Campinas" (Costa 1997a), and that "lack of security, religious preaching, irregular selling, yelling, begging and pamphleting" could soon be stamped out if the Association were formed (Piloni 1997).

Thus, the different types of appropriation and modes of use of the centre of Campinas by vulnerable groups, that relied on it for their survival, socializing or work, clashed directly with the symbolic and market interests of the Association, so that the driving out of this population, seen as "unbefitting", was a precondition for obtaining the expected aesthetic results and the appreciation of the space.

Again, the proposals to recover the centre were not facilitated by the municipal administration or by the private finance initiative. However, we should highlight the position of these strategic actors in the formulation of proposals for the centre. When aiming at improving its image, they placed the above-mentioned social groups in a vulnerable situation, as the latter were not included in proposals of a social nature or in discussions on policies for the centre of Campinas.

3.3 *Remodel to reoccupy the shopping streets*

In 2001, under a new municipal administration called "Governo Democrático e Popular de Campinas" (Democratic and Popular Government of Campinas), an instrument of democratic administration called Orçamento Participativo (Participatory Budget) was implemented. This enabled the street vendors to have funding approved to part-finance the construction of an official space for informal activities in the centre. The occupation, implementation and regularisation of this area was called the Popular Shopping Centre, and its construction was made official in 2003 in the vicinity of the Central Terminal (Figure 2).

The organisation of informal vendors was part of the Projeto Centro (Centre Project), under the municipal administration of Izalene Tiene (2001-2004). It consisted of a set of 14 emergency actions coordinated by the Municipal Department of Culture to yield quick results for the recovery of the centre of Campinas over a period of four years. Projeto Centro presented proposals for improving urban equipment, promoting cultural activities, visually depolluting facades, and refurbishing and repurposing historical, artistic and cultural heritage constructions such as the buildings and warehouses located in the *Fepasa* Railway Complex and the former mansion of the Baron of Itatiba, better known as Palácio dos Azulejos (Palace of Mosaic Tiles) – a historic building listed by *Iphan* – Instituto do Patrimônio Histórico e Artístico Nacional (National Institute of Historic and Artistic Heritage).

Although the implementation of the Centro Popular de Compras legalised the activity of street vendors, in a way it also resulted in their segregation from a specific area, as they were prevented from occupying certain avenues and public spaces in the centre, and it was the remit of the municipal council that manages public land in Campinas, *Setec* – Serviços Técnicos Gerais (General Technical Services) to inspect and charge a fee for the use of public land (Sakai 2011).

As well as this measure to indirectly meet the interests of the formal retail trade in the centre, another action by Projeto Centro was the previous proposal to remodel Rua 13 de Maio. According to information published in the Diário Official do Município (Official Municipal Gazette) at the time, unlike the previous project, the implementation cost would be the City Government's responsibility (Campinas 2004).

The renovation project for Rua 13 de Maio, which had been a pedestrianized street since the 1970s, was only concluded during the administration that followed and included changing the paving, replacing the water and drainage network, undergrounding the energy, telephony and data networks and cleaning up and adapting the facades of commercial establishments in order to reduce visual pollution (Figure 3).

Figure 2. Centro popular de Compras. Source: Authors' photo, 2019.

Figure 3. Rua 13 de Maio. Source: Authors' photo, 2019.

Therefore, it was a project that proposed a remodelling of the street so that it could be reoccupied in an "orderly" manner.

3.4 *Expel, construct, appreciate – the nostalgia of a centre for the few*

In 2008, the newspaper Correio Popular published a special article titled "Apagão Urbano" (Urban Blackout), in which it exposed the lack of political action and urban planning in the centre of Campinas, stating that "for almost two decades the urban renewal of the centre of Campinas has been on the agenda of successive municipal governments" (Busch & Villa 2008) and pointing out, in a nostalgic tone, what the centre used to represent for the population of Campinas:

> A symbol of economic thriving in the last century, until the early 1980s the region was the best that a metropolis can offer in the shopping, services and leisure sectors. Cinemas, well-structured squares frequented by families and large retail chains made the central area the destination of choice for shopping and outings. It was a time that has remained in the memory of the older residents and even arouses feelings of nostalgia in those who witnessed the heyday of this now disfigured place.
>
> (Busch & Villa 2008)

Although the article denounced the lack of urban planning in the Campinas city centre, the terms "degradation" and "deterioration" were used when referring to the presence of street vendors, prostitutes, beggars, homeless people and high-turnover hotels that were home to prostitution. In the same article, the Secretary of Urbanism at the time, Hélio Jarreta, expressed that "there is no [such thing as] ever-present and all-powerful inspection; the best form [of inspection] that can exist is the appropriate occupation of the region" (Busch & Villa 2008).

It should be noted that the use of the terms "deterioration", "decadence" and "degradation" mentioned so far generally refer to the appropriation of the centre of Campinas by lower socioeconomic classes. However, far from being an isolated case in the centre of Campinas, this kind of stigmatisation disseminated by public and private actors, and often by the media, has previously been mentioned by authors such as Villaça (2001) and Carlos (2007).

Villaça (2001: 344) stated that the ideology of "deterioration" serves to naturalise social processes. Considering that urban centres are highly strategic points for the exercise of domination, the ruling class is unable to own the fact that at a given moment it abandoned the main centres in favour of new areas, thus disseminating a version of events that does not compromise its position as the dominant class, with the mainstream media serving as an efficient vehicle for these versions.

According to Carlos (2007: 89), the process of renewing urban centres in Brazil was based on a rationale that proposed the demolition of places, furthering processes of segregation and expulsion of certain populations considered "unbefitting", all due to the need to produce a "new space":

> Thus, urban renewal is first and foremost a process of revalorising urban land, which changes the use of a space through the imposition of an exchange value, expelling those unable to pay for it, as seen, for example, in São Paulo, Salvador and even Paris. Urban renewal also produces an asepsis of places, as what is considered "degraded" is always what appears in the landscape as the poor, the dirty, the ugly, and calls for replacement with the rich, the clean, the beautiful – characteristics that do not befit poverty.
>
> (Carlos 2007: 89)

Thus, it was no coincidence that, only two years after the publication of the special article "Apagão Urbano" by Correio Popular – which exposed, among other aspects, the "degradation" of the area of the old bus station in the centre and highlighted the presence of "slums, guest houses that enable prostitution, drug trafficking and areas with poor lighting" (Busch & Villa 2008) – the Hélio de Oliveira Santos municipal administration (2005-2011) launched what was known as the Pacote de Revitalização do Centro, (Centre Urban Renewal Package). Among other measures, the package created a mechanism to encourage investment in the region of the old and new bus stations, through the almost simultaneous enactment of two laws, Complementary Law No. 29 of January 11, 2010 and Complementary Law No. 30 of January 13, 2010, both of which were authored by the Executive Branch.

The first law established specific conditions of urbanisation for the area of the old bus station – made up of the block surrounded by the important avenues of the centre –, which was demolished in 2010 and since became under disuse, with some ruins remaining (Figure 4). It was about the possibility of adopting new construction parameters and establishing land use and occupation criteria in order to leverage private investment in the area. In other words, the degradation resulting from demolition carried out by the municipality itself actually provided an opportunity to change the construction potential and appreciate of the area.

According to an online article published on the Campinas City Government website in 2015, Rede D'Or São Luiz [an upmarket hospital network] proposed to build a private hospital on the area with tax incentives for the City Government such as exemptions from fees, emoluments, public prices from technical bodies, exemption from the Urban Land Tax (IPTU) for ten years and a reduction in the rate

Figure 4. View of the site after demolition of the old bus station. Source: Authors' photo, 2019.

of the Tax on Services of Any Nature (ISSQN) (Prefeitura Municipal de Campinas 2020).

The second Law created the so-called "Área Especial de Reurbanização do Entorno do Terminal Multimodal de Campinas" (Special Reurbanisation Area of the Region of the Campinas Multimodal Terminal) which, besides including the old bus station area, also focused on the surroundings of the newly built Multimodal Passenger Terminal – areas relatively close to one another.

Complementary Law No. 30 of 13 January, 2010 established new criteria for land use and occupation, with a view to promoting the urban renewal of the area. It prohibited certain activities such as the sale of scrap iron and the purchase and sale of recycled materials, as well as the activity of street vendors and waste pickers. In addition, with the requirement of minimum standards for the establishments such as private toilets, accessibility and parking spaces, the law directly affected properties that were used as collective dwellings, hotels and boarding houses. According to a survey conducted by Camargo (2010), this region was home to most of the centre's slums.

Therefore, it is evident that whatever the name used, the proposals to "regenerate", "renew" or "redevelop" the centre of Campinas prioritized, above all, proposals aimed at resuming private investment and revalorising its services and commercial nature. To this end, a consensus on the degradation and devaluation of the area had to be created, to then justify actions and strategies of urban and social hygiene and cleansing.

3.5 *Beautify, re-valorise: Resuming the discourse*

In this retrospective, it is evident that an urban policy that would include the central area and aim to improve the centre of Campinas as a long-term plan – as opposed to a plan based on topical and selective actions by respective municipal administrations from the 1990s onwards – has so far been lacking. In actual fact, the plans, programmes and projects proposed have either been mere tools for negotiating the areas that were the object of the interventions, or have served as marketing for the administrations themselves. The recent administration of Jonas Donizette (2013-2020) contributed to the continuity of this situation with the launch of a new project in 2014 – the Projeto de Requalificação da Avenida Francisco Glicério (Redevelopment Project of Avenida Francisco Glicério).

Once again, the project – and in this case the actual urban intervention – prioritized the tertiary and financial nature of the centre, since Avenida Francisco Glicério Avenue (Figure 5) is considered by many to be the "heart of Campinas", as it houses commercial and service buildings important to the economic dynamics of the city as a whole, as well as being home to historical public institutions, buildings and spaces such as Largo do Rosário, the Palace of Justice and the Campinas Metropolitan Cathedral.

The intervention included direct participation by class entities, companies and segments of the real estate, commercial, telecommunications and civil construction sectors (Prefeitura Municipal de Campinas 2014). Additionally, the project was donated to the City Government by *IAB-Campinas* (Institute of Architects of Brazil-Campinas) with the support of the Associação Brasileira de Cimento Portland – *ABCP* (Brazilian Portland Cement Association) and the Sindicato da Construção de Campinas (Campinas Construction Union), bodies that bring together real estate and construction entrepreneurs (Reunião Ordinária 2014). In other words, a project that was not drawn up within the municipal administration and which involved strategic segments with vested interests in economic, construction and real estate benefits resulting from this intervention was made viable.

The project proposed a clear improvement of only the aesthetic aspect of Avenida Francisco Glicério. In 2015, the Avenue received standardized paving, underground infrastructure and telecommunications cabling, improved lighting and new street furniture, and its pavements were widened.

Figure 5. Avenida Francisco Glicério. Source: Authors' photo, 2019.

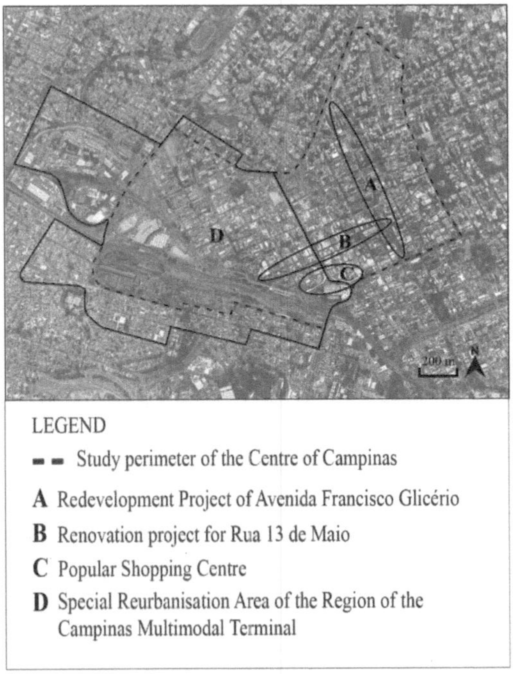

Figure 6. Overlaying of the perimeters of the different intervention proposals for the centre. Source: Authors' own, based on an aerial photo from Google Earth, 2018.

4 CONCLUSIONS

For almost three decades and over successive municipal administrations, the centre of Campinas has been the target of plans, programmes and projects (Figure 6). Although these have gone under different names, always striving to be a novel policy for the centre, the assumptions behind these processes have remained the same: to carry out interventions in favour of a clean city centre, without the presence of lower-class groups, providing a favourable environment for new economic and real estate investments.

It is evident that the emphasis on the use of the terms "deterioration" and "degradation" associated with the centre of Campinas has been recurrent in the discourse of municipal administrators and other strategic actors, and disseminated by the local media, which has sought, above all, to establish a general consensus that the cause of the deterioration of the centre has been the presence of lower-class and vulnerable groups considered "unbefitting" or "decadent".

However, as this article has shown, this reiteration of the stance of public and private agents working in the centre over successive administrations leads to an understanding of the opposite – that the cause of the deterioration and degradation of the centre lies not in the fact that the area was appropriated by these groups considered "unbefitting", but rather, in a set of factors. Among these factors is the interest in reclaiming the centre not for its use, but for investment by the ruling classes, and the absence of an effective and long-term urban planning system to promote new spaces for socialising, interacting, leisure and housing, that is, one whose principle is the diversity of uses required by the contemporary city. However, the main issue is not the causes, but the permanence of the discourse of "deterioration and devaluation" mentioned, which remain as a pretext for the creation of a stock of areas of opportunity for future real estate investments.

The way in which plans, programmes and projects were drawn up – as well as the underlying discourses –, without any kind of survey of the needs of other sectors or social actors that use the centre, has revealed a recurrent attitude by public administration of relinquishing its central role in proposals, which are meant to prioritise the public and collective interest.

The issue is not a lack of proposals, discussions or actions for the area, but rather, a break with this *modus operandi* for the centre of Campinas, where specific actions, emphasis on promoting tertiary activities, and the selectivity of actors in urban policy-making disregard the different rights and needs of the population as a whole. Above all, it is about deconstructing the nostalgic and stigmatising discourses by public administration and several political and social segments, which insist on the notion that only a few "deserve" to live and enjoy the centre of Campinas.

REFERENCES

Alvim, A.T.B., Castro, L.G.V.& Zioni, S. 2010. Avaliação de políticas urbanas. In Alvim, A. T. B. & Castro, L.G. V (Eds.), *Avaliação de políticas urbanas: contextos e perspectivas*: 13–41. São Paulo: Universidade Presbiteriana Mackenzie, Mackpesquisa and Romano Guerra Editora.

Busch, G. & Villa, M. 2008. Apagão Urbano. *Correio Popular*, August 24, Campinas.

Camargo, C. M. 2010. *Habitação coletiva popular na área central de Campinas: tendências e caracterização*. PhD diss., Universidade de São Paulo, São Carlos.

Campinas 1993. Lei No. 7.850, August 16, 1993. Autoriza o Poder Executivo a conceder o uso especial da Rua 13 de Maio à Associação de Lojistas, a ser constituída para a implantação de "Shopping de Rua", e dá outras providências. *Diário Official do Município de Campinas*, August 17, 1993: 1.Accessed February 28, 2020. https://bibliotecajuridica.campinas.sp.gov.br/index/visualizaroriginal/id/108978

Campinas 2004. Reforma da 13 de Maio vai custar R$ 366 mil menos do que o previsto. *Diário Official do Município de Campinas* No 8.404, April 1, 2004: 2. Accessed February 27, 2020. http://www.campinas.sp.gov.br/uploads/pdf/811927328.pdf

Campinas 2010a. Lei Complementar No. 29, January 11, 2010. Estabelece condições específicas de urbanização para a área do antigo Terminal Rodoviário e dá outras providências. *Diário Official do Município de Campinas* No 9.811, January 12, 2010: 1-2. Accessed February 27, 2020. http://www.campinas.sp.gov.br/uploads/pdf/319903744.pdf

Campinas 2010b. Lei Complementar No. 30, January 13, 2010. Cria a Área Especial de Reurbanização do entorno do Terminal e implementação de restrições de uso e ocupação do solo nesta área. *Diário Official do Município de Campinas* No 9.813, January 14, 2010: 1-2. Accessed February 27, 2020. https://bibliotecajuridica.campinas.sp.gov.br/index/visualizaratualizada/id/86106

Carlos, A. F. 2007. *O Espaço Urbano: Novos escritos sobre a cidade*. São Paulo: FFLCH/USP

Carrión, F. 2009. El centro histórico como objeto de deseo. In *Sesión Inaugural del Seminario Permanente Centro Histórico de la Ciudad de México*. Programa Universitario de Estudios sobre la Ciudad – UNAN, Ciudad de México. Accessed October 17, 2019. https://puec.unam.mx/pdf/seminarioschcm/spponencias/03.pdf

Costa, M. T. 1997a. Campinas se une para revitalizar Centro. *Correio Popular*, May 28, Campinas.

Costa, M. T. 1997b. Plano de intervenção revitalizará Centro. *Correio Popular*, July 11, Campinas.

Piloni, D. 1997. Centro de Campinas está um verdadeiro mercado persa. *Diário do Povo*, July 12, Campinas.

Prefeitura Municipal de Campinas 1993a. Programa de Revitalização da Área Central do Município de Campinas. Campinas: *Secretaria de Planejamento e Coordenação* (Seplan).

Prefeitura Municipal de Campinas 1993b. Novo Centro de Campinas: Revitalização da Rua 13 de Maio. Campinas: *Serviços Técnicos Gerais* (Setec).

Prefeitura Municipal de Campinas 2014. Proposta de Requalificação – Avenida Francisco Glicério. Campinas: *Secretaria Municipal de Desenvolvimento Econômico, Social e de Turismo*. Accessed February 28, 2020. http://www.campinas.sp.gov.br/arquivos/comunicacao/revitalizacao_glicerio.pdf

Prefeitura Municipal de Campinas 2020. Rede D'Or lança pedra fundamental de Hospital São Luiz em Campinas. Campinas: *News*. Accessed February 22, 2020. http://www.campinas.sp.gov.br/noticias-integra.php?id=38085

Reunião Ordinária 2014. Ata No. 439, November 27, 2014. Campinas: *Secretaria Municipal de Cultura*. Conselho de Defesa do Patrimônio Cultural de Campinas (Condepacc). Accessed February 29, 2020. http://www.campinas.sp.gov.br/uploads/atas/155451_Ata%20439.pdf

Rojas, E. 2004. *Volver al Centro: la recuperación de áreas urbanas centrales*. Collaboration of Eduardo Rodríguez Villaescusa and Emiel Wegelin. Banco Interamericano de Desarrollo: Departamento de Desarollo Sotenible. https://publications.iadb.org/publications/spanish/document/Volver-al-centro-La-recuperaci%C3%B3n-de-%C3%A1reas-urbanas-centrales.pdf

Sakai, R. Y. 2011. *A ocupação da área central pelo comércio ambulante: negociações e produção do espaço urbano*. PhD diss., Universidade de São Paulo, São Carlos.

Vainer, C. 2002. Pátria, empresa e mercadoria. Notas sobre a estratégia discursiva do planejamento estratégico urbano. In Arantes, O.; Vainer, C. & Maricato, E. *A cidade do pensamento único: desmanchando consensos*: 75–103. Petrópolis: Editora Vozes.

Verzignasse, R. 1995. Planos de Revitalização Urbanística envelhecem na gaveta. *Correio Popular*, June 25, Campinas.

Villaça, F. 2001. *Espaço intra-urbano no Brasil*. São Paulo: Studio Nobel, Fapesp.

Research Tracks in Urbanism: Dynamics, Planning and Design in Contemporary Urban Territories – Allegri et al (eds)

Slum urbanization: Impacts of interruption on the lives of residents

C.S. Souza & P.R. Samora
POSURBARQ, CEATEC, Pontifícia Universidade Católica de Campinas, São Paulo, Brazil

ABSTRACT: Slum upgrading policies require long-term actions, continuous financial support, and the government's commitment to accomplish the goals, seldom seen in Brazilian social housing experiences. This paper aims to discuss the effects on the lives of the residents of "Vila Parolin", a precarious settlement in the city centre of Curitiba, Brazil, of the repeated interruptions in social housing programs over the last few decades. The present investigation was based on official data and interviews with residents, mapping the finished works and their outcomes. We observed failings as a result of limited actions that did not consider overall quality and effectiveness. This ineffectiveness impacts householders′ decisions regarding their housing rights, creating the illusion that decent housing can only be achieved by their own means, without the State's commitment. Consequently, this weak dialogue between residents and public authorities contributes to promoting further urban segregation and spaces of social vulnerability.

1 SLUM UPGRADING: IMPACTS OF HOUSING POLICY INTERRUPTIONS ON THE LIVES OF RESIDENTS OF INFORMAL SETTLEMENTS

2 INTRODUCTION

In the past two decades, Brazilian housing policies have built a large number of housing units and started the urbanisation of numerous favelas. From 2007 to 2017, the Brazilian government dedicated about 30 billion reais to 1,058 cities located in different regions of the country for investment in slum upgrading programs (CARDOSO, DENALDI, 2018). However, many housing developments were affected by interruptions. This issue, plus a level of housing demand which outstripped the previously established estimates of the public authorities in charge, has significantly impacted the lives of the residents of informal urban areas included in social housing schemes.

As actions to improve informal areas require long-term commitment, misconceptions can occur during the upgrading of informal settlements. The regular succession of political groups in public management impacts the previous line of action. Other errors can arise from attempts to implement urban services without considering their effectiveness or outcomes on urban settlement quality. Some of these mistakes are recurrent, addressed, and flagged by experts, yet, established public management practices are failing to consider these.

During the slum upgrading process, routine changes discourage dialogue between residents and the government representatives toward solving residents' needs and issues. This ineffectiveness of public housing policy impacts householders` decisions regarding their housing rights. This scenario forces slum-dwellers to seek decent housing through their own means, often leading to more precarious housing conditions. It also creates the illusion that this issue can be addressed without the State's commitment.

This study presents an analysis of the Vila Parolin case, conducted by the city of Curitiba, Brazil. This settlement received investment from both federal and local governments through different housing policies over many years. However, the process of upgrading was interrupted many times, affecting the residents′ expectations and often causing frustrations as the quality of the built environment did not improve as expected.

Vila Parolin is an interesting case, a well-located slum area that faces constant harassment from public authorities seeking to evict residents. The community is subject to an intense urban segregation process that fuels social vulnerability due to unconcluded upgrading interventions over the decades. This article discusses the impacts of housing program discontinuity on the affected population's lives, expectations, and frustrations. It also brings to the debate the solutions sought by the Vila Parolin community in response to the ineffectiveness of government actions.

The study is based on records provided by the City of Curitiba on programs implemented in the area. We assessed how the upgrading projects have

DOI: 10.1201/9781003220855-21

changed, the unfinished works and their impact on the built environment. The process shows how the government prioritizes solutions that transfer former dwellers to new housing developments far from the original settlement. In addition to this, we draw on material from interviews with householders and community leaders conducted in 2015 (SOUZA, 2016), when the last interruption occurred. As a result, we discuss how discontinuations affected the community and how they express their desire to see the works restart.

The social and urban complexity of precarious settlements requires comprehensive prospect studies before proposing solutions to improve these areas, to avoid negative impacts on the life of the community'. In Vila Parolin, the intervention indicated the relocation of many dwellers due to overcrowded housing conditions and the need to lay roads and promote better infrastructure networks. In 2018, however, when the city authorities restarted the works, they focused on finishing the housing developments outside the settlement, while the infrastructure works within the Vila Parolin perimeter were neglected.

Discussing housing policies that promote the relocation of slum dwellers in Buenos Aires, Lacarrieu (2018) stated this move was not the wish of the affected population, which prefers to remain in place, while the government insists on relocating them to other areas. According to the author, "el futuro de los asentamientos precarios no es uno solo, sino diversos y desiguales futuros constituidos a partir de contextos y procesos históricos diferenciados" (2018: 387). In other words, the different context and process that constitute the slum settlements impact its future outcome.

The main objective of this article is to discuss how the interrupted or partially implemented slum upgrading policies impact the lives of the local population and how public policy dynamics can contribute to a fast return to the previous precarious condition, which exacerbates urban segregation and social vulnerability.

3 EFFECTIVENESS OF HOUSING PUBLIC POLICIES

In the second half of the 20th Century, Federal and State investment in housing and urban development, albeit significant, was smaller than that needed due to address lack of affordable housing for Brazilian families. Davis (2006: 71) clarifies that "the minimalist role of the National government in housing provision has been reinforced by the current neoliberal economic orthodoxy defined by the IMF and the World Bank". Due to this, many of the programs implemented have a more palliative than structural nature, not reaching the core of the problem nor substantially changing the lives of the low-income inhabitants.

Also, implementing housing policies without effective integration with other public policies, such as those aimed at generating jobs and income, produces new adversities such as households struggling to sustain themselves in unaffordable housing units. As a result, moving to another place does not produce a qualitative change in those transferred to new housing schemes, making these neighbourhoods spaces of social disintegration. Often, these settlements suffer a lack of basic infrastructure and urban services. When the urban upgrading works are not concluded nor legalized, it is not fully incorporated into the regular city (CALDEIRA, 2017).

The way public managers face the slum dweller's housing needs does not consider project quality or community expectations. The decision-making process ignores communities' demands by not listening to households, and partly explains the challenge developing more effective slum upgrade programs.

Some resident input can improve public policies and create awareness of how to develop successful projects. The main features are the quality of housing design and urban infrastructure, aiming to keep most households on the site after upgrading it. Matters such as the sense of belonging and individual needs do not feature in ordinary housing programs. However, this information could change the transient condition that often affects low-income families living in informal areas, promoting the concept of permanence (Lopes, 2002).

The incompleteness of housing and urban projects and other social policies such as education, work, and income generation, keep low-income families in poverty. Here lies a perverse dichotomy: (i) the public authorities do not consider solving the social housing deficit a priority of the State, leaving the poorest to shoulder the responsibility of overcoming the problem; (ii) the insufficiency in social policies such as education, labour, and income does not confer citizens sufficient autonomy to solve their condition of precariousness in the event of a partially implemented policy.

The poor tend to blame themselves for their social precariousness and vulnerability, and bring it upon themselves to solve their material problems (Veras, 2001). This individualization stems from a social ascension perspective in which low-income individuals see it as their responsibility to improve their station and become middle class or achieve wellness. However, according to Souza (2017: 80), this is an unrealistic aspiration, because "The theory of modernization has sophisticated the fashion to euphemize reality to deny forms of domination which tend to be eternal. Thus, poverty and maladjustment are transient, often resulting from transitory situations, such as the passage from the countryside to the city".

However, the possibility of social ascension, the dream of owning a house and of better living conditions involve structural issues that go beyond individual effort in the daily life of the poorest families.

Therefore, the responsibility of the State to provide its population with minimal living conditions that go beyond survival cannot be lifted.

Brulon and Peci (2018: 79) conclude, after studying the slums of Rio de Janeiro, that the residents "[...] went on to assume by themselves the responsibility of 'fighting' for their demands to be satisfied, whether by doing it themselves or by means of a great effort to call for responses from the State. There seems to be a generalized belief that the residents of the slums themselves must fight".

This individualization puts the poor in charge of their living conditions (akin to the middle class), as opposed to them mobilizing to demand collective solutions from the State and thereby protecting their rights.

According to Kowarick (2017: 23), one way of solving the impasse is by strengthening "the capacity building of several groups and social movements to put pressure and obtain from the State the essential elements for their survival in the cities". The elements to which the author refers are "access to land, housing and collective goods". Pressure from the poor is fundamental for public policies to offer better living conditions.

Social inequality is in the essence of Brazilian formation (Veras, 2001; Souza, 2017) and in the development of the capitalist State (Harvey, 2006), which produces a process of urban spoliation (Kowarick, 2017) and social vulnerability (Veras, 2001).

Given this, the housing and urban development policies must be associated with other public policies, such as education, labour, and income-generating to modify the unequal social and urban condition. It should be rooted in more effective proposals able to overcome the complicated and comprehensive situations like those affecting households in precarious settlements.

4 UNFINISHED CITY: VILA PAROLIN, CURITIBA/PR

Regarding Curitiba, research often shows that this capital city follows the same exclusionary pattern seen in other Brazilian cities. Thus, the informal settlements are being consolidated, while new housing schemes are established on the outskirts or within poorer cities of Greater Curitiba (Silva, 2012; Stroher, 2017).

Social-spatial inequality constitutes the city's urban fabric, and is concentrated in lower-class peripheral neighbourhoods, while middle and upper-class neighbourhoods are well-equipped (Silva, 2012; Carvalho, 2014; Stroher, 2017).

COHAB-CT, a municipal body, is in charge of the social housing policy, and historically focuses on new housing developments located in outlying neighbourhoods. (Boletim, 2006; Carvalho, 2014). In the last few years, the policy has been changing, with investment in informal settlement regularisation and transferring at-risk households to surrounding areas. (Curitiba; COHAB-CT; IPPUC, 2007).

The Vila Parolin settlement has occupied the banks of the Vila Guaira river since the 1950s, in a privately-owned area of 240,000 square meters. It is located in the city centre, among middle and upper middle-class neighbourhoods, a poor enclave that suffers urban segregation and in conflict with its neighbours. COHAB-CT registered 1,507 families and 5,385 inhabitants in 2006. Although the public authorities have made several attempts to remove the settlement between 1960 and 1980, the residents resisted, and pressured the government to improve the neighbourhood (COHAB-CT; Curitiba, 2007).

In 2007, the settlement was selected to receive resources from the PAC Program (PAC-PPI), a federal government slum upgrading program. Before this program, the area had received only a few investments for sporadic actions.

The PAC guidelines required a full urban design project and the COHAB-CT staff developed this. The Upgrading Master Plan envisaged the removal of 40% of the dwellings, which was then justified by the high-density land occupation level, which required the laying of roads to improve the settlement's infrastructure. The removed dwellers would be transferred partly to surrounding areas in new developments composed of two-story buildings. Another group would be relocated to the Santa Cândida neighbourhood, far from the original settlement (Fig. 01). The Plan also included urban infrastructure improvement, public facilities implementation, and recovery of the riverbanks as a permanent preservation area (Table 1). In addition to the funds from the PAC/PPI, the overall scheme would be financed by resources from the Federal Program *Minha Casa Minha Vida* (MCMV) and the Social Housing National Funding (FNHIS), as well as municipal funds (Curitiba; COHAB-CT; IPPUC, 2007).

The upgrading plan proposed a comprehensive intervention in the settlement and was carried out according to a process that recognizes the needs of the population by allowing them to participate in the decision-making. In 2010, it was awarded the "Merit Seal" by the Brazilian Association of COHABs in the category of projects with a regional approach (COHAB-CT).

However, the field works started in 2007, but suffered cuts in 2014 due to downsizing of the Federal Government Programs, and was later halted altogether by the municipal government. Given the lack of resources, many of the planned phases of work went unfinished. The precarious dwellings previously located along the Vila Guaira riverbanks were removed; however, its green areas were not recovered nor was the proposed park implemented.

Although part of the intended plan was implemented, the previous informal condition of the settlement remains. Since some residents generate income from collecting and selling recycling materials, the lack of proper storage led it to being kept inside the

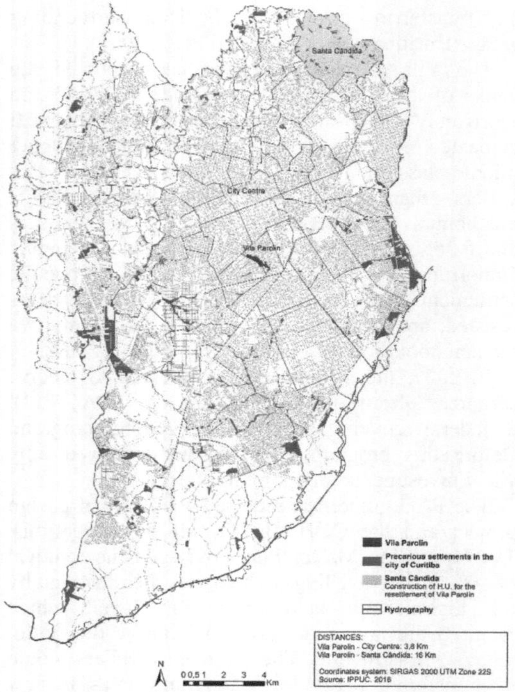

Figure 1. City of Curitiba and Vila Parolin. Produced by the authors (2020) based on IPPUC data (2016).

Table 1. Summary of interventions carried out in Vila Parolin. Produced by author (2020) based on COHAB-CT.

Phase	Upgrading action	Resources
Started in 2006; halted in 2014	To improve the housing conditions; to promote the incorporation of the area into the formal city; to improve urban infrastructure; promote the recovery of the banks of the Guaira river and creation of sports and leisure area around the river banks; land regularisation (COHAB-CT; CURITIBA, 2007; COHAB; CURITIBA, s.d.)	Federal and Municipal
2010	Community in Colours: improvement of precarious housing (COHAB-CT; CURITIBA, s.d)	Municipal (FMHIS and COHAB-CT)
2011	Complementary actions: building of a Social Assistance Reference Centre (CRAS); Childhood Education Municipal Centre (CMEI) and reconstruction of the Health Clinic (COHAB-CT; CURITIBA, s.d.)	Municipal
Started in 2007, halted in 2014 and restarted in 2017.	Federal Program PAC I/PPI –Slum upgrading: 491 new housing units (COHAB-CT; CURITIBA, s.d.)	Federal (PAC I/ PPI)
Started in 2007; halted in 2014	Action (resources of FNHIS): building of 150 housing units to substitute precarious housing (COHAB-CT; CURITIBA, s.d.)	Federal (FNHIS)
Started in 2007; halted in 2014 and restarted in 2017	PMCMV: building of 176 housing units in surrounding new developments.	Federal (PMCMV)
Started in 2011; halted in 2014 and restarted in 2017	PMCMV: building of 49 housing units in Santa Cândida neighbourhood	Federal (PMCMV)

houses and alleyways, fostering social and environmental vulnerability. Although the plan outlined environmental awareness training for residents and construction of three triage warehouses in the surrounding area, the problem was not resolved.

In addition to Vila Parolin, another 20 informal settlements in Curitiba received PAC resources between 2007 and 2014 (Silva, Vasco and Teixeira, 2018). According to Silva, Vasco and Teixeira (2018), these actions focused on developing new housing units in areas far from the intervention areas without engaging in preliminary stages. This is the opposite approach to that taken under the PAC guidelines, which recommend full intervention in these areas and the recovery of the permanent preservation areas. Thus, there have been similarities in the policies implemented in other city settlements with those carried out in Vila Parolin.

When Vila Parolin's upgrading works restarted, the authorities prioritized developing the housing schemes 16 kilometres from the settlement, in the Santa Cândida neighbourhood. Even though it was previously planned and, according to COHAB-CT staff, the households agreed to leave the former dwellings, the fact is that most of works needed to upgrade the area, then addressing a larger number of residents, remained incomplete. Also, the city authorities did not finish the 150 housing units located within the settlement's perimeter to relocate households, and also suspended works to upgrade infrastructure networks in the area.

Interviews with community leaders and some residents carried out in 2015 (SOUZA, 2016) revealed a mix of hope and discontent. The discontinuation of the works was not perceived as an unchangeable fact, although there was great apprehension that the city would not be complete it. Some residents praised the finished works and believed their lives would be changed when the project was complete. However, others contested the results when they complained that "they sold a dream and did not deliver even half of it".

This population has struggled with subsequent eviction attempts in the past decade and those removed always returned to the area, due to its privileged location, a transportation-rich area near job opportunities and surrounded by well-equipped neighbourhoods. Until the start of the upgrading project, the conflict with the middle-upper class neighbourhood gained momentum. The neighbours' representatives complained to the public authorities, demanding they demolish Vila Parolin dwellings instead of regularising them. However, the community protested and manifested their desire to remain, pushing the COHAB-CT to continue the project.

Although the city authorities restarted the works in 2017 under a different government mandate, they prioritized solely transfer of residents to the new housing schemes located far from the settlement and finished in 2018 and 2019. Thus, the core characteristics of urban segregation and social vulnerability were unchanged. The settlement returned to its previous precarious condition, and new dwellings are occupying the riverbanks and at risk from the constant floods. Solid waste remains spread throughout the common areas and the feeling of abandonment affects those who remain.

However, even though there are finished works, the fact is that the authorities failed to conclude the settlement regularisation process required to accomplish the ultimate goal of its integration with the regular city. Thus, having been only partially implemented, the PAC project ended up not substantially changing the residents' reality.

5 CONCLUSIONS

The case of Vila Parolin, as seen in other precarious settlements, reinforces what was discussed by Lacarrieu (2018), demonstrating that the public authorities still prioritise the implementation of new housing schemes over regularising established informal settlements. Analysis shows the city of Curitiba conducted the works for this settlement, in tune with other informal areas benefited by Federal Government resources (Silva, Vasco and Teixeira, 2018). It reveals that Vila Parolin is no exception, but representative of Curitiba's housing and urban policy.

For the community, it was an accomplishment to remain in the location after struggle and resistance. However, according to interviews carried out in 2015, the incomplete regularisation process promotes the perception that the struggle must go on, or worst, that each household must continue their struggle alone, as the State will not grant them better housing conditions.

All the investment made, whether material or human, is affected, thereby impacting the settlement's future. The discontinuation of housing policies stokes frustration among residents. It casts doubt over the public authorities' effectiveness in solving their housing problems, fostering individual determination to change their own reality by themselves.

The solution to social and urban problems is the State's ultimate responsibility, and this assumption prompts the poor to demand more effective public policies. The State must be sensitive to the poor's demands and needs, acting to prevent the establishment of other informal settlements and avoid future social housing schemes that are not fully integrated into the city's urban fabric.

Understanding the processes that lead to intensifying urban segregation and social vulnerability is fundamental to devising effective public policies. Successful policies must guarantee the affected population's involvement and be associated with other public policies on education, labour and income. The State must be committed to avoiding discontinuity in housing policy actions, minimising the adverse outcomes that directly impact the community's everyday life, and to improving precarious settlement conditions. Finally, we highlight the need for integration of the various different public policies, as housing and urban development policies alone cannot change the situation of vulnerable low-income families.

REFERENCES

BOLETIM CASA ROMÁRIO MARTINS. 2006. COHAB-CT: 41 anos de planejamento e realizações. Curitiba: Boletim.

BRULON, V.; PECI, A. 2018. Quando processos de organizar se chocam: hibridismos no espaço social de favelas. *O&S* - Salvador, v. 25, n. 84, p. 068–086.

CALDEIRA, T. P. R. 2017. Peripheral urbanization: Autoconstruction, transversal logics, and politics in cities of the global south. Environment and Planning D: Society and Space, v.35, n.1, p.3–20.

CARDOSO, A. L.; DENALDI, R. 2018. Urbanização de favelas no Brasil: um balanço preliminar do PAC. 1ª edição. Rio de Janeiro: Letra Capital.

CARVALHO, A. de S. 2014. Vivendo às margens: habitação de interesse social e o processo de segregação socioespacial em Curitiba. Dissertação de mestrado. Florianópolis: UFSC.

COHAB-CT; CURITIBA. 2007. Projetos Prioritários de Investimentos – PPI. Intervenções em Favelas/

Saneamentos Integrados. Projeto de Trabalho Técnico Social. Bacia do Rio Belém: Vila Parolin. Curitiba.

COHAB-CT. – (s.d.) Comunidade em Cores. Prefeitura, COHAB e Comunidade Juntas Revitalizando a Vila Torres. – (s.d.) Apresentação fornecida pelos técnicos da COHAB-CT.

COHAB-CT. Companhia de Habitação Popular de Curitiba. http://www.cohabct.com.br/ (consulta: 25/02/2020)

CURITIBA; COHAB-CT; IPPUC. 2007. Plano Municipal de Regularização Fundiária em Áreas de Preservação Permanente. Curitiba: IPPUC.

DAVIS, M. Planeta favela. 2006. Tradução Beatriz Medina. São Paulo: Boitempo.

HARVEY, D. 2003. Acumulação via espoliação. En *D. Harvey. O novo imperialismo* (115-148). São Paulo: Edições Loyola.

KOWARICK, L. 2017. Escritos urbanos. São Paulo: Editora 34.

LACARRIEU, M. 2018. Futuros inciertos y precarios: asentamientos precarizados en la ciudad de Buenos Aires. *Oculum ens. Campinas*. 15(3). 377–398.

LOPES, D. 2002. Casa, despejo e cultura do habitar. *Cadernos CERU*, série 2, no 13.

SILVA, M. N. da. 2012. A dinâmica de produção dos espaços informais de moradia e o processo de metropolização em Curitiba. Tese de doutorado. Curitiba: UFPR.

SILVA, M. N. da; VASCO, K. M. C. M. e TEIXEIRA, A. G. 2018. O arcabouço institucional e normativo da política municipal de habitação e as contradições na execução do PAC em Curitiba. En A. L. Cardoso e R. Denaldi. *Urbanização de favelas no Brasil: um balanço preliminar do PAC*. 1ª edição. Rio de Janeiro: Letra Capital.

SOUZA, C. de S. 2016. Resistência, justiça ambiental e política pública: urbanização de áreas de vulnerabilidade socioambiental consolidadas em Curitiba, PR – Vila Torres e Vila Parolin (1980-2014). Dissertação de mestrado. Curitiba: UFPR.

SOUZA, J. de. 2017. A elite do atraso: da escravidão a lava jato. 1ª edição. São Paulo: Editora Leya Brasil.

STROHER, L. 2017. Reestruturação da metrópole periférica e o impasse da reforma urbana em Curitiba. *EURE*, Santiago, v. 43, n. 128, p. 273–294.

VERAS. M. P. B. 2001. Exclusão social: um problema de 500 anos: notas preliminares. Em B. Sawaia (org.). *As artimanhas da exclusão- análise psicossocial e ética da desigualdade social*. Petrópolis: Vozes. Petrópolis.

Research Tracks in Urbanism: Dynamics, Planning and Design in Contemporary Urban Territories – Allegri et al (eds)
 ISBN 978-1-032-10788-2

Subjective and symbolic disputes in the fight for the right to the city: Actions of the collective Stop Despejos for resistance of the 6 de Maio neighbourhood in Lisbon

D. Bogado
University of Architecture and Urbanism, Universidade do Planalto Central Aparecido dos Santos, Brasília, Brazil
Universidade de Brasília, Brasília, Brazil
Centre for Social Studies, University of Coimbra, Coimbra, Portugal

S-M. Saaristo
Global Development Studies, Department of Political and Economic Studies, Helsinki, Finland
Centre for Social Studies, University of Coimbra, Coimbra, Portugal
Urbaria - Helsinki Institute of Urban and Regional Studies, University of Helsinki, Helsinki, Finland

ABSTRACT: This article draws from action research carried out with the collective Stop Despejos in the city of Lisbon from 2018 to the beginning of 2019. This period marks a time of growth in the real estate market, associated with international visibility of the metropolis. Cities are promoted globally in a picture-postcard manner, in line with the strategies of international financial capitalism, driving a rapid increase in the price of land, housing, and living costs. Stop Despejos was formed during this period, with the purpose of exposing the increasing evictions in the city. The article explores the actions carried out with the collective in the neighbourhood of 6 de Maio, Greater Lisbon, and analyses the strategies that consisted of adapting methods of community mobilisation previously used in the Vila Autódromo neighbourhood of Rio de Janeiro. These strategies focused on the subjective impacts of displacement.

1 SEGREGATION AND EVICTIONS AT THE HEART OF NEOLIBERAL MANAGEMENT

1.1 *Spectacular picture-postcard city*

In recent years, land speculation in Lisbon has produced socio-territorial transformations that cannot be conceived as particular feature of the Portuguese capital. On the contrary, these shifts reflect a behavioural pattern of real-estate on a global scale, which currently involves Lisbon with "speculative effects": cruises, residential renovations, proliferation of hotels and apartments for temporary leasing, scooters and *tuk tuks* - in addition to other transport innovations –, in order to increase the circulation of tourists on the sidewalks. All these elements result from the tourist and real-estate boom in the city.

Such changes form a part of the behavioural dynamics that are considered characteristic of the contemporary capitalist city. The present capitalist base articulates with the speculative financial movement, contributing to the diffusion of ideologies through the media and attaining significant transformations of a symbolic and subjective nature in society, which shape consumption according to business concerns, according to Milton Santos (2013). Consequently, transformations are not restricted to changes in the physical space of the city, but encompass the functioning of social life.

In articulation, financial capital and the media create individual needs founded on the invention of a lifestyle based on constant consumerism. This entire global financial structure is ensured by the violence inherent in the capitalist economic system, which manifests through institutional and police coercion and repression that support the reproduction of consumerist behaviour (Santos, 2013). Hence, Santos (2013) argues that all social life processes are intermediated by consumerism, which penetrates numerous spheres of life. As a result, social behaviour is standardised and spectacularised in merchandise, tourism, housing, and other sectors. Relationships between individuals become competitive and are reduced to appearances (Debord, 1964). Such competitiveness is not restricted to the sphere of individual relationships, but becomes the predominant logic justified by itself in the consumer society (Santos 2013).

DOI: 10.1201/9781003220855-22

The same logic applies to cities, whose management aims primarily at the production of globalised urban spaces, a kind of international picture-postcard to compete with other metropolises. The race for the production of spectacular landscapes thus becomes a goal of urban management: converting the city into an attractive, visible, stunning venue standardised to the taste of the cosmopolitan public becomes the model to be followed by public administrations. As a result, the idea of appropriate urban management, previously understood as one that promotes basic infrastructure and access to the city for all, is replaced by the transformation of spaces into international window-dressed picture-postcards that attract tourists by offering the perfect backgrounds for selfies. The fundamental problem in this transformation is that, as a by-product of these urban operations, violations of the right to the city and the right to housing befall the local population. Indeed, the urban project implemented in the metropolises for the transformation of places into spectacle venues supports land speculation, gentrification, and the replacement of the local population by a volatile public: users of local accommodation with a cosmopolitan and consumerist profile.

It is worth mentioning that such transformations often occur faster when major international events are held. This is what took place in Barcelona before the Olympic Games of 1992, and in Rio de Janeiro before the World Cup of 2014 and the Olympic Games of 2016. The territorial transformations in Rio de Janeiro at the time of the mega-events reflected the new ways of conducting urban policy, adopted in several cities around the world in line with neoliberal paradigms. Pérez de Lama (2002) proposes that neoliberal urbanism has an atmosphere that is favourable to privatisation and the massive entry of global capital into the city, resulting in the proliferation of disintegrations and segregations.

In years leading up to these mega sports events, 20,299 families were evicted in Rio de Janeiro (Azevedo & Faulhaber 2015). The sheer number illustrates the high price to be paid for the policy of "urban entrepreneurship". According to David Harvey (2005), urban entrepreneurship is a characteristic of the current model of urban management that seeks to incorporate the city into capitalist economic dynamics. Harvey argues that this adaptation changes the relationship between economy and power, with the state taking on new roles and neglecting its conventional functions. Mega-events create an environment that is favourable to mega-projects and to the logic of urban entrepreneurship, in record time and in a discretionary manner. Changes in bureaucratic parameters, implementation of exceptional procedures, and enablement of unilateral trade agreements are customary measures adopted during mega-events. However, they are capable of restructuring the territory with the objective of materialising business and elite interests. This situation can be explored using Giorgio Agamben's (2005) idea of the "state of exception", in which an established authority is not overthrown, as occurs in a coup d'état, but the power is exercised in a way that allows the "right to suspend rights" (Agamben 2005: 39).

1.2 *Neoliberal city*

The megaprojects - and also the mega-events - are fundamental components of neoliberal city management. Neoliberalism emerged as a paradigm of political-economic guidance between the 1970s and 1980s, with its implementation in Latin American cities occurring in the from the 90s onwards (Harvey, 2005). The marriage of marketing with urban management, in order to make market interests feasible, to the detriment of collective interests, is a phenomenon not peculiar to Rio de Janeiro. The adaptation of the territory to the fluidity demanded by global capital has had a considerable impact on the geography of many cities. In this scenario, the State concentrates resources primarily to serve business interests, leaving no resources for social investment. In addition, privatisations, state financing of private companies, and sale of share capital are carried out indiscriminately (Santos 2013).

Therefore, the situation is characterized by an inversion in which the business sector becomes the author of the political process and the public administration converts its activities into marketing, since negotiations to meet market demands occur in the political sphere, and the State exempts itself from the function of social regulation to conclude large contracts associated with urban projects.

Although the repercussions of mega-events are disastrous - when analysed from the perspective of the loss of social rights, since the burden falls on the population in the socio-cultural fringes - mega-events are not the only factor responsible for the processes of gentrification and displacement of communities in cities. When cities are advertised as global tourist destinations, great socio-spatial transformations are also promoted. Indeed, this is what took place in Portugal, elected as the best tourist destination in the world by Veja magazine in 2017: "Country is the first European country to receive the top prize at the World Travel Awards" (Veja online magazine 2017).

The territories affected by urban entrepreneurship are gradually deformed: affective, socio-cultural and identity references are dismantled, together with the destruction of houses and neighbourhoods in the processes of urban reform. The wiping out of places and disappearance of spatial structures caused by gentrification, and the removal of buildings and neighbourhoods, contribute to social disintegration, territorial expropriation and the erasing of local history (Garnier 2014).

In a study on urban transformations that occurred in the port area of Marseille, Jean-Pierre Garnier (2014) identified the unfolding of a "global process of de-civilisation". However, population dispersion is not a characteristic of Marseille, but typical of the processes triggered by the execution of major urban projects in the metropolises. The appreciation of the land and the consequent population replacement causes a rupture between territory and memory, and consequently, a loss of local identity, a phenomenon observed in Marseille, as well as in Vila Autódromo of Rio de Janeiro, the 6 de Maio of Greater Lisbon, and likewise in many other places.

2 OBJECTIVE AND SUBJECTIVE RESISTANCE

2.1 *Erasing of communities*

Over the years, studies of displacement, eviction and gentrification processes triggered by urban reforms have led to an understanding of the strategies used by public administrations to clear popular neighbourhoods. Alex Magalhães (2013) shows that cutting off public electricity and water supply, as well as use of personal threats, lies and seeding doubt with residents make up the repertoire of "biopolitical eviction strategies", explaining that these tactics penetrate various aspects of life, in addition to legal issues and house negotiation policies. The forms of action of municipal administrations involve resources that transcend the legal and economic spheres of negotiation, permeating aspects of life, and can thus be defined as "biopolitical" tactics according to the meaning used by Foucault (1999, 2006). The biopolitical actions of the State have the purpose of controlling the marginalized population (Zibechi 2011: 16). Furthermore, Mendiola Gonzalo (2016) elaborates on the biopolitical ontology of habitability, which is based on the relationship between inhabitant and inhabited place, as a way of being in the world that is projected through the production of subjectivities (Mendiola Gonzalo 2017: 226). This conception gains further depth when linked with Judith Butler's (2006) notion on the production of the uninhabitable, which occurs to the extent that life is careless, and of precariousness, which is exacerbated under circumstances of violence.

Among the countless consequences of forced population displacement, we observe an erasing of local history and community memory, severing of social ties, and psychological traumas that mark families who go through episodes of state violence. The erasing of a community has subjective and symbolic impact. In the case of Vila Autódromo, a community located in the West zone of Rio de Janeiro, the eviction of the community was associated with the construction of the Olympic Park, the largest sports structure to be built for the Rio 2016 Olympic Games. The origin of Vila Autódromo dates back to the 1960s, when fishermen settled on the Itapeba Peninsula, on the banks of the Jacarepaguá Lagoon, in the West zone of the city of Rio de Janeiro. The occupation began with the construction of houses and small shops. Most of the residents earned their living by fishing in the Jacarepaguá lagoon (Bogado 2017).

At the time of announcing the construction of the Olympic Park, the City authorities of Rio de Janeiro not only intended to demolish and remove Vila Autódromo, as it had done with the other *favelas* at the time (O Globo online 2010); it sought to disseminate the narrative that the Rio 2016 Olympic Games would leave a legacy for the city. Nonetheless, the residents saw the issue in a different light: according to Sandra Maria, "The Olympics brought with it an eviction process that is much bigger than the ones we had seen here before. The Olympics gave the government a reason for the eviction that society accepts, and additionally, they [the government] now also had the capital they did not have before the Olympics. So, it was then that the demolition of Vila Autódromo indeed happened. There was a greater need for people to take the lead in the struggle. It was that need that made me embrace the cause of the fight against eviction." (Sandra Maria, interview in August 2016. See also Bogado 2017).

2.2 *Eviction Museum*

The supposed legacy of the Olympics is thus not a narrative shared by all *Cariocas* (residents of the city of Rio de Janeiro). The Eviction Museum, a community museum built in Vila Autódromo, promotes a counter-narrative about the Olympics, challenging the state power and refusing to keep quiet in the face of the endless violence suffered by favela residents in the eviction processes preceding the sports events, undertaken by the city council. In Vila Autódromo, only 20 families remain of the 700 families that originally lived there. Out of this process of resistance, loss, violence, revolt, but also community unity, the Eviction's Museum was born. The museum is the result of the resistance struggle of Vila Autódromo and an extension project carried out by the Faculty of Architecture and Urbanism of the Anhanguera University, headed by Professor Diana Bogado, who took a stand in support of the community and against the violations of the right to housing and of human rights that the city council was perpetrating. The museum is the result of a process of struggle against the commodification of the city.

The assumed political position in relation to the municipal urban policy and the action research were central features of the extension project that gave rise to the Eviction Museum. The actors involved emphasise that the Universities of Architecture and Urbanism should not exempt themselves from criticizing state actions in the field of urban planning, especially within the framework of neoliberal or "market-oriented" urbanism (Mascarenhas 2016), due which the right to the city fails to be secured.

The community, already self-organized, received support from the university, developing several territorial intervention projects. The experience of social museology aimed at rescuing and communicating the history of the community and the counter narrative about the Olympics. The construction of the Eviction Museum included "Memory workshops". The Memory workshop is a strategy to achieve a dialogue with the community. This enabled students to have contact with the history of the residents. Finally, the students designed and built the sculptures that marked the location of the demolished houses. The sculptures, built from the rubble of the destroyed homes, mark the route to be followed by the visitors.

On the Museum´s inauguration, seven sculptures were presented that led the public through the first exhibition of the Eviction Museum in the neighbourhood now reduced to rubble. "All Saints' Village" was a sculpture paying tribute to the spiritualist centre of African origin situated in the community. "The Many Faces of Penha" was a sculpture in tribute to Dona Maria da Penha Macena's home and "The Holder of Evils" a sculpture in tribute to Dona Jane Nascimento. "The Light that Does Not Go Out" was the only sculpture that relates to one of the two buildings that have not been removed: the São José Operário Church. The church had welcomed residents whose homes were demolished, and it was thus decided to honour the church because of its importance to the community. "The Sweet Childhood" was a sculpture in tribute to the local playground; "The Ocupa Space and Dona Conceição's Home", a sculpture that referred to the festivals of Ocupa Vila Autódromo; and lastly, "I am the Association" was a tribute to the Residents' Association of Vila Autódromo, relating to the campaign organised by the community when the building was demolished.

The Memory workshops were designed collectively by the team of community supporters together with the residents. These were devised for the creation of the Eviction Museum, but when carried out, their effects acquired proportions that were more significant than their pedagogical purpose or the communication of the local history to students for creating the sculptures. The workshops revived stories, reconnected residents, dampened conflicts, reconstructed narratives about the violent actions of the government, comforted anguished families, and communicated the violence endured by Vila Autódromo and the other favelas during the preparations of Rio de Janeiro for the Olympics.

Hence, the process of building the Eviction Museum was able to touch the subjective encroachment of the ongoing violent eviction operation. The dynamics in the neighbourhood were clearly in direct contradiction with the project for society that had been designed by the hegemonic actors and institutions. The solidarity and sense of community and cooperation that exists in the poor suburbs breaks with the notion of individualism, fear and competition that underlies the capitalist consumerist society. Resistance to eviction, in turn, challenges public administration plans for the community. Such resistance, creatively developed during oral workshops that stimulated coexistence and persistence in the struggle, in the face of attacks and duress from the public authorities involving Vila Autódromo, placed the Eviction Museum on a solid footing based on bonds, while galvanizing other communities to fight.

One of the main objectives of the Museum was to contribute to the community's permanence in the territory. This objective was achieved to some degree, promoting further advancement and globalisation of the struggle of the Eviction Museum, exceeding the material sphere and reaching the domain of subjectivity, symbolism and memory, going beyond the city ´s borders. The Eviction Museum acquired an unexpected local, national and international visibility within the spaces of struggle for upholding urban social rights. In the context of speculation and gentrification in Lisbon, the experience of the Eviction Museum resulted in the introduction of a new approach to the dynamics and activities adopted by a Portuguese social movement. The Memory workshops reapplied in Lisbon highlighted the perspective of the immaterial impact caused by the evictions and the importance of seeking creative forms of struggle, capable of accessing subjective consequences in the wake of mobilisation and popular articulation activities.

3 POPULAR MEMORY AS AN INSTRUMENT IN THE STRUGGLE FOR DECENT HOUSING

After the construction of the Eviction Museum in Rio de Janeiro and securing the right of the Vila Autódromo community to stay put, the residents, along with the other members of the Museum, remained engaged in continual political and cultural activity. The purpose was to disseminate information about the struggle of Vila Autódromo and to denounce the state violence in several communities that have suffered from, or were threatened with, eviction in Rio de Janeiro, Brazil and the world. During these constant activities, the Eviction Muse, with the goal of articulating the fight for the right to housing, the city and to memory in a network, is in line with the mission of the collective Stop Despejos in Lisbon. Stop Despejos is a collective formed at the end of 2017 with the objective of exposing the countless evictions that had been taking place in the city due to the international promotion of the Portuguese capital as a "trendy city", as described earlier. The Stop Despejos collective is a group derived from the Association Habita in Lisbon, since the collective was formed with the idea of working together with Habita. Habita works in many domains in the area of right to housing and to the city, providing support to families at risk of eviction, backing their mobilisation and helping them present their cases to the City Council. Although their actions are articulated, the two collectives are independent (Bogado 2019). Together with Stop Despejos and Habita, we carried out various activities during 2018 and early 2019, to support communities in their struggle for the right to housing, establishing direct and indirect dialogues between residents of Lisbon neighbourhoods and Vila Autódromo.

Workshops were held with Stop Despejos in the 6 de Maio neighbourhood, Amadora, Greater Lisbon. The neighbourhood is situated in Damaia, bounded by the roads Rua Militar, Rua das Fontainhas and Rua Francisco Simões Carneiro. The first houses of the neighbourhood were wooden, but were later replaced by brick structures. The residents of 6 de Maio are predominantly migrants from Cape Verde who came in search of work in civil construction in the 1970s, as well as their children and grandchildren. The name of the neighbourhood derives from the first meeting of residents that took place on 6 May 1976 (Bogado 2019). The activities run at the 6 de Maio were aimed at prompting social memory as a means for supporting resistance during the ongoing eviction and eviction process (more in Bogado 2019). During the action research, the methodology of research-action-participation (IAP) was employed. The IAP methods use participation from a methodological perspective to outline actions that form part of the study and allude to practice and life. They are particularly important as tools to develop more appropriate models of intervention and of housing management, especially in the contexts of social crisis that call for new spatial intervention strategies (Medina 2010). The IAP methodology prioritises praxis in promoting strategies of biopolitical resistance, which are defined here as resistance strategies which articulate popular resistance with subjective aspects.

During the process of fighting against the demolitions in Vila Autódromo in the previous years, it was concluded that not only did the State's actions have deeper implications than the loss of homes, of a symbolic and subjective nature, but that there is a *modus operandi* of eviction involving a repertoire of violent and diversified actions to demobilize popular resistance. In this sense, Vila Autódromo's creative and diversified form of struggle demonstrated its capacity to act in subtle spheres that mitigated the destructive impacts of ongoing evictions. Strategies to encourage dialogue about history, memory, commonality, identity, and feelings stimulated residents to continue to struggle and support each other.

During the study, we observed that Amadora City Council was carrying out the eviction and destruction of 6 de Maio using the same strategies employed by the Municipality of Rio de Janeiro to destroy Vila Autódromo. Furthermore, it was possible to conclude that the organisation of Memory Workshops in 6 de Maio also had an effect on alleviating trauma, reversing impacts and mobilising the community.

"By not informing the day of eviction, including only some of the residents for resettlement, negotiating individually with each family, making threats asserting that "either the residents accept the situation or they will not receive anything", were tactics used both by the City Council of Rio de Janeiro during the Olympics evictions, as well as by the Amadora City Council to demobilise the social struggle in the execution of demolitions and evictions in 2018 and 2019. Such practices discourage residents, lead to personal disagreements and compromise collective resistance. Due to the use of these biopolitical strategies by the State, it becomes necessary to work on resistance in the spheres of life: community bonds, identity and social memory" (Bogado 2019).

At the time, the neighbourhood resembled a war zone, with a few isolated buildings still standing. It was going through a process of obliteration caused by the destruction of its territory and the residents were drained. We decided to organise the workshops as a mobilisation strategy. Several memory workshops were thus held in 6 de Maio, with one occurring during an occupation that residents and activists carried out at the Lisbon Housing and Urban Rehabilitation Institute, *IHRU*, in September 2018, to demand that all residents threatened by evictions would be added to the IHRU list to receive social housing, given that only a few residents had been included.

During the "Hands Off Action Camp" the annual meeting of the European Action Coalition for the Right to Housing and to the City, held in Lisbon in 2018, we organized a visit to the 6 de Maio, with the residents hosting the activists. A few weeks later, we held another workshop in the neighbourhood, in which we addressed the theme of "women in struggle", using the Amnesty International's "Write for Rights" Campaign materials.

"The Eviction Museum, together with the collective Stop Despejos and Habita, started running strategic activities in the 6 de Maio community. The "Memory Workshops", held regularly in the community, are an adaptation of the processes carried out during the Vila Autódromo resistance struggle. We are making use of the material made available by the Amnesty International Write for Rights Campaign, and addressing issues such as democracy, participation, bonds, women in struggle, and love. Because it is love that weaves the entire web of resistance and resilience in communities affected by evictions." (Text prepared by the activists).

The text above refers to the dissemination of the Memory Workshop held in 6 de Maio, linked to the visibility of women's leadership in the popular struggle. We began the workshop by sharing the story of Marielle Franco, murdered on March 14 that year. We told the story of Marielle who was born in the favela of Maré and dedicated her life to fighting for

Figure 1. A Memory Workshop in 6 de Maio, coordinated by Diana Bogado and Saila-Maria Saaristo together with the collective Stop Despejos. A community resident is holding the photograph of Marielle Franco, a City Councillor of Rio de Janeiro, who was assassinated the same year. Photo: Diana Bogado.

the rights of favela residents. The aim was to draw attention to the fact that the violation of basic rights is recurrent in different parts of the world, as well as to argue that community union is the only path to resistance (Bogado 2019).

Other workshops were held in 6 de Maio together with Stop Despejos and Habita (Figure 1). It is not possible to prove that the work of subjectivities was decisive in promoting greater mobilisation of the neighbourhood. However, as participants in the process, we observed and talked with the residents, leading us to conclude that the organisation of the workshops had spawned other resistance activities, through the community´s own initiative, demonstrating that, in some way, the workshops encouraged residents to remain united and continue the struggle for their homes. The 6 de Maio continues to fight against evictions, while Amadora city carries on with the accelerated demolition process.

4 DISCUSSION

The objective of the action research discussed in this article was the interrelation of urban planning in a neoliberal framework with the mobilisation of social movements for the right to housing and the city. The central theme of analysis drew on the impacts of contemporary territorial transformations on the non-material subjective realms of low-income communities - made invisible and torn apart by the public authorities in the context of neoliberal urban management - as well as their reflections on the history and memory of the city.

We also addressed the particularities of gentrification and eviction processes, which in the context of urban changes, are associated with a specific eviction methodology used by municipal administrations in different contexts and countries. Different states reproduce similar violent and basic procedures, as was identified in the field research in Rio de Janeiro, 2015 and 2016, and in Lisbon, 2018 and 2019.

The similarity between the so-called biopolitical eviction strategies (Magalhães 2013) in both cities studied justified the replication of the resistance strategies used in Vila Autódromo (Rio de Janeiro) in 2015 and 2016, in 6 de Maio (Amadora) in 2018 and 2019. The Memory Workshops were oriented towards working on subjective aspects of memory and social history. They were adapted to the Portuguese neighbourhood to restore ties, revert trauma, and regenerate lives that had met with the institutional machine of city destruction. The methodology adopted was Research-Action-Participation (IAP), due to its focus on praxis in the promotion of biopolitical resistance strategies. The "Memory workshops", in turn, touch upon these same subtle spheres, thus acquiring the potential to regenerate memory, as well as psychological and community bonds and, as such, can be considered biopolitical resistance strategies.

However, the novelty brought to the fore in this research is the finding that first, the actions of the State are strategically articulated to interfere in the symbolisms and social psychology of the community, exhausting mobilisation possibilities; and second, that the movement for the right to housing must, in response, act consciously to reverse these subjective impacts. This is made possible through the promotion of dynamics that give communities a voice and further the dissemination of popular narrative. The constant encouragement to continue the fight, as well as the actions of everyday life, make it possible to rebuild lives after episodes of institutional violence, as pointed out by Veena Das (1999). It is this stimulus based on affection, bonds and ordinary life that we propose as a fundamental ingredient in social struggles.

REFERENCES

Aguilara, F. e Naredo, J. 2009. *Interés y contexto del tema tratado. Economía, poder y megaproyectos*. Madrid: Cromoimagen S.L.

Azevedo, L. e Faulhaber, L. 2015. *SMH 2016: Remoções no Rio de Janeiro Olímpico*. Rio de Janeiro: Editorial Mórula.

Bogado, D. 2017. *O Museu das Remoções da Vila Autódromo: Potência de resistência criativa e afetiva como resposta sociocultural ao Rio de Janeiro dos megaeventos*. Sevilha: Universidade Sevilha.

Bogado, D. 2019. O uso da dimensão social da memória como instrumento emancipatório em comunidades em situação de vulnerabilidade sociocultural. *Caderno sociomuseologia* 58(14): 61–106.

Butler, J. 2006. *Vida precária: el poder del duelo y la violência*. Buenos Aires: Paidós.

Das, V. 1999. Fronteiras, violência e o trabalho do tempo: alguns temas wittgensteinianos. *Revista Brasileira de Ciências Sociais*, 14(40): 31–42.

Debord, G. 1969. *A sociedade do espetáculo*. Rio de Janeiro: Ed. Contraponto.
Foucault, M. 1999. *Vigiar e punir: nascimento da prisão*. Petrópolis: Vozes.
Foucault, M. 2006. *A hermenêutica do sujeito. Curso dado no Collège de France (1981-1982)*. São Paulo: Martins Fontes.
Garnier, J.P 2014. Marsella 2013: el urbanismo como arma de destrucción masiva. *GeocritiQ* 24 [without page numbers].
Harvey, D. 1985. *The Urbanization of capital*. Oxford: Basil Blackwell.
Harvey, D. 2005. A *produção capitalista do espaço*. São Paulo: Annablume.
Lefebvre, H. 1974. *Le droit à la ville (suivi de) Espace et Politique*. Paris: Editions Anthropos.
Magalhães, A. 2013. *Transformações no "problema favela" e reatualização da "remoção" no Rio de Janeiro*. Doctoral thesis. Instituto de Estudos Sociais e Políticos: Universidade do Estado do Rio de Janeiro.
Mascarenhas, G. 2016. Rio de Janeiro 2016: a cidade em movimento. *Revista USP* 108: 49–56.
Mendiola Gonzalo, I. 2016. El dispositivo de la captura: espacios y cuerpos bajo el signo de la excepcionalidade. *Athenea Digital*, 16(1): 83–111.
Mendiola Gonzalo, I. 2017. De la biopolítica a la necropolítica: la vida expuesta a la muerte. *Eikasia: Revista de Filosofia* 75: 219–248.
Medina, M. 2010. Metodologías participativas para la gestión social del hábitat. *Hábitat y sociedad* 1: 83–103.
Pérez de Lama, J. 2002. Ciudades y globalización capitalista durante la IV Guerra Mundial. Conference presentation at the Social Forum of Sevilla.
Rolnik, R. 2016. *A Guerra dos lugares: A colonização da terra e da moradia na era das finanças*. São Paulo: Ed. Boitempo.
Santos, M. 2013. *Por uma outra globalização: do pensamento único à consciência universal*. Rio de Janeiro: Record.
Tripp, D. 2005. Pesquisa-ação: uma introdução metodológica. *Educação e Pesquisa* 31(3): 443–466.
Zibechi, R. 2011. *Territorios en resistencia. Cartografía política de las periferias urbanas latinoamericanas*. Málaga: Editora Zambra.

Research Tracks in Urbanism: Dynamics, Planning and Design in Contemporary Urban Territories – Allegri et al (eds)

Live centre Active centre - a practical and proposal-based experience for the people of Suzano

E.J. Vieira, C.A.G. Gallego, R.S.H. Lú & S.Y. Takahashi
Suzano City Hall, Brazil

ABSTRACT: The aim of this study is to present the investigative path that guided the proposal for redeveloping Suzano city center streets, whose intense and conflicting flow prioritized cars over people. Thus, how can an intervention take place in consolidated areas given the present situation? And what is the population's perception of the presence of cars and the impact of removal of parking spaces? Based on these issues, and others, the assessment of the urban space and resultant data provided support for devising a proposal for redeveloping and redesigning the network of central streets to balance the presence of cars and people, the protagonists of public space. The use of tools such as the Active Design: Shaping the Sidewalk Experience, Walkability Index (iCam) and Active Design Guidelines has shown it is possible to intervene in consolidated areas and provide an active territory where spaces are just, pleasant and safe for all users.

Keywords: walkability, urban redevelopment, city for the people

1 INTRODUCTION

The benefits of walking are clear, and the long-recognized importance of staying active means that cities should be designed to promote easy, pleasant and safe walking.

Since the late 1980s, the urban debate has guided issues of active mobility and comfort in walking, which has a direct relationship with urban environment accessibility and the ease of travelling around the city. However, the importance given to cars in the city streets persists, impacting the urban space, the environment and health of the local population. Therefore, the concept of walkability addressed in this study proposes that the city design is focused on pedestrians and the ease of walking the streets. It has been suggested that the urban design provides a sense of well-being and security, whereby walking can be a pleasant activity and the streets a place for meeting as well as transit.

According to the available data from the General Report by the National Public Transport Association – Simob/ANTP (2007), 28 billion people travel by bicycle or on foot, while 19.6 billion travel by individual means of transport – cars and motorbikes, and 18.5 billion people by public transport. This indicates that the majority of the population does not drive daily and that most trips occur actively (by bicycle or on foot).

There is a need to rethink the urban dynamic of the cities, transforming the "walking citizen" or urban space protagonist who, according to Jeff Speck (2016), is directly linked to the vitality of urban centres, because the streets are not spaces exclusively for cars, but also convey non-motorised vehicles, such as bicycles and public transport (SPECK, 2016). In addition, the pavement is an integral part of the street, and should allow people to walk around the city safely and comfortably.

In the United States, New York City has undergone a real transformation with the use of the "Plazas" project which closed numerous streets to vehicles. This move enabled pedestrians to occupy these spaces, with significant gains in return of public space, economic gains for commerce and services, and a reduction in traffic accidents in those areas covered by the scheme. Examples in Brazil, such as pedestrianised pavements in Curitiba, São Paulo, Porto Alegre, among others, are important models that can be replicated in other cities, promoting more vitality and an active and safe environment, especially in city centres.

2 GENERAL LAYOUT OF SUZANO AND WALKING

Urban designs stems from a need for mobility and space occupation and can only be a product of social relationships. These designs may possess characteristics that interfere with the quality of life of residents or those simply travelling through these

DOI: 10.1201/9781003220855-23

spaces. Similarly, the relative width of the streets and pavements can indicate priorities of urban space use.

Suzano is an exception to this logic. The city center is the focus of this urban study and dates from the city's initial occupation. It is made up of a network of central streets designed like a chessboard. This layout, allied with its location in Greater São Paulo and easy access to the railway line, transformed the city center into an important trading hub. Consequently, the central area sees heavy traffic of motor vehicles and pedestrians, creating a conflicting flow that prioritizes cars over people.

Many problems are evident and the proposal to mitigate this situation entails creating a logical design on a human scale within which the pedestrian is the protagonist. This design is aimed at providing pedestrian routes that connect up with other key modes of transportation (train and bus), in addition to promoting safe, welcoming and inviting pavements. Cars shall travel on a well-marked signposted road at safe speed, so that the street can become the stage for the various actors that make up the city.

2.1 *Intervention area*

Located between São Paulo City, the Paraíba Valley and Baixada Santista, Suzano is situated 45km east of the capital, with connecting links to Guarulhos International Airport (GRU-AEROPORT) and Santos Port. Suzano is one of the 39 cities comprising Greater São Paulo.

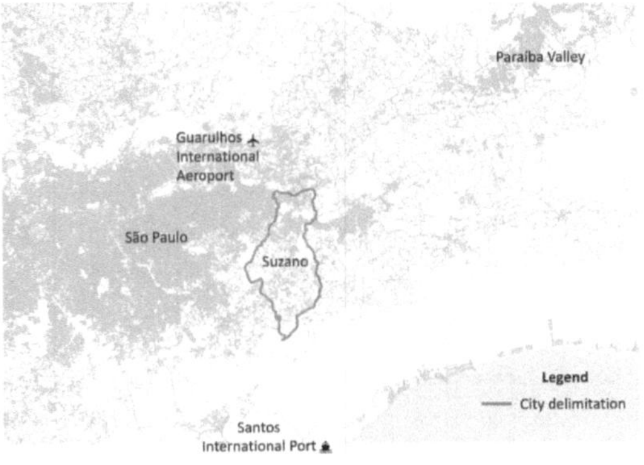

Figure 1. Location of Suzano/ SP/ BR.
Source: Map produced by author from Google Earth satellite image (2019)

The main mobility links in the Suzano region are major roads, including the Mario Covas Jr. Ring Road (east side), Ayrton Senna da Silva motorway (SP21) and Presidente Dutra motorway providing access to GRU-AEROPORT and Paraíba Valley, and Índio Tibiriçá motorway (SP 31) providing access to Santos, where Brazil's main port is located. This places Suzano in a strategic location due to its proximity to Santos and São Sebastião ports, Congonhas and Cumbica airports, as well as the economically strong cities and regions, such as Paraíba Valley, Grande ABC, São Paulo Capita and Guarulhos. The two railways (for passenger transport – CPTM connects to São Paulo city, and for cargo transport – MRS connects the region to Santos Port) contribute toward reinforcing the economically important role that Suzano has been playing for the development of the Alto-Tietê Cabeceiras. This region also becomes a passage and flow route of local, industrial and farm production between the coast, Greater São Paulo and the interior of the state.

The project involves the central area of the city, which supports predominantly commercial and service activities, attracting a large number of people from the city and surrounding area, thus generating a large flow of pedestrians and demand for transport.

The focus of the intervention is two streets of great importance in the city – Benjamin Constant Street and General Francisco Glicério Street. The first street carries heavy bus traffic and has intense private vehicle flow that competes for space with public transport; the second, receives a large flow of people due to the large commercial activity and access to rail transport in the area.

2.2 *Understanding urban space*

Freedom of movement is a constitutional right enshrined in section 05 of the Federal Constitution, which provides for freedom of movement in national territory, entitling all individuals by law, to leave and return freely.

In Brazil, public transport and non-motorized vehicles (travel by bicycle or on foot) were never a priority in promoting urban mobility, nor was the use of private vehicles restricted (OLIVEIRA et al, 2011). In recent years, however, issues related to transit within cities have become increasingly relevant, while growth in urban sprawl has been accompanied by disputes over the use of space on roads and streets, which are insufficient to accommodate everyone. This is a result of the urban mobility model adopted, which is mirrored on the American model – *transport on wheels* – which favoured the use of private cars over public transport or non-motorized transport. Consequently, the resultant excess of cars causes pedestrians discomfort and negatively impacts the urban environment. It thus follows that the design of the current urban space should consider people's accessibility and ease of travel within the city.

According to Jan Gehl (2013), areas for pedestrians and cyclists in the cities should be perceived as part of integrated urban policy in order to enable the creation of living, safe, sustainable and healthy cities. Therefore, it is necessary to observe people's needs in the use of public space, to stimulate the use of non-motorized transport and promote improvement in transit and environmental quality of the city.

In this regard, with the aim of understanding the space and carrying out an urban study to transform the central area into a walkable and pleasant space for people, a strategy was used based on the concept Active Design[1], developed by New York City that has led to interventions in Times Square, among others. This methodology has proven a powerful tool in developing plans and projects in many cities around the world and has helped toward establishing the general objectives of interventions and to understand the role of public space in supporting people's transit and leisure.

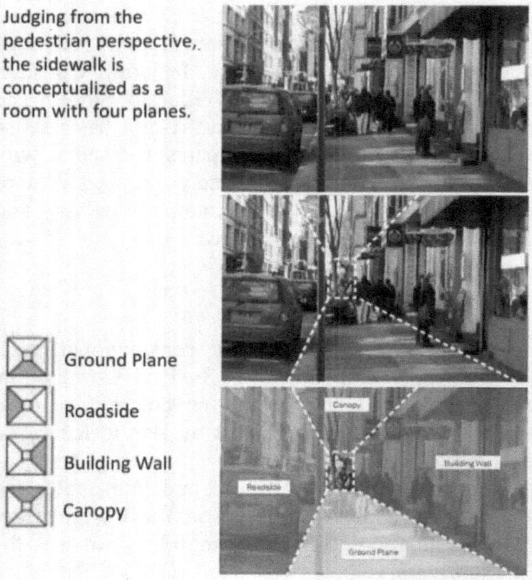

Figure 2. Pavement Analysis Methodology on four planes (ground plane, roadside, canopy, and building wall).

Source: Active Design

The method developed by the local Transport and Development Policies Institute (ITDP) was also applied, the Walkability Index (iCam) which measures the characteristics of the urban environment determined for pedestrian transit, producing recommendations based on the results obtained in the evaluation. Thus, this process enabled devising of a proposal for safe and comfortable transit to promote public space use, leisure and appropriation, yielding a concept which can be replicated elsewhere in the city, providing a more walkable city that is less dependent on individual vehicular transport.

In the study conducted in Suzano, both methodologies were applied and compared to verify whether results were similar. The results, in turn, were also compared against studies conducted for the formulation of Suzano´s Mobility Plan whose subsequent analysis guided the development of the proposal presented to the Ministry of Development in the Cities Progress Program – Urban Mobility.

For the development of the proposal to redesign Central Suzano, a timetable of actions was established that would allow analysis of the urban space and provide an understanding of the population's expectations, in addition to the possibilities of improving quality, urban security and walkability in the city centre. As a result, a 26-block cutout was defined in the central area for the data field survey, measurements and urban space analysis. Within this cutout, 06 points along General Francisco Glicério Street and 03 points along Benjamin Constant Street were marked for application of the methodology abased on the subject research and flow measurements, and urban space observation drawings. Two (02) of these points were located on João Pessoa and Expedicinários Squares.

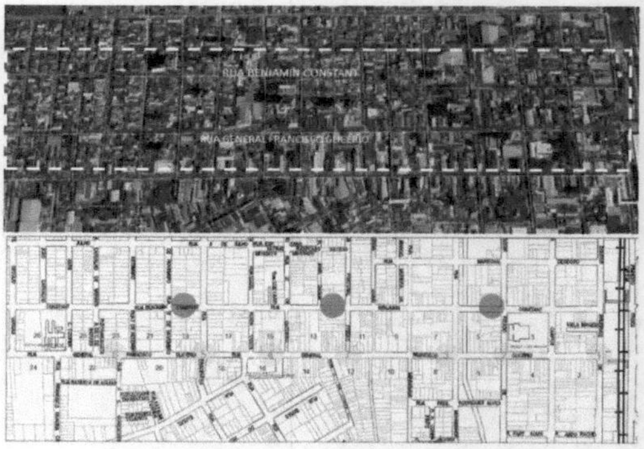

Figure 3. Demarcation of the urban intervention area in Suzano/ SP.

Source: Produced by author from Google Earth satellite image (2019) and Municipal Cartography Base.

The on-site survey and urban recognition were based on the field visit and application of

1 The *Active Design Concept* is based on how cities and locations are planned and produced, where places with easy pedestrian access, comfortable stairways and bicycle park area foster healthier habits among regular users.

methodologies with the participation of twenty students from the Braz Cubas University Architecture and Urbanism Faculty overseen by the Suzano City Hall technical team.

Figure 4. Perception drawings and measurements of urban space conducted in partnership with FAU-UBC students based on the applied methodology.
Source: Suzano City Hall (2019)

For urban space recognition, the *Active Design: Shaping the Sidewalk Experience* tool was used from the perception of the four planes of each block of General Francisco Glicério Street. Measurements of the pavements were then taken in order to determine the dimensions of the public urban space.

Figure 5. Drawings produced from the observation plane.
Source: Suzano City Hall (2019)

This assessment produced graphs and materials that led to the development of the proposal for redeveloping the urban space in central Suzano.

The people flow count from General Francisco Glicério Street was performed on alternate days and times over a 2-week period, while the sample on Benjamin Constant Street involved| 60 people interviewed on a single day, mainly shopkeepers and local business workers. The count was performed every 10 minutes broken down into travel modality (on foot, by bicycle, car, bus, motorbike, etc.). Thus, an estimate of the number of pedestrians, bicyclists, motorbikes, cars, lorries and buses that travelled in pre-established points every hour was determined.

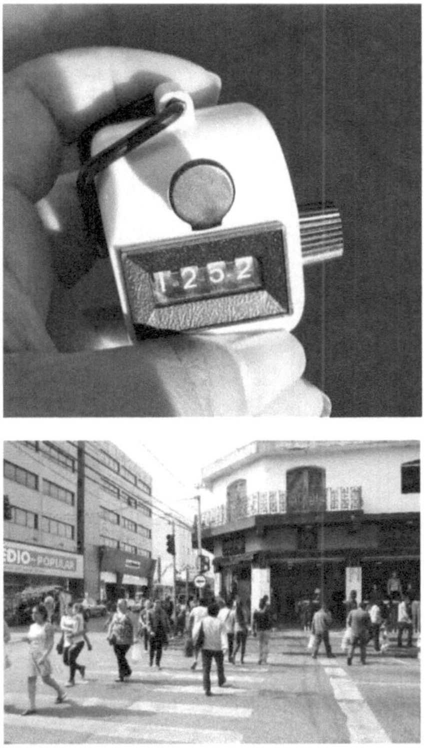

Figure 6. People flow count on section 03 of General Francisco Glicério Street which at 5pm reached 1,252 people.
Source: Suzano City Hall (2019)

The graph produced from the count performed on General Francisco Glicério Street, in general, presented the following results:

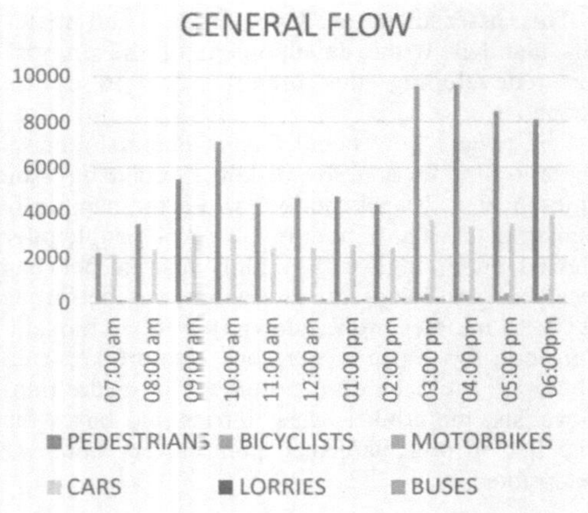

Graphic 1. General People and Transport Flow on General Francisco Glicério Street.

Source: Produced by Suzano City Hall (2019)

The results of the sampling conducted on Benjamin Constant Street are depicted below

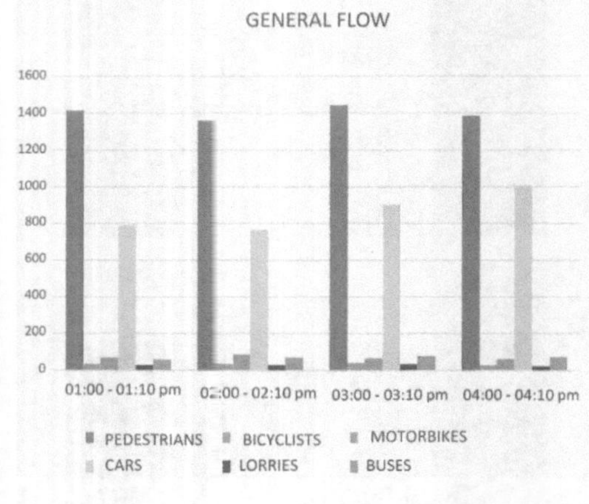

Graphic 2. General People and Transport Flow on Benjamin Constant Street.

Source: Produced by Suzano City Hall (2019)

Although individual graphs were produced, the overall count was also verified, as evidenced in Graphs 01 and 02.

The results showed a similarity in the evaluation of the street relative to General Francisco Glicério Street, where the worst evaluations were for accessibility, pavement maintenance, landscaping, lighting, and safety. However, it is noteworthy that the street had lower pedestrian flow and a greater amount of public transport.

Nevertheless, at point 01 the number of cars exceeded the number of pedestrians from 3pm onwards, And likewise at points 05 and 06 from 5pm onwards. Motorbikes were observed mainly between 9am and 3pm. and at point 01 between 3pm and 6pm.

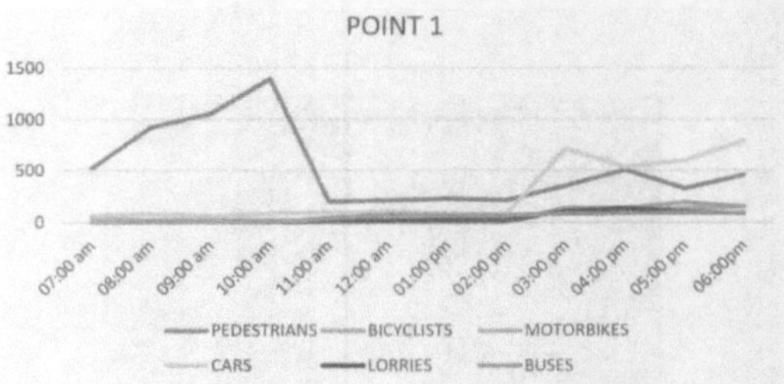

Graphic 3. People and Transport Flow on General Francisco Glicério Street at Point 1.

Source: Prepared by Suzano City Hall (2019)

Despite the results cited above, for the questions surveying proposed solutions, most answers were against road closure to cars, yet favoured improving walkability.

This contradiction between a desire to improve walking conditions and car dependence is latent, despite considering that the number of pedestrians travelling in the central area exceeds the number of individual motor vehicles.

2.2.1 *Pavement evaluation criteria*

In order to measure the urban space quality in central Suzano, the questions and criteria for answers adopted in the research and field survey drew on the work developed by GEHL Architects, 2003, Active Design Guidelines, NYC and Active City, which are based on a score obtained on existing elements in place. The topics covered were: Connectivity, Accessibility, Safety, Diversity, Scale, and Sustainability.

According to the criteria indicated by the methodology, it was found that the existing situation in terms of connectivity was rated regular-to-good, given the presence of linear continuous pavements, traffic signals at crossroads, few driveway ramps for cars, proximity to public transport and squares within a 500m radius. In relation to accessibility, this parameter was rated bad-to-terrible, owing to lack of tactile or audible signals, road humps, obstacle-free road surfaces or continuous free crossings. With regard to safety, the evaluation rated this parameter as regular-to-good, owing to public lighting, commercial use on the ground floor, a large number of people on pavements and windows on pavements with multiple entrances. Diversity was rated as bad-to-terrible, owing to the absence of urban furniture, inadequate width and limited variety of uses at the ground floor. The parameter pedestrian scale was rated as good given the multiple entrances, compatible ground floor heights, façades using various materials, windows facing the pavement and absence of garages in the front setback. Sustainability was rated as bad because of lack of afforestation, flowerbeds, gardens, shade, and inefficient drainage.

The results for students' perceived experiences proved consistent with the measurements obtained using the criteria. However, students tended to rate more negatively due to a harsher critique; explained by the architecture students′ quantitative plus qualitative analyses, where experience and local appropriation influenced the views expressed.

With regard to the Walkability Index (iCam) developed by the Transport and Development Policies Institute (ITDP)[2], walkability worksheet Final Result and measured score was 1.45, this parameter rating of SUFFICIENT is a low-to-medium score (on the evaluation scale). This low rating was attributable to the fact that most items existed, but failed to meet requirements ore be fit for purpose.

Table 1. Walkability Index evaluation results and score using iCam 2.0 methodology.

Indicators and categories	Final score (0-3)	Evaluation and scoring criteria (Insufficient, Sufficient, Good, Excellent)
Paving	1.54	Sufficient
Width	2.00	Good
Pavement	1.77	Sufficient
Block size	2.42	Good
Walking distance to transport	1.22	Sufficient
Mobility	1.82	Sufficient
Physically permeable façades	3.00	Excellent
Visually permeable façades	2.71	Good
Day and night public use	0.42	Insufficient
Mixed uses	1.25	Sufficient
Attraction	1.85	Sufficient
Street typology	1.00	Sufficient
Crossings	0.00	Insufficient
Road safety	0.50	Insufficient
Lighting	1.00	Sufficient
Day and night pedestrian flow	1.82	Sufficient
Public safety	1.41	Sufficient
Shadow and Shelter	0.70	Insufficient
Noise pollution	1.28	Sufficient
Garbage collection and cleaning	2.00	Good
Environment	1.33	Sufficient
iCam Final Result	1.45	Sufficient

Source: Results compiled by author using methodology applied (2019)

2.2.2 *Space analysis after applying methodologies*

The analysis of the surveys conducted on site using the applied methodologies and cross-comparison of results, revealed that walking in Suzano central area was rated as not bad, but also not suitable, especially for those with reduced mobility. It was not pleasant as an experience, because of visual pollution, high interference on all planes and the lack of landscaping/green area. In addition, it was deemed unsafe for pedestrians and bicyclists, because they vied for room with cars which were the dominant agents of the space, occupying more than 70% of it.

Figure 7. Current vehicle dominance in city center – General Francisco Glicério Street (1) and Benjamin Constant Street (2).

Source: Suzano City Hall (2019)

Although the ground floor properties were practically all occupied by shops, there was no life dynamic after business hours. There are no spaces for rest and contemplation, or places encouraging interaction. Also, waste management and incentives for renewable technologies were lacking.

Taken together, these results point to a need to review the urban space design. The solution may not imply a complete ban on cars, but striking a balance of forces that recognizes the importance of walking and respect for pedestrians.

3 A PRACTICAL AND PROPOSAL-BASED EXPERIENCE FOR THE PEOPLE OF SUZANO

From the applied methodology implementation and urban space assessment, Suzano City Hall technical team focused on the material produced to translate this methodological experience into a proposal for adapting and redesigning the urban space.

2 ITDP Brasil designed the iCam tool to support walking studies in Brazilian cities.

In this regard, the proposed solution was to implement the shared street concept for the first three blocks, near the railway station and bus terminal, where there is the highest density of people; rebuilding the street, expanding the pedestrian crossing significantly, allowing cars access only at specific times and ensuring full accessibility.

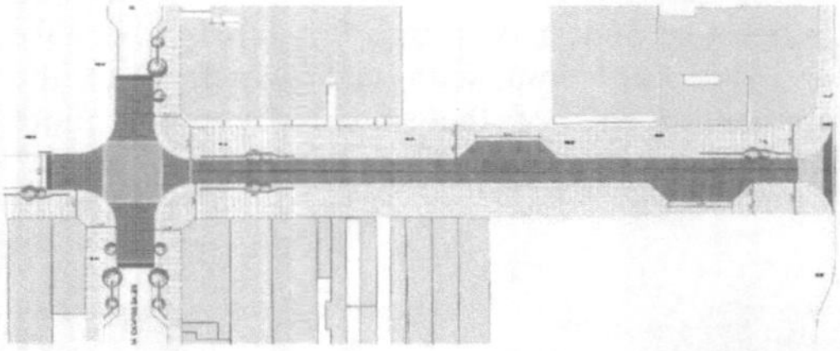

Figure 8. Proposal to implement concept of "shared street" on first block of General Francisco Glicério Street.
Source: Suzano City Hall (2019)

Figure 9. Computer generated view of "shared street" implementation proposal on first block of General Francisco Glicério Street.
Source: Suzano City Hall (2019)

Figure 10. Computer generated view of "shared street" implementation proposal on first block of General Francisco Glicério Street from Railway Station (CPTM).
Source: Suzano City Hall (2019)

A proposal to landscape an area was put forward, using the concept of rain gardens with urban furniture and living areas. These would be installed in pre-identified points so as to create areas for catching breath and resting, aimed at greater thermal comfort for walkers and improved efficiency of urban drainage.

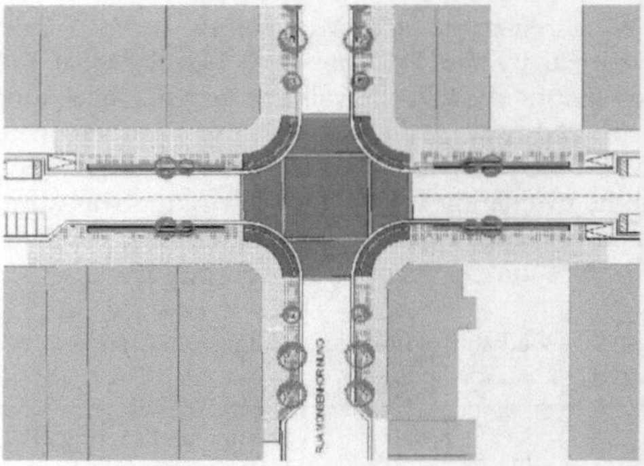

Figure 11. Proposal for pavement adaptation and urban furniture and landscaping.
Source: Suzano City Hall (2019)

The proposal also included the reconstruction of the street from the sub-base, enabling the implementation of new public transport lines, and building a drainage system that meets actual needs.

The proposal also envisaged increasing the number of streetlamps and replacing the current technology with LED lamps, improving efficiency; and erecting appropriate signposting, making the streets accessible to all travelling through the city centre.

However, drawing on the results obtained after the application of the methodologies, there was a collective understanding that, although the overriding wish would be to ban cars from this part of Suzano´s territory, implementation would be unfeasible due to the prevailing city´s urban mobility configuration.

Thus, the main intervention entailed the broadening of the pavements at their extremes (corners) ensuring shorter crossing times. This required a reduction in parking spaces, but not their total removal.

Pavements in the city center area vary in size, but the stretch facing General Francisco Glicério Street and Benjamin Constant Street have an average width of 2.50 metres.

By broadening the pavements at corners, it was also possible to ensure accessibility by implementing ramps and tactile floor or by elevating the level of crossings so as to promote walking comfort. The possibility of ensuring a safer and more comfortable walk for pedestrians was addressed and, therefore, the option of raising the crossing contributed greatly.

This reclaims areas for interaction, where the urban space serves as a place of encounter and passage where people's presence plays a key role, as protagonists of this space.

The shortening of crossing time is fundamental in this process, because it highlights that people's transit must occur respectfully and not in a way that competes with cars. The simple fact of raising the crossing point to the level of the pavement contributes to making vehicles slow down and respect the pedestrians' passage.

It is important to emphasise that the urban design envisaged in the proposal for redeveloping the city centre constituted a mix of alternatives that sought to promote a greater sense of security and well-being when walking through the central streets to develop space and urban life.

4 FINAL CONSIDERATIONS

The methodology of urban space assessment validates public actions. Cities need a methodological urban space analysis tool so that actions can be delineated based on a real-life perception, verified according to recognised technical parameters.

Consequently, in addition to the technical issues of public space evaluation, through application of the methodologies, it was also possible to identify the profile of the people who travel through Suzano city centre. In this regard, it was found that 60% of the interviewees were classified as medium socioeconomic class, according to FGV parameters, and 65% were aged between 20 and 50 years.

The applied methodology showed that 62% of the people interviewed in the city centre used the public transport system. This is a consequence to the fact that the central area connects to all other areas of the city through public transport and affects the range of mobility on foot. Thus, the number of people walking in this area was increased to 73% of those interviewed. The application of the methodology allowed us to verify people's reason for visiting the city centre. Surprisingly, the proportion of people who walked through the center for the purpose of shopping was low at only 23%. If workers (shopkeepers interviewed) are excluded then this rate increases to 38%. However, the figures in the other categories suggest that there is a great opportunity to boost trade, in view of the high number of people travelling on the site.

The results obtained according to the research showed that people are in favour of interventions, especially toward improving the walking experience, either through adequate width, specific road closures or decreased car presence. Much of this perception of the need for presence of cars is due to the fact that mobility in the city is difficult and also because the central area is an important route to access other areas of the city or neighboring towns.

People are still favourable to the notion of creating space for bicyclists, advertising and paving standardisation, and more efficient streetlamps.

The desire to introduce landscaping, urban furniture, technology availability (internet and surveillance cameras) is valid.

However, all changes cause an impact and a fear of the unexpected, but urban life plays a fundamental role in people's lives, in which the presence of a suitable space promotes pleasant sensations good for human life. Conversely, problematic spaces increase the stresses of daily life.

REFERENCES

ANTP Associação Nacional de Transportes Público (2017). *Relatório Geral da Associação Nacional de Transportes Público* - Simob/ANTP. Available at: http://files.antp.org.br/simob/sistema-de-informacao-de-mobilidade-urbana-da-antp–2017.pdf. Accessed: 02 February 2020.

CIDADE ATIVA. (2018a). *Diagnóstico áreas 40: São Miguel Paulista. São Paulo: Cidade Ativa*. Recovered: 01 June 2019, from www.cidadeativa.org

GEHL, J. (2013). *Cidades para Pessoas*. Ed. Perspectiva, São Paulo.

GAETE, C. M. *"'A arrogância do espaço': A distribuição desigual do espaço público em relação aos pedestres, ciclistas e automóveis"* 24 Out 2014. ArchDaily Brasil. Accessed: 04 March 2020. <https://www.archdaily.com.br/br/755838/a-arrogancia-do-espaco-a-distribuicao-desigual-do-espaco-publico-em-relacao-aos-pedestres-ciclistas-e-automoveis> ISSN 0719-8906

ITDP Brasil. *Índice de caminhabilidade versão 2.0*. (2018). Available at: <http://itdpbrasil.org.br/icam2/> Accessed: 08 May 2019.

NEW YORK CITY DEPARTMENT OF CITY PLANNING. et. al. *Active Design: Shaping the Sidewalk Experience 1. New York: 2013.* Available at: http://centerforactivedesign.org/sidewalks. Accessed: 08 May 2019.

NEW YORK CITY DEPARTMENT OF CITY PLANNING. *Active Design: Shaping the Sidewalk Experience 2 – Tools and Resources. New York: 2013.* Available at: http://centerforactivedesign.org/sidewalks. Accessed: 08 May 2019.

OLIVEIRA, L. (2018). *Forma e função no projeto da praça urbana. Paisagem E Ambiente*, (41), 35–55. https://doi.org/10.11606/issn.2359-5361.v0i41p35-55

OLIVEIRA, A. C. A. et al. (2011). *A promoção da mobilidade urbana sustentável em detrimento do uso indiscriminado do automóvel.* Anais 18° Congresso Brasileiro de Transporte e Trânsito, Rio de Janeiro.

PREFEITURA DE SUZANO. *Plano de Mobilidade Urbana. Lei Municipal nº 5190*, 31 May 2019. Instituted the Plano de Mobilidade de Suzano. Suzano: 2019.

SPECK, J. (2012) *Cidade caminhável*. São Paulo: Perspectiva.

Research Tracks in Urbanism: Dynamics, Planning and Design in Contemporary Urban Territories – Allegri et al (eds)

Collaborative Land Model. Centralities with rights in the West Municipality of Montevideo

E. Leicht, A. Quintans, A. Matos & C. Centurión
ITU, School of Architecture, Universidad de la República, Uruguay

ABSTRACT: The Urban Centralities research team has the goal of contributing to the strengthening of centralities as urban scenarios of universal access, promoting the sense of belonging of local communities and consequently contributing to the attractiveness and vitality of urban space. This article provides a summary of the results of research based on a case study reported under five sections. The first section focuses on the research problem and places this in context with Montevideo. The second explores the conceptual theoretical corpus. The third explains the work methodology through the "laboratory", so-called due to its experimental nature. The fourth defines aspects and centrality attributes to quantitatively and qualitatively quantify the entities studied. The fifth section concludes by proposing a collaborative model for the centralities of Municipality A.

1 THE DECLINE OF CENTRALITIES IN THE CONTEXT OF MONTEVIDEO

1.1 *The role of traditional centralities today*

The worldwide decline in neighbourhood centres known as *centralities* is a frequent topic in current urban planning literature.

Figure 1. Paso Molino high street (2017, authors).

As an alternative, some strategies have been proposed, such as reclaiming the street and public space as a vital place for the citizen, the management of commercial areas addressing the mix of uses and activities, as well as morphological and heritage aspects. Some urban design principles can be highlighted, such as the permeability of the public-private interface, and the legibility, variety, robustness and visual appropriateness of the built environment (Bentley et al, 1985). On a broader scale, the recovery of centralities is highlighted as one of the main strategies toward achieving the *compact city*, which is the most accepted model for sustainable urban development.

At an international level, the concern for neighbourhood centres, which are a peculiar combination of public and private space, is a recurrent topic mentioned by scholars of contemporary urbanism, from the morphologist Philippe Panerai (1983) to Jane Jacobs (1967) and their followers such as Jan Gehl. From a broader perspective, the paradigm of sustainability, Richard Rogers (2000) introduces the concept of the compact city and its neighbourhood centres.

The concept of *compact city* is mainly associated with the medieval city model: densely built cities with important civic spaces, surrounded by defensive walls. Rogers reinterprets the *dense city* model —unfortunately this last term has also been associated with the overcrowding of the industrial cities of Victorian England with their problems of congestion, poverty and illnesses, and calls for elaboration of the idea of the compact city: *a dense and socially diverse city where social and economic activities overlap and where communities can integrate into their neighbourhood.* According to Rogers, a sustainable city is, among other things, a compact and polycentric city that protects the surrounding countryside and centres, and integrates the communities within their neighbourhoods and optimizes their proximity. In addition, it attaches great importance to public space as a meeting place for diversity and its potential as a factor for social

DOI: 10.1201/9781003220855-24

inclusion. The neighbourhood centres, so unique in London, are for Rogers a key factor in promoting the vitality and densification of the city. Although the aforementioned reference is a seminal study, we complemented it with specific information regarding Latin American cities and informal trade. The approach in this case requires an interpretation which focuses not only on the mechanisms of economic circuits, but also on deep social fragmentation processes and cultural phenomena linked to survival strategies, as well as to mechanisms of relationship, belonging and visibility (Almeida dos Santos, 2015).

In Montevideo (Uruguay), traditional centralities are identified with commercial areas and neighbourhood centres in which facilities and services such as shops, public offices, transportation, social and cultural institutions, and so on, historically converge. Despite its limited scale, proximity allowed its consolidation as references for these local communities, who contributed significantly to the territorial balance of the city and to neighbourhood identity, conferring vitality and character.

According to the relationship with the consolidation process of these neighbourhoods and to an ingrained logic, these centralities were organized along streets or crossroads, which defined their public and urban nature. In addition to supporting the fundamental structuring of centrality, the street constitutes, due to its nature as a public space of proximity, a territory of particular symbiosis of public and private worlds in which its responsibilities and powers become blurred; this aspect leads to one of the main complexities in the approach to these areas. Despite the relevance that these urban centres acquired throughout the twentieth century, the scenario posed by contemporary consumer spaces (shopping centres, large footage) in the last three decades —which also involve urban changes, new social behaviours and new consumer trends—, erodes its urban and socio cultural role and causes, among other aspects, a decline in commercial and cultural activity with the consequent loss of vitality in both meeting and socializing environments.

Since the late 20th century, the commercial distribution sector has undergone major transformations involving profound changes that have taken place at the level of the economy, society, technology and culture worldwide. This new context has led to a complex specialization of roles within the distribution channels that have transformed not only the types and formats of shops, but also the association and distribution systems.

Thus, following the development of the so-called vertical distribution systems (particularly corporate) or within the scheme of associated commerce (franchised chains) large commercial firms have formed branch chains on a global scale, found in the main areas and commercial sectors worldwide. Within the system of associated trade —and completing this complex scenario— there has been a rapid expansion in the closed shopping centre format (horizontal spatial systems), whose development worldwide has not only radically modified the logic of competition with traditional retail trade, but also given rise to new places of centrality. These centralities are characterized by hybrid complex spaces, accommodating container buildings and free spaces managed by private parties (Paris, 2013) catering to new consumer habits and cultural models.

The traditional local shops have been replaced or displaced in some cases by expanding multinational firms in search of strategic locations. These approaches have hegemonized and standardized the shopping options available in these areas, shed any identity and impoverished the shopping experience. The complexity and diversity of consumer goods available, dictated by the new logic of world trade, generate new controlled spaces for consumerism. To make things more complex, electronic commerce contributes to the deterritorialization of consumption and changes the traditional role of centralities.

In summary, the loss of vitality and visual appropriateness experienced by the traditional urban centres of Montevideo has been identified as a central problem. This process, which affects both commercial and socio-cultural activities, as well as environmental aspects, has eroded their traditional symbolic function and role as reference areas for local communities. Despite the apparent consensus regarding its importance for the city and of international experience on the issue, in the case of Montevideo there have been few recent studies addressing the subject. The centrality system is seen as an opportunity to densify and revitalize a city that continues to expand without an increase in population. It is known that successful centralities contribute to the urban development of neighbourhoods and stimulate settlement of the population, according to the principles of responsive environments (Bentley et al, 1985).There are few but notable examples of good practices involving the implementation of co-management experiences (public-private), and promotion of social appropriation that warrant revisiting.

1.2 *The centralities in the Montevideo land use plan*

In Montevideo, the Land Use Plan - POT (1998) recognises the centralities as a hierarchical system. Subsequently, guidelines for centralities were updated in the Departmental Directives (2014).Certain centralities have been updated in the scope of partial plans such as Goes, Casavalle, or more recently Pantanoso.

The centralities identified in the POT are classified into four categories according to the process of historical conformation and to different qualitative and quantitative data. This classification of centralities denotes the intention to address the issue and respond to the transformations experienced in the 90s linked to the generation of new commercial areas and their impact.

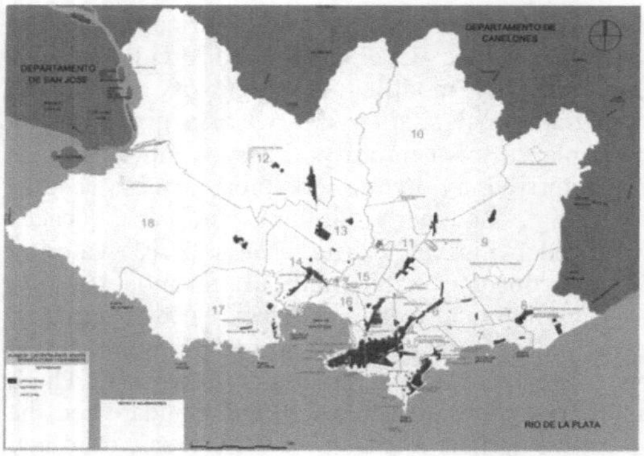

Figure 2. Centralities in the Montevideo land use plan (IMM, 1998).

- The *main centre*, with an axis along *18 de Julio Avenue*, identified as a project of strategic value in the POT and ratified in the Departmental Guidelines.
- The *urban and metropolitan centralities*, which encompass those considered historical (Goes, Unión and Paso Molino) and others that are of metropolitan importance due to their scale (Colón, Portones, Pocitos). Six centralities are identified in this category
- *Zonal centralities*, which cater to the need for commercial areas associated with the growth of the city after the emergence of the historical centralities. There are seven zonal centralities: Paso de la Arena, Curva de Tabárez, Sayago, Piedras Blancas, Cerrito, Curva de Maroñas, Carrasco.
- *Local centralities*, which follow the logic of urban organization whose basic cultural foundation lies in the idea of the neighbourhood. The local centres are, more than venues providing services, areas in which society manifests self-representation. The daily supply type of commerce predominates and 42 local centralities are registered.

At the Institute of Theory and Urbanism –(ITU), Professor Antonio Cravotto put forward a proposal for the revitalization of the main Centre through the development of the 18 de Julio Avenue, its spaces and commercial galleries. The impact of the first generation of shopping centres in Montevideo was identified as a problem to address early on (Urruzola, 1994)

There are now more than ten articles in the urban regulations of Montevideo with significant consequences directly related to the centralities. These rulings, among other issues, allow for the enclosure of the front yard of properties for commercial uses within the centrality area, or more parking lots are required for the same condition. Nevertheless, the preference in centralities is mixed use, which enables more flexibility and stimulates 24-hour activity. The importance assigned to the system of centralities as a land structure is of such significance that the execution of a centrality may serve as a basis for expropriation.

On the other hand, some centralities resulting from the initiative of civil society or spontaneous commercial groups do not have a correlate in the Plan. As an example, the Barrio de las Artes, the Bajo de Ciudad Vieja or the structuring of Av. Italia in certain sections —with some specialization in the automotive market, decoration and household items— do not have a registry in the "institutional" (according to Plan) centrality system, nor are those sectors specialized in specific lines of business, such as car spare parts along Galicia Street, or the wholesale of clothing and accessories on Colón Street, registered at an instrumental level.

Figure 3. Sketch of 18 de Julio Avenue by Daniel Venturini (ITU, 1995).

2 THE DEVELOPMENT OF THE CONCEPT OF CENTRALITY AND ITS POTENTIAL TO SECURE THE RIGHT TO THE CITY

The review of the Centrality System in the city of Montevideo, proposed as one of the aims of this study, called for the need to focus on the meaning of the term *centrality*, a term commonly coined in the jargon of the urban discipline, but which needs, in the face of contemporary processes, to be redefined, as a tool and support for action in the territory. Starting from the theory of central places (Christaller, 1933), which holds that land is organized into hierarchical spaces according to the "surplus of services" they offer, various concepts of centrality have been developed where the common focus is centred on the definition as a place of attraction for people where a high level of exchange takes place.

Starting in the seventies, this definition based on criteria with a predominance of economic flows was revised with the contributions of authors from the social sciences, such as Henri Lefebvre (1969), among others, who placed focus on social relations that give meaning and coherence to these spaces within the fabric of the city.

According to Lefebvre, centrality, for those who study the territory, is the very essence of the urban phenomenon. Lefebvre did not define it as static in nature, but quite the opposite, that is, as a dialectical movement that, over time, creates and changes. Centrality is that aspect of places that allows each point in the territory to be a centre, characterize the environment and fill it with meaning. (Paris, 2013)

In the last few decades, positions have emerged that make explicit the richness of centralities as "multifunctional spaces of different scales, with a defined role, which attract people and goods where intense collective exchanges take place" (HYDEA - Target Euro, 2008). They identify: "i) a high concentration of services and activities of different scales that determine their degree of attraction to people; ii) good internal accessibility and from the rest of the city (usually have transport nodes); and iii) recognition of the community as the centre of the area of which they are part, constituting a point of reference and symbolic expression of the living conditions of its inhabitants" (Cuenin and Silva, 2010). Faced with the phenomenon of globalization and urban expansion in Latin America, new places of consumption arise that have the capacity to attract the population, defining new temporal-spatial relationships.

To understand the current System of Centralities, we refer to the study by Paris (2013) re-examines the classification of central functions of other authors and identifies: traditional centralities, emerging centralities and what he defines as «places of centrality». The traditional centres or centralities are those corresponding to symbolic spaces which have been consolidated over time, linked to the original layout or with the subsequent growth of the city, where towns or villages that were outside are integrated in the plan of the original layouts and that reproduce the same scheme of the centre within their area of influence, where their identity value and the predominance of public space (squares in some cases or the street itself) are a distinctive characteristic where intense exchanges are condensed.

Emerging centralities are found throughout the territory, characterized by a "high degree of accessibility that attracts and stimulates their own users/ consumers" (Paris, 2013) where functions are grouped yet without being consolidated in the collective imagination, constituting spaces designed in contemporary architecture with the potential to generate new notions of spaces for interaction.

The places of centrality exacerbate a phenomenon which Lefebvre had first noted in the 1970s. These are hybrid complex spaces accommodating buildings —containers and free spaces— managed by private parties where contemporary interactions occur, commonly linked to consumption. Therefore, space and action are controlled, thereby undermining the essence of public space for socialization as a meeting place and to manifest the contradictions the community has, and exacerbates the processes of socio-spatial segregation.

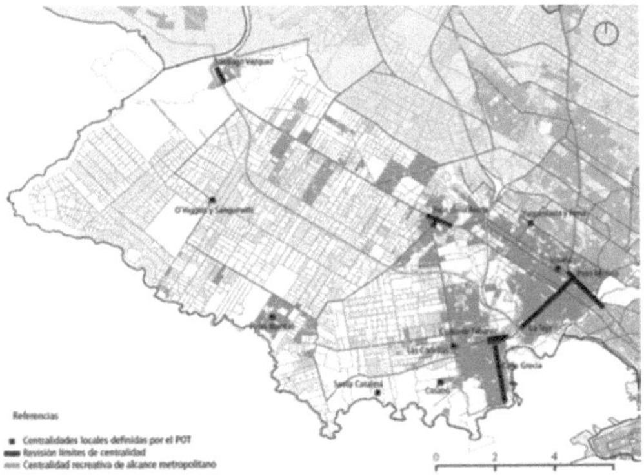

Figure 4. Centralities in the west of Montevideo (2018, authors).

The key to public space, mobility, and co-management between actors seems to present a challenge in re-examining the centrality system in the metropolitan context by investigating the roles and relationships of the centres of consumption, the traditional centralities, and the possible projects of new emerging centralities in the territory to contribute to the right of all citizens.

3 THE LABORATORY OF URBAN CENTRALITIES AND THE RIGHT TO THE CITY. A PILOT EXPERIENCE IN MUNICIPALITY A

Although the research team[1] understands that a comprehensive approach to the metropolitan system of centralities of Montevideo is necessary for a full update of the configuration of centralities proposed by the POT, it considered the approach of centralities located in the western part of Montevideo a priority. In this sector, centralities have a relevant role to play within a perspective of balanced development of the city and the national territory. In all national level statistics that indicate values regarding quality of life (education, food, housing, security), the West of the city is flagged "in red" (Director of PTI Cerro, G. Gonsalves, 2017).

1 The research team is identified as 883032 at Comisión Sectorial de Investigación Científica, Universidad de la República: Eleonora Leicht, Leonardo Gómez, Amancay Matos, Andrés Quintans, Carola Rabellino, Carlos Musso, Camila Centurión, Estefanía Mannise, Natalie Sobot, Lucía Álvarez

Municipality A or *del Oeste* (West) is one of the municipal regions of Montevideo that is part of the third level of government according to the 2009 decentralization law. Therefore, it constitutes a “young territory” which, although included in earlier extensive urban studies, none had the specificity of scope or territorial area of interest of the present investigation. It is a region in urgent need of answers to the many problems it faces where centralities are relevant as part of its land structures and systems. The starting point is the group of thirteen centralities defined in the POT located in Municipality A:

3.1 *Objectives and strategies of the laboratory*

The Urban Centralities Laboratory is a space built from collective exchange involving people with different degrees of knowledge and specialization. Through various forms of experimentation and learning, ideas and experiences are shared, with the aim of generating new knowledge and producing synergistic exchanges. The «laboratory» format aims to promote collaborative environments and to experiment with new methodologies for analysis and action, encouraging rehearsal, disciplinary transfer in the search for attributes of centrality and solutions to the problems that these spaces face.

The right to the city, as a basic human right, requires the care and development of these meeting places where various activities converge —economic, recreational, cultural, commercial— and where people are involved in the capacity of not only consumers but citizens who can make full use and appropriation of these spaces.

<u>Objectives of the Laboratory of Urban Centralities</u>

General:

- To focus on the concept of urban centralities as an object of study, as an essential territorial system of urban planning
- To visualize the concept of centralities and its problems through research, teaching and extension
- To promote interdiscipline and comprehensiveness of university functions through the generation of a work group adopting the «laboratory» format

<u>Specific:</u>

- To generate new knowledge regarding the concept of centralities through the casuistic corpus of Municipality A as the first stage of a study that in the future could be replicated in other contexts
- To cooperate with the socio-urban balance of the Montevideo territory by addressing the centralities of Municipality A, both as public spaces and potential “engines” of local economic development
- To identify attributes, potentialities and singularities of the various centralities located in Municipality A
- To propose a conceptual model for the centralities of Municipality A

4 ASPECTS OF ANALYSIS AND ATTRIBUTES OF CENTRALITY

By way of systematization, a registry model was developed that synthesizes the most relevant aspects which define and characterize the centralities in our context based on various attributes, grouped according to the following aspects:

4.1 *Urban environment and public space*

One of the keys for the study of these areas is the development of a diagnosis of their urban context in terms of their location, morphological and architectural characteristics and the aspects involving the attributes of public space (urban design, trees, sidewalks, equipment, etc.). Other factors are also involved, such as traffic volume, hygiene, noise pollution, public safety, and so on. The unique articulation and incidence of each of these aspects defines the identity of each urban area and determines the environmental quality of its public spaces and, thus, of the centrality as a whole.

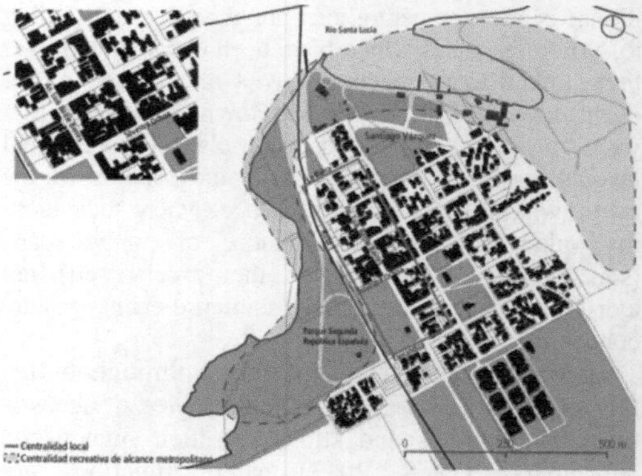

Figure 5. *Pueblo* of Santiago Vázquez. Metropolitan recreational centrality (2018, authors).

4.2 *Commercial activity and its management*

In each of the cases, the quantity, characteristics and category of the commercial establishments included are identified and indicators related to commercial density and relative weight of the different commercial formats (retail, franchises, chains, large stores, etc.) are proposed. This also establishes the degree of commercial association and the existence or otherwise of joint initiatives by the businesses. To complement this perspective, we seek to determine the main economic activities of each sector and their degree of link with commercial activity and with the centrality in general terms. We investigate the link between the centralities and the industrial and logistics activities that take place in Municipality A.

4.3 *Social facilities and public services*

According to the historical makeup of these urban areas, the presence or absence of cultural and leisure services available, as well as the existence of public services and facilities, play a leading role in their development and attractiveness. A process of identifying, mapping and evaluating their current weight is carried out, as well as determining their potential roles in recovery and revitalization processes of these areas.

4.4 *Mobility*

From the perspective of accessibility, a global view is promoted regarding the forms of access and movement towards centrality, whether these are collective or individual wheeled transport systems (motorized or non-motorized)or pedestrianized areas. The role and quality of mobility systems and the spaces intended for them are evaluated (location of public transport stops, walkability, provision of parking, accessibility from logistics areas, etc.).

4.5 *Identity and symbolic role*

Given the relevant —and historical— role that these areas played as reference centres for the city's neighbourhoods, it is important to recognize the value that these local communities assign to them today in light of the transformations the city has undergone on this ground. Investigating this field allows us to establish the degree of identification and the sense of belonging that these centralities generate both among the local population and in the group of actors directly involved. From the analysis of the defined centrality aspects and attributes, the following matrix is structured, which synthesizes the study carried out in each of the centralities.

5 CONCLUSIONS: TOWARDS A COLLABORATIVE MODEL AND INTERVENTION STRATEGIES FOR THE CENTRALITIES OF MUNICIPALITY A

5.1 *Intervention strategies to work in a varied polycentricity*

In the updated model of centralities of Municipality an according to hierarchy, specificity, delimitation and area of influence, the various modalities of interrelation, complementarities and links between them are made explicit. Different intervention strategies emerge from this model.

Among the centralities in the continuum of the urban fabric (Paso Molino, La Teja, Curva de Tabárez, Calle Grecia and Las Cadenas), there is greater relative dynamism and interaction expressed asa greater presence of commercial activity and services, generating a process of conurbation of centralities on the Avda. Agraciada and Carlos María Ramírez axes (main road connectors in the area). Mapping the shops in these centralities, reveals that the density of commercial establishments on the high streets varies from 60 shops per block (Paso Molino)to 10 shops (Carlos María Ramírez or Grecia).

The densities, profiles and items of the shopping areas, as well as the presence of equipment and services, continue to establish hierarchical relationships in the network of centralities that define the type of links they maintain with each other.

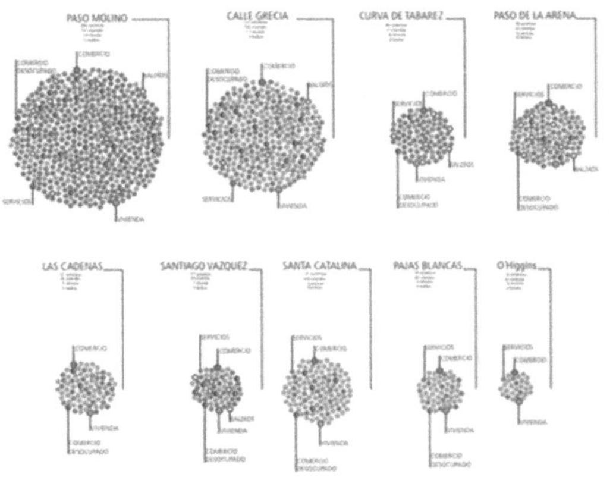

Figure 6. Relation between shops and residences in centralities of *Municipality* A (2018, by the authors).

5.2 *Sub system of centralities of the west municipality*

Based on this study, a reconfiguration of the subsystem of centralities of the West Municipality is proposed based on complementarities and links marked by territorial accessibility, with the following guidelines:

In cases where valuable pre-existing elements are found (urban and historical aspects, consolidated commercial area, equipped public spaces, facilities), the proposed strategy is to explore comprehensive projects with the aim of strengthening identity in each of the centralities, giving value to the imprint that gives it character and interest on the metropolitan scale.

Examples of this are the potential of Santiago Vázquez linked to the various quality public spaces around the Santa Lucía River as a leisure space at the

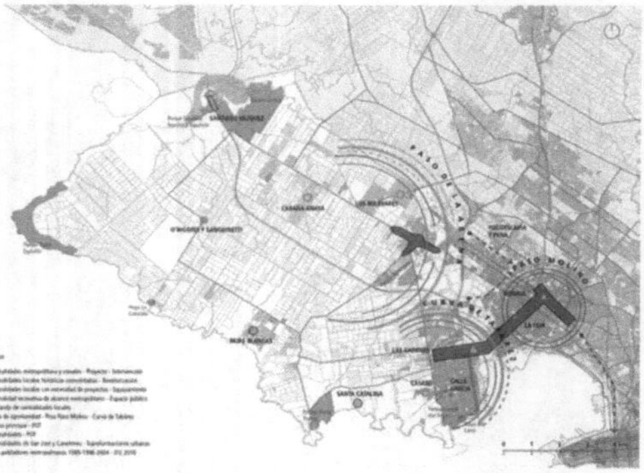

Figure 7. *Municipality A*. Collaborative land model (2019, by the authors).

metropolitan level; and Santa Catalina with the possibility of exploring links with the Punta Yeguas Park.

In addition to local attractions, as in the case of PajasBlancas, its link to beach access; and the importance of the enhancement of historical local centres such as Calle Grecia, O Higgins and Sanguinetti, and Vitoria are of note.

The relevance of promoting and projecting centralities of local hierarchy in areas of consolidated housing with poor services, such as the cases of Casabó, Cabaña Anaya and Los Bulevares, is evaluated.

For the conurbation of Centralidades del Oeste on the Avda. Agraciada and Carlos María Ramírez axes, a comprehensive project is proposed, which must include the centrality of Paso de la Arena with its role as a rural-urban centrality; with the objective of carrying out interventions to guarantee the right to the city where the relevance that this territorial area has for the West of Montevideo is also visualized.

REFERENCES

BENTLEY, I et al. (1985). Responsive Environments. A manual for designers. London, Routledge

BEUF, A. (2011). Nuevas centralidades y acceso a la ciudad en las periferias bogotanas. *Boletín del Instituto Francés de Estudios Andinos* (en línea), (*40*)1, 147–78. DOI: 10.4000/bifea.1663

CAPOROSSI, C. (2017). *Las centralidades barriales en la planificación urbana. Barrios pericentrales de Córdoba: el caso de San Vicente*. Buenos Aires: Editorial Café de las ciudades.

CUENIN, F. y SILVA, M. (2010). Identificación y fortalecimiento de Centralidades Urbanas: El Caso de Quito. Banco Interamericano de Desarrollo. Notas Técnicas IDB- TN – 156.

FACULTAD DE ARQUITECTURA, UNIVERSIDAD DE LA REPÚBLICA (1997). Insumos para la elaboración del Plan de Ordenamiento Territorial. Memorias de Información y Memoria de Ordenación. Capítulo Centralidades urbanas. Montevideo. Multicopiado, ITU.

HYDEA-TARGET EURO (2008) Programa de Fortalecimiento del Sistema de Centralidades Urbanas de Quito (EC-L1041). Informe Final de Consultoría.

GEHL, J. (2013). *La humanización del espacio urbano (EUA09). La vida social entre los edificios*. Barcelona: Editorial Reverté.

HARVEY, D. (1977). *Urbanismo y desigualdad social*. España: Editorial Siglo XXI.

LEFEBVRE, H. (1969). *El derecho a la ciudad*. Barcelona: Ediciones Península. Publicación original: *Le detroit a la ville* (1968). París: Editions Anthropos.

LÓPEZ DE LUCIO, R. (coord.) (1996). Centros urbanos frente a nuevas centralidades comerciales. Un análisis del sur metropolitano de Madrid. *Cuadernos de Investigación Urbanística*, 14. Madrid: Departamento de Urbanística y Ordenación del Territorio, Universidad Politécnica de Madrid.

PARIS, M. (2013). De los centros urbanos consolidados a los lugares de centralidad: una propuesta metodológica para su estudio. *Dossier Monográfico Ciudades*, (*16*). Recuperado de http://uvadoc.uva.es/bitstream/10324/10345/1/CIUDADES-2013-16-DELOSCENTROS.pdf

ROVIRA LARA, A. (2000). Propuesta y modelos para la gestión estratégica de los centros urbanos desde la cooperación pública y privada. *Revista Gestión y Análisis de Políticas Públicas* (GAPP), 19, 19–34, España.

URRUZOLA, J. P. (1994). Nuevas centralidades o viejos problemas. *Revista Elarqa*, *(4)*12. Montevideo: Editorial Dos Puntos SRL.

VECSLIR, L., et al. (2018). Un corredor sin autopista. Centralidad y movilidad cotidiana en el sur de la Región Metropolitana. Buenos Aires: Café de las Ciudades

Minor cities and development. Correlations between morphology and urban planning in Arica between 1950 and 1980

H. Torrent
Escuela de Arquitectura, Pontificia Universidad Católica de Chile, Chile

ABSTRACT: Between 1950 and 1980, many intermediate and minor cities underwent similar phenomena to those experienced by large metropolises, such as accelerated population growth, urban plan expansion, land markets and the adoption of new denser architectural typologies. This paper reviews the case of Arica, a northern Chilean town, as an example of urban morphology produced according to the notion of developmentalism for intermediate and minor cities. It considers exploiting urban structural analysis to revise current phenomena and smaller urban configurations that have emerged under developmentalism ideas. It proposes to exploit structural urban analysis for the revision of more current phenomena and to analyze minor urban configurations that emerged under developmental ideas, where the relationship between architectural typologies and urban morphologies is not strictly corresponding and can assume characteristics that are difficult to define in traditional urban morphological studies.

1 INTRODUCTION

Between 1950 and 1980, many intermediate and minor cities underwent similar phenomena to those experienced by large metropolises, such as accelerated population growth, hasty expansion of urban plans and land markets and the adoption of new architectural typologies that tended toward a denser habitat. These phenomena have already been relevant regarding the so-called *primate cities* of Latin America, cities which, due to their size and scale, have been associated with significant urban problems and dominated the focus of studies in recent decades. However, little progress has been made in characterising these phenomena in minor cities that have undergone expansive growth promoted by special regimes to promote economic and social development, such as the case presented here.[1]

Today, it seems timely to review these cities' scale when it becomes clear that they are approaching a more balanced and sustainable development of urban communities, and their territorial role is on the rise. Therefore, this paper presents a new stage of Arica's case study and progress in knowledge, as an example of urban morphologies proposed under the notion of developmentalism for minor and intermediate cities.

The previous study emphasised the urban form configured from the periphery's large, autonomous and discontinuous portions. In this paper, growth modes are the focus of investigation, presenting these along a timeline of implementation. The correlations with the urban plan proposals aimed at governing development are examined, along with how these correlations had positive or negative outcomes according to different timepoints. Likewise, further decomposition of the expansion is proposed in order to deepen the analysis of turban fabrics, its continuities and discontinuities, with an emphasis on the relationship between plot dimensions and occupation surfaces, given that many of the operations centered on the configuration of extended peripheries.

This essay also asserts the importance of urban structural analysis as a tool to examine current phenomena and smaller urban configurations which emerged under developmentalism ideas, where the relationship between architectural typologies and urban morphologies is not strictly corresponding and can adopt characteristics that are challenging to define in the tradition of urban morphological studies. The mutations in these cities' growth and occupation modes seem to play a more substantial role in urban space configuration and the living and inhabitation approaches they propose. Arica's case can help shed light on how urban instruments should assume a major role according to the situation of the city at

1 This work is a partial advance of the research Fondecyt N° 1181290 "Modern Architecture and the City: works, plans and projects in the development laboratory. Chile 1930–1980". The author is grateful to Fondecyt for the support given.

DOI: 10.1201/9781003220855-25

the time, in pursuing sustainable growth for cities in the future.

2 ARICA'S TRANSFORMATION BETWEEN 1950 AND 1980

Arica may still be considered a minor or intermediate city. Its current population is around 230,000. In 1953, however, at the beginning of the process we are studying, it had about 23,000 inhabitants. Founded on the Pacific coast during the 16th century, its layout took on characteristics of the Spanish colonisation cities in America: blocks arranged in a grid-like structure. It also registered some typical deformations in its proximity to the main surrounding landform, the Morro de Arica, a hill of about one hundred meters high that emerges as the last expression of the coastal foothills before a flat depression starts, which corresponds to the elbow that South America makes on the Pacific Ocean. It was on this continuous plane that the city first took root, near San Jose de Azapa River.

Having previously been part of Peru, after the War of the Pacific in 1883, it came under Chilean rule. Despite its strategic border position, both in relation to Peru and Bolivia, it received little investment after its definitive integration into Chilean territory with the 1929 treaty. Perhaps the most significant achievement was the building of the infrastructure that dominates the city, the Port and the railroad lines to Tacna and La Paz, both products of international agreements.

In 1953, a development project that sought to awaken the city from its lethargic state and transform its economic base was proposed: the port's liberalisation. The Free Port regime, coinciding with its configuration as the port of Tacna and La Paz, established a series of tax exemptions for the region, encouraging a free-trade zone and an area for promoting industrial settlement. Faced with the limited success the regime had in its early years, in 1958 a new strategy was introduced: the creation of a decentralised regional body that would manage the port resources and reinvest them both in the city and the region to foster its development.

The Junta de Adelanto de Arica, *JAA* (Arica Advancement Board) was founded to lead regional and mainly urban progress. Expectations generated by the boost for development led to significant migration that doubled the population during the ensuing eight years, repeated in the following decade. Thus, the urban population grew from 23,033 inhabitants in 1952 to 46,686 in 1960, reaching 87,726 by 1970.

This process produced expansive growth in the city plan. The initial 260 hectares core in 1953 expanded to more than 580 ha by 1960, and then to over 820 ha by 1970. Although the expansive growth seems to have gone hand in hand with increases in the population, the first major increase (1953–1960) occurred after incorporating a large industrial area. Afterwards, housing demand drove the expansion, principally because there were large land areas available – flat swathes of desert, mostly state-owned. The state was the central factor in promoting growth since the absence of real estate and materials markets, and the limitations of self-construction due to the city's position in the desert, left very few possibilities for large housing investment by the inhabitants themselves.

Migration and rapid population growth put public authorities on alert, and they repeatedly drew attention to the problem. Housing construction during the 1950s was initially carried out by Corporación de la Vivienda, *Corvi* (Housing Corporation) and Sociedad Modernizadora de Arica (Arica's Modernising Society). From 1958 onwards, *JAA* was responsible for coordinating with state corporations and entities, such as *Corvi*, Corporación del Mejoramiento Urbano, *Cormu* (Urban Improvement Corporation) and Caja de Empleados Particulares, *Empart* (Private Employee Welfare Fund), among others. Housing construction constituted an unprecedented experience for the region, using various types and formats, both in layouts and occupations.

JAA also promoted urban planning, road and infrastructure works, public facilities (educational, hospital, security and recreation), parks, promenades and housing estates, all with a strong modern urban design imprint and formal identity based on modern architecture. Urban development during the three decades of Arica's experience can serve as a magnificent laboratory for analysis.

3 GROWTH IN THE CITY: EXPANSION AND DEVELOPMENT

The study of city growth offers 'a global glimpse of the urban agglomeration in a dynamic perspective, followed by precarious stabilities based on the city's internal cohesion to which periods of rupture and leaps in development follow, the current state representing a mere snapshot of this evolution' (Panerai 1983:28). It is thus possible to observe the most stable and dynamic periods of the city's configuration, and we can also note in each of these the recurrent logic of the urban form.

A sequence of graphs has been drawn up for this paper, allowing us to account for the successive periods of growth in the city plan, and we have also studied the most apparent forms of this growth, moving forward in some of the recurrent logic.

It seems essential to situate the beginning of the process around 1932 when the population started out on the course of transforming demands that would be seen only in the decade of the 50s. Up to that point, Arica had occupied a small portion of territory, with an initial orthogonal layout somewhat deformed by the presence of the Morro and bay, and in which the blocks became misshaped as they encountered critical geographical features, mainly owing to differences in topography. The centre appears as a configured area that disintegrates towards the northeast and the southeast, where there were no boundaries and areas which remained desert. There was not yet any advance beyond the limit imposed by the San Jose River, and the city emerges protected by the bay and the Morro. The main growth points were found along the roads to the north, mainly the route to Peru and the Azapa Valley, whose occupation was separate from the city.

By 1956, the city was contained by two fronts and began its expansion to the northeast and southeast, crossing the barrier of the San Jose River. In this period, growth led to a doubling in size of the urban plan, essentially through large sectors, which involved the military settlements, the stadium site and some blocks facing the river. Towards the south, on the slopes of the Morro and following the road's path, a series of blocks can be seen in the form of smaller constructions resulting from settling by inhabitants. There was no beachfront occupation, and the growth seems to be quite erratic: no clear plan is evident, but growth is rather organic in the direction of available free land. There were only a few cases of planned housing projects (such as the Chinchorro estate) located quite far from the intensely occupied areas.

By 1960, the city had grown mostly towards the Northeast, the banks of the San Jose River were already built on, regarded as a break in the urban fabric. To the north of the river, a planned programmed configuration can be clearly seen, which extends to the side of the route to Peru and corresponds to the industrial zone established from the different studies of zoning plans, due to the demands imposed by the free port policy and developmental industrialisation. The growth towards the southeast is much more orderly than in the previous plan, where blocks are regular in size and streets have continuity. The blocks are smaller and suggest a higher density of housing than in the city centre.

In 1970, the city grew mostly toward the northeast, occupying the desert's flatter areas. The industrial zone, previously easily distinguishable, was then completely absorbed by the urban configuration. Minor housing developments can be distinguished, which continued to grow northward, with the Pan-American Highway as an axis.

Towards the beach, large blocks can be seen, while the granularity changes as one moves southwards. Two macro blocks can be observed to the southeast on the road to Azapa Valley. If we read the urban fabric carefully, different grouping modes related to the concept of a laboratory are evident, like various ideas being tested in the territory. A route to the south appears, bordering the beach and connecting to new urbanisation in the Morro's southern border, between two beaches. By 1980, the growth of the city extended to the north following the road, instead of heading east or southeast. A coastal border can be recognised, and the urbanised surface in vicinity of the road and the

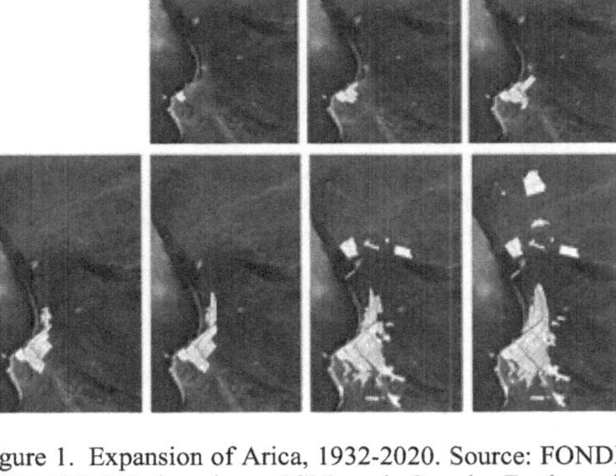

Figure 1. Expansion of Arica, 1932-2020. Source: FONDECYT 1181290 based on IGM and Google Earth aerial photographs.

perimeter ring of the city is increased. The new layout in the north suggests a high building density.

In 2000, the city had grown mostly towards the southeast and was connected in continuity with Azapa Valley. Some growth arises from the plan and appear linked to different routes that head toward the north, south and the interior. The beach road leads to new urbanisation that connects with the route to Lluta Valley, accompanied by large industrial blocks, and some exempt urbanisation emerged.

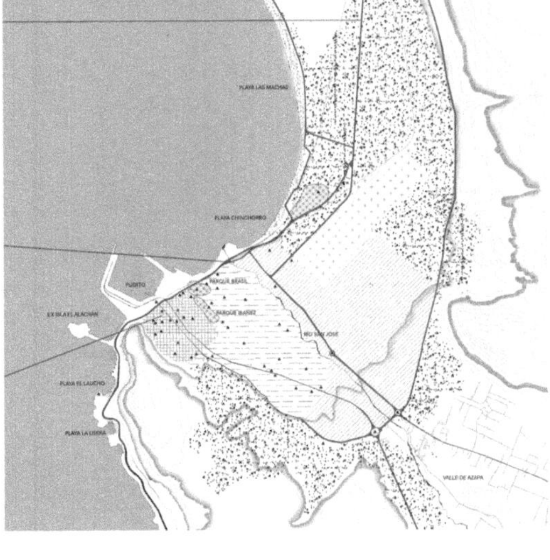

Figure 2. Study of the areas of growth and organization of the landscape in Arica. Source: FONDECYT 1181290 based on local study with notation according to Lynch and on current plan, 2020.

By 2020, some of the plots previously detached from the urban layout are now part of the conurbation, as parts of a more cohesive city. The city's growth continues to the north in the form of small urban constructions detached from the Pan-American Highway, and a new nucleus appears next to the airport, near the border with Peru. Growth towards the southeast are also evident, on the slopes of the Morro and beyond, with new forms of land division, in some cases as macro blocks and large estates.

4 URBAN PLANS: ZONING AND URBAN FORM

The intense urban growth was a severe problem in a still relatively small town, mainly due to the doubling in size of the city plan and the expectations of housing complex installation. The lack of growth foresight surpassed the urban plan carried out around the 1940s by the Sección Urbanismo of the Dirección General de Obras Públicas (Urban Planning Section of the General Division of Public Works), and of which very little is known. Although this could have contained some preliminary urban regulation, it was indeed confined to the Morro and San Jose River's limits. The expectation of both port and industrial settlement growth generated by the Free Port required specific areas not considered.

The debate on the need for a new urban plan marked the period between 1953 and 1958. Proposals for regulation took on three challenges: the pressure on the centre caused by equipment and services increase, the situation of land areas destined for industry and the residential fabric expansion.

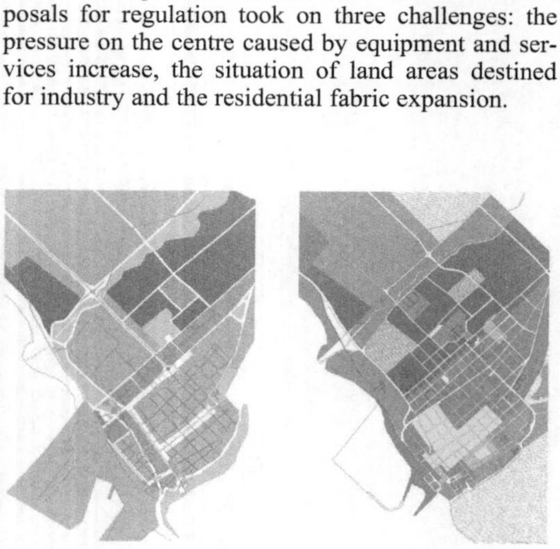

Figure 3. Studies and proposals of zoning plans of Arica: 1955, Jorge Poblete and 1957, Osvaldo Hufe. Source: FONDECYT 1181290 based on information provided by the Documentation Center of the MINVU.

In 1954, a year after establishing the Free Port, Jorge Poblete Grez carried out studies for the urban plan that determined zoning and road management, two crucial problems to face in the urban future concerning its new regional and national status. These studies, which did not become a definitive plan, established the industrial location to the north of the city plan, and a road connecting it to the port. The residential area's planned expansion was limited to that between the existing city and the industrial zone, which soon proved insufficient.

In 1957, a new study by Osvaldo Hufe revised the proposal within the Planning Department of the Ministry of Public Works, enlarging the residential expansion area towards the east, but maintaining the limitation imposed by the industrial zone location. Up to this point, discussions were mainly dominated by required zoning for industrial and residential extension purposes.

In 1958, Amador Brieva and Ventura Gonzalez, continued the study of the plan and delimited the industrial zone, conserved the road hierarchy without major changes and established residential areas according to socio-economic levels. This plan took on a definitive form, was approved and remained in effect during the 1960s, along with the Local Ordinance that specified it. The plan included, with greater detail, aspects of the existing urban form, identified types of residential fabric and extended the layout's characteristics, incorporating accurate detailing of the city's road hierarchy and setting up a clear urban border.

Figure 4. Zoning Plan 1958, Amador Brieva and Ventura González. Source: FONDECYT 1181290 based on information provided by the MINVU Documentation Center.

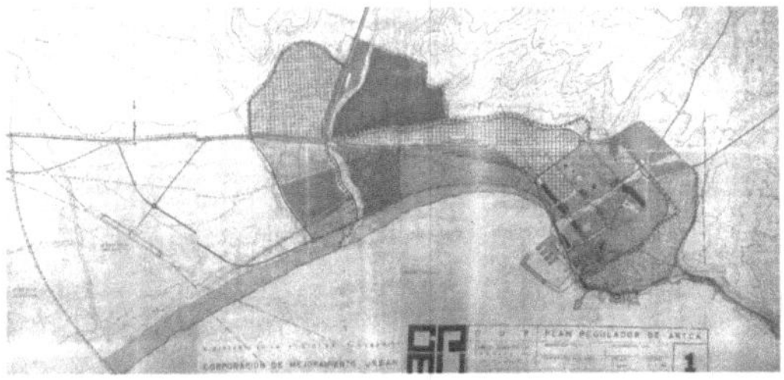

Figure 5. Emilio Duhart's proposal for a Zoning Plan for Arica, 1969. Urban Improvement Corporation. Documentation Center of the MINVU.

Later, attempts at revision were successively constituted by the Federico Oherens diagnosis, the Guillermo Jones Odriozola proposal, produced as a consultant to the United Nations in 1964, and the plan by Emilio Duhart in 1969, none of which were definitively approved.

In 1971, a new urban plan was developed and approved by *Cormu*. The planning team consisted of Patricio Arancibia, Anselmo Trepiana, Pedro Salas, Sergio Roman, Nelson Berthelon, with Jorge Elton and Karin von Buch as collaborating architects. It established new limits for the urban extension: the industrial area and its housing from the pre-existing urban border to the Lluta River. In addition, it provided for a section for tourism, housing facilities and a reserve between the road and the sea, up to the Gallinazos estate in the north. Beyond the urbanised zone, a satellite extension was planned on the south bank of the Lluta River, another reserve area to the north of this to absorb housing, equipment or services growth (new studies centre and the regiment) and a sector for heavy industry in the river valley, thus confirming the new productive role of the city and its growth projections for the future. It also recognised the old city conditions and gave it a preponderant services character, proposing growth only through harmonious complexes. The plan allowed for the configuration of the neighbourhoods close to the centre, maintaining high-density conditions, and the more distant districts with medium and low density according to their distance from the centre. It also suggested some particular conditions for the growth in the slopes of the Morro and La Lisera, like staggered three floors houses. Some specific sectors supported combined densities, with some indirect typological references. It is important to highlight that the 1971 plan recognised the sectional plans' pre-existence that *Cormu* had carried out facing the lack of anticipation and the demand for localised transformation.

Although the present analysis does not examine the different proposals registered in any depth, it is clear that the city grew beyond that envisaged by the planning instruments. The growth in the urban plan that have been determined – even partly – based on documentation of operations constituted by lots and housing estates, show that the need for housing and the particular projects to cater to this demand far exceeded the provision that the planning proposed in the urban plans or its studies. Private housing projects surpassed that envisaged in the urban plans in force and were mostly achieved through partial extensions of the urban limit and according to the figure of the sectional plan. But how it was done in this morphology can be understood by presenting several cases, both in the generation of modern urban fabric and the extension of residential fabric, mainly formed by the repetition of units.

5 FREQUENT ARCHITECTURAL TYPOLOGIES AND RESULTING URBAN MORPHOLOGIES

The city's expansion took shape through successive housing projects of various types, although adopting two common forms: blocks, houses – paired or isolated –, and a combination of both. This was initially significant, not so much in number as in quality. The Ex-Estadio estate and the Lastarria complex assumed this form in an effort to constitute a paradigm that proposed it as a model of urban growth, where density and the more traditional ways of living in isolated units on a single floor with a garden were equated.

The Ex-Estadio estate (Bresciani, Valdes, Castillo & Huidobro, 1956-57) is quite well known and was the first in the city to establish a combination between blocks that were arranged independently of the topography slopes and a series of associated houses established as a tapestry that followed the slope. The Lastarria complex (Saint Jean, Moraga, Vallejo, 1961-67) has been recognised as a paradigmatic work and was declared as a National Monument in 2016. Built to house the staff of *JAA*, it combined a series of 21 houses with a superblock.

The blocks instead were a present option, whose use grew from the mid-sixties with the series of constructions that *Corvi* carried out, first with 1010 blocks (Gonzalez & Moreno c.1967) and 1020 blocks (Sepulveda & Perelman, c.1968), and some private *Empart* projects. Later, *Cormu* introduced a particular type developed for the city: a square floor block with flats at half levels and circulation systems that allowed the differences in the levels of the land to be bridged, and which were applied in the complexes Pucara (Galleguillos, Rodriguez & Toro, *Cormu*, 1968) and Chungara (Trepiana & Roman, *Cormu*, 1971), respectively.

However, the bigger housing production resulted in estates with isolated or paired housing units, and these shaped the vaster urban plan extension. As analysed in the previous paper (Torrent, 2019), most settlements had an extensive horizontal layout and occupation by serial house buildings.

The isolated house estates were organised according to urban units such as blocks, derived from longitudinal groups of lots, with few spatial qualities. In some cases, they had continuity with pre-existing layouts, as in San Jose (Saint Jean, Moraga & Vallejo, 1961) and Juan Noe (Rebolledo & Rivera, 1959-60) estates. But these were projected with autonomy and probably guided more by their constructive possibilities than by their urban conditions. Some were conceived independently of the minor urban layout, using the concept of a superblock that had subdivisions in its interior, according to the type of occupation, such as Radio El Morro estate (Moreno & Gonzalez, *Corvi*, 1971).

Proposals of grouped houses sharing basic services should be included, such as those built in the Paula Jaraquemada estate (Perez & Roman, *Cormu*, 1969), executed under the Operation Site.

In the 70s, the block became a more viable option, developing particular alternatives, such as those of the Parinacota complex (Trepiana, Roman & Berthelon, *Cormu*, 1971), with 12 blocks built out of 25 projected for the Pampa Nueva section, and which explored the possibilities of staggering flats and duplexes in relation to elevated streets, with wide typological variations of the unit.

At the end of that decade, several groups addressed the possibility of combining blocks again with extended housing units. Perhaps the exemplary case in this context is the Citroën complex (Roi, 1974), with 19 blocks and almost 700 dwellings. In 1975, the 11 de Septiembre housing project was developed by *Empart*, consisting of 9 blocks, with 260 flats.

From the late 1960s to the early 1970s, the relationship between urban plans and architectural definition was deepened in an unprecedented way. It is symptomatic that Emilio Duhart's 1968 plan drew on the same author's idea for the Pre-investment Study, in which urban densification was proposed through larger blocks, very similar to that of the Lastarria complex. Also, the maximum three-floor staggered houses design, was recommended in 1971 for the Morro and La Lisera foothills and seemed to resemble some schemes that *Cormu* had used for collective housing in the Parinacota complex.

6 CORRELATIONS BETWEEN GROWTH AND URBAN PLANNING

Arica's size, as a minor or intermediate city, and the development processes proposed over thirty years (1950–1980), allow us to verify some correlations between urban growth and urban plans.

The phenomenon of city growth, reflected in the succession of building plans, bear witness to processes that the large primate cities also underwent, at least in population size, in their capacity to envisage industrial installations and the expansive growth of the urban plan.

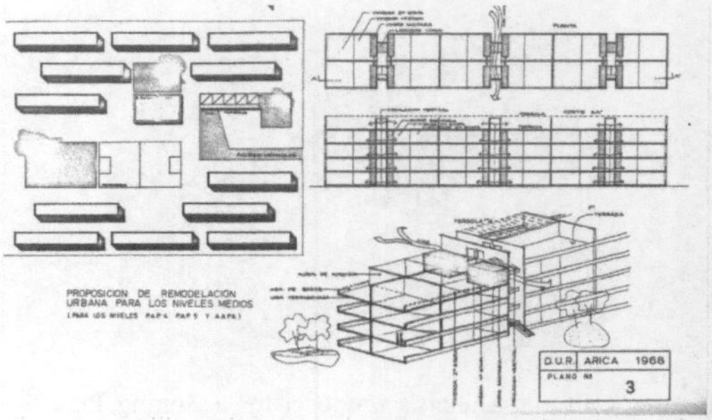

Figure 6. Emilio Duhart. Housing types proposal. Source: Ministry of Housing and Urbanism. Pre-investment Study, Arica (Santiago: DUR, 1968).

It is clear that urban plans, as instruments designed to guide urban development, only lagged behind the city's real growth. The city's study and planning alternatives accelerated as demands increased. The instruments were not in the advanced stages of forward planning, but rather responding to the conditions imposed by the impact of population growth and urban dynamics. Perhaps only the foresight of the hierarchical road system can be considered an exception.

The relationship between urban growth and planning seems to have had some ups and downs that were not resolved effectively, even at the level of the typological choices made. Although initially favoured at the architectural level, the use of blocks and superblocks was widely superseded by the extensive isolated housing alternatives, built on the basis of land availability and with the prefiguration given in each case by a sectional plan which extended an urban plan restricted in its possibilities. The incorporation of these sectional plans would not occur until 1971 when the majority had already expanded the urban plan in a discontinuous manner. However, it seems that planners were aware of the expansion problem, given the need to extend the proposals to typological levels, which were not present in traditional regulations. Thus, Arica became a laboratory where the correspondence between urban development and planning strategies can be read, even at the level of the proposed architectural typologies.

REFERENCES

Chile, Corporación de la Vivienda. 1963. *Plan Habitacional de Chile*. Santiago: CORVI.

Mangin, D., Panerai, Ph. 2006. *Projet urbain*. Marseille: Éditions Parenthèses.

Lang, A. 1962. *Puerto libre de Arica*. Santiago de Chile: Tesis de Licenciatura, Universidad de Chile.

Panerai Ph. 1983. Crecimientos. En: Panerai, Ph., Depaule, J-C., Demorgón M., y Veyrenche, M. 1983. *Elementos de Análisis Urbano: 25-56*. Madrid: Instituto de Estudios de Administración Local.

Pérez, F. 2006. *Bresciani Valdés Castillo Huidobro*. Santiago, Chile: ARQ.
Pérez, H. 1967. El Plano Regulador y el Desarrollo Urbano de Arica. En Colegio de Arquitectos. IV *Convención Nacional del Colegio de Arquitectos de Chile*. Santiago: Colegio de Arquitectos.
Portas, N. 2012. *Os Tempos das Formas - Livro 2: A Cidade Imperfeita e a Fazer*. Minho: Universidade do Minho.
Ruz, R; Galdames, L; Díaz Araya, A. 2015. *Junta de Adelanto de Arica (1958-1976). Experiencia, Documentos e Historia Regional*. Arica: Ediciones Universidad de Tarapacá.
Solá-Morales, M. 1974. *Las formas de crecimiento urbano: urbanística I: curso 1973-74*. Barcelona: Edicions UPC.
Solá-Morales, M. 1989. The Culture of Description. in *Perspecta* Vol. 25, 16–25.
Solá-Morales, M. 1997. *Las formas del crecimiento urbano*. Barcelona: Edicions UPC.
Torrent, H. 2015. El patrimonio moderno en la construcción del país urbano: Dialécticas entre arquitectura y ciudad en Chile 1930-1970, *AUS (Valdivia)*, *17*, 11–17.
Torrent, H. 2018. Modern Heritage and the Challenge of Urban Conservation: Between Singular Buindings and the Metamorphosis of the Urban Fabric. In: Tostões, A. y Koselj, N. *Metamorphosis. The Continuity of Change*: 140–146. Lisboa-Ljubljana: Docomomo International-Docomomo Slovenia.
Torrent, H.; Ruz, R.; Morán B. 2018. Arquitecturas para la institucionalización del desarrollo: tres dimensiones en la obra de la Junta de Adelanto de Arica. In Torrent, Barría et al. (eds.), *Patrimonio Moderno y sustentabilidad: de la ciudad al territorio*: 124–128. Valdivia: Docomomo Chile - Universidad Austral.
Torrent, H. 2019. *Arica, 1950-80: la forma urbana del desarrollo. Estructura y lógica de partes en el crecimiento extensivo*. XI SIIU - Seminario Internacional de Investigación en Urbanismo, Junio 2019; Santiago de Chile.

Research Tracks in Urbanism: Dynamics, Planning and Design in Contemporary Urban Territories – Allegri et al (eds)

Narratives under construction: Negotiating belonging in the city[i]

I. Sancovschi & C.R. Duarte
Architecture Subjectivity and Culture Laboratory (LASC), Postgraduate Program in Architecture - PROARQ/ Federal University of Rio de Janeiro UFRJ, Brazil

ABSTRACT: Addressing the subject of migration and refuge, this article reflects on belonging to place within the city. Through a case study, we analyzed narratives constructed by a Venezuelan asylum seeker, based on her memories and everyday experiences in the city. The study is based on an approach to urban space that adopts the concept of sensitive ambience, focusing on the social and cultural aspects of space. We believe that the imposed conditions for a legal and socially acceptable existence transform the way in which asylum seekers inhabit the city. As we demonstrate at the end of the article, life narratives are constructed daily, reflecting a negotiation of how one belongs that is permeated by factors such as hospitality, suffering and segregation.

1 INTRODUCTION

To think about the city today is to take on the need to move through multiple (social, cultural) layers within an increasingly complex space that manifests through material and subjective dimensions. When addressing this complex space, it is necessary to understand that these layers are often intertwined through relationships between the self (individual) and the Other (space, society) in the processes of (re) construction of identities and the structuring of subjectivities (albeit those of space or of the individuals present within it).

From our perspective, knowledge borne by the subjectivities of the various cultural groups that inhabit large urban centers today is capable of radically transforming the city project (s) that we seek for the future. In fact, research on the sensitive aspects of space has been growing in recent times. By looking at subjectivities, the study of ambiences, on which we base our work, focuses on the sensitive and dynamic aspects of places, as well as the cultural and social aspects of the users. This allows an understanding of the sensitive experience of cities, thus bringing new ways of thinking and acting within the urban environment (Duarte, 2013). In this context, our research has been focusing mainly on the relationship between culture and space, intermediated by subjects, with the main objective of understanding the inner dynamics of the city in relation to the multiple cultures that coexist within it.

The phenomenon of mass migration and the question of refuge are not new to the dynamics of cities. However, as the sociologist Bauman (2017) points out, "the modern way of life" serves as an impulse to this phenomenon, causing it to emerge as an important point for reflection in studies about the contemporary city. The research that forms the basis for this article is dedicated to the theme of migration and refuge, with a focus on asylum seekers and narratives they have built from memories and daily experiences of the city. We argue that these narratives support a territory in constant transformation, playing an important role in the (de) construction (Derrida, 2013) of the dwelling in exile. In this article we reflect on the question of belonging to place under the conditions of instability and fragility which asylum seekers find themselves.

Our hypothesis is that the imposed conditions for a legal and socially acceptable existence transform the way in which asylum seekers inhabit the city, bearing considerably on their life narratives, and, in turn, taking part in a dialectical process with the (de) construction of dwelling in exile.

In the first part of the article, we propose a theoretical debate about the notion of Dwelling,

i Translator's note: This title makes use of the complex Romance copula, split in Portuguese into 'Ser/Estar' from the Latin verbs 'ESSERE'/'STARE,' which could be translated literally as 'to be' vs. 'to stand.' In Portuguese, as in other Romance languages, the split copula allows for conceptual separation between an existential experience of being (or a being in some way imbued with consciousness) and the mere occupation of physical space — that which merely 'is'. The 'ESSERE'/'STARE,' split lends itself with ease, or one might say even invites considerations of the subjective dimensions of space.

DOI: 10.1201/9781003220855-26

where we contrast the idea of an essential dwelling to the idea of dwelling through everyday experience, which is constituted in a plural form. In the second part of the article, we will discuss urban space in terms of its sensitive dimension, based on the concept of Sensitive Ambience. It is in this context that we explore the notions of hospitality, suffering and of segregation, notions which encounter the subjectivities inscribed in the spaces in which asylum seekers are integrated within cities. Based on this theoretical foundation, we present our work methodology where, according to ethnotopography, we provide a critical analysis of spaces through tools borrowed from the human sciences, notably psychology and cultural anthropology.

This work is illustrated by the case study of Christina[1], a Venezuelan asylum seeker[2], who we followed throughout the second semester of 2019 in lectures given in public and private schools in Rio de Janeiro, as part of the "Refugees in Schools" program offered by the PARES-Caritas-RJ Refugees and Asylum Seekers Program. This case illustrates a qualitative study that seeks meaning and values attributed to the spaces experienced by asylum seekers.

2 TO BE IN SPACE X TO OCCUPY SPACE

From a psychosocial perspective, Fischer states that space is not defined as an external property, but as a set of matrices that provide a "concrete existence for individuals" (1994: 15). We can say, then, that the relationship between the individual and their place of origin, or the place they consider "home" will define part of their identity and way of life. We affirm this, because as we work with the concept of displacement, it is essential to understand dwelling from the perspective of everyday experience, to understand that living in exile is not limited to being in space but is (de) constructed in the relation between being and occupying space.

Heidegger defines dwelling as a fundamental trait of the individual. Turning to the interpretation of language, the philosopher sought the essential vigor of dwelling by stating that "The way you are and I am, the way in which we are men on this land is the Buan, the dwelling. Man being says: to be like a mortal on this land" (1954: s/p) and explains to us that being on this land involves: staying, lingering; to live in order to maintain daily habits; to build, in the sense of relating to the land, cultivating, protecting. In this sense, we understand that dwelling emerges from an essential notion of the individual's relationship with the land. This space where the individual is, where he dwells, is a space perceived as fixed, rigid, and permanent, formed by very clear borders. In Bachelard's philosophy, we also find an idea of dwelling as an essence of being. "The house is our corner in the world, it is as we often say, our first universe" (Bachelard, [1957] 2008: 24), human existence, then, would begin with this connection to the universe. The philosopher also stated that, once the house exists, the universe can no longer be comprehended by the dialectical game of self and non-self. The true value of inhabited space would come from the non-self that protects the self.

We must not put aside this notion of being in space, of dwelling based on an original place of reference, however, when studying groups of refugees, we are working with displaced subjects, and more than that, they are subjects forced into displacement. In this sense, the relationships between individual and space will occur in a more fluid and dynamic way. The rigidity and permanence of/in space become volatile. This individual who exists in space begins (also) to occupy space; the daily experience becomes important to the constitution of their identity. They can no longer only identify with their place of origin, with their essence. This subject in displacement (de) constructs his/her dwelling from a plurality of belongings that accumulate within this displacement (Said, 2003; Brah, 2011). From the case presented in this paper, we will see that these belongings which accumulate in the paths of individuals seeking asylum are negotiated daily.

3 THE SUBJECTIVITIES OF URBAN SPACE

As stated in the introduction, to understand the complexity of the city, it is necessary to address subjectivities inscribed within the dialectic relationship between the individual and space. When addressing the idea of dwelling based on everyday experience, we cannot ignore the sensitive, dynamic, social and cultural characteristics of space. Our proposed approach moves us somewhat away from concrete, tactile and physical space, seeking instead a fluid, dynamic and imaginary space. To discuss this urban space that coexists, and modifies (itself) daily along with the concrete city, we work with the concept of sensitive ambience from the perception of another, in the case of this study, from the perception and narrative of an asylum seeker.

3.1 *Sensitive ambiences*

"The ambience is born from the encounter between the physical properties of the environment, the corporeality and capacity for movement and affective tone" (Augoyard, 2007/2008: 60).

1 Name changed to protect the informant's identity.

2 Christina and her family were granted refugee status at the end of 2019, along with other Venezuelans in an unprecedented action by the Brazilian government that granted refugee status to approximately 21,000 Venezuelans. In this article, we use the title of asylum seeker because throughout our fieldwork, Christina had not yet been granted refugee status.

The concept of sensitive ambiences allows, within urban studies, the possibility of seeing a city as revealed between the lines, bringing subjectivities into the objective aspects of its spaces. Pallasmaa (2009) talks about a paradigm of vision that has been built in the Western world since the ancient Greeks and how this paradigm deprived the comprehension of spaces, images, architecture and urbanism of its own complexity. The concept of sensitive ambiences breaks precisely with this paradigm, placing the other sensitive and perceptual aspects alongside vision (in a completely horizontal manner), adding the psychological, social and cultural values found in them to the physical spaces. It is in this context that we decided to examine the spaces experienced in the daily lives of asylum seekers, based on the notion of ambiences.

"Ambience evokes our subjective interpretation of the collective experience, the awareness of being part of an urban place whose sensations have meanings shared by its users. Thus, there is a clear "incarnation" of subjectivity, which gives rise to the idea that an ambience is recognized in the community, despite representing itself in individuality." (Duarte, 2013: 3).

Thus, the study reported by this article shows the tension existing in the encounter of the environment experienced collectively in the city of Rio de Janeiro with the individual daily experience of the asylum seeker whose story was heard. From this tension, small anchors begin to emerge to a space created between the material and the imaginary, enabling identification of this asylum seeker with her "provisional" living space, which is the city of Rio de Janeiro.

3.2 *Hospitality, suffering and segregation*

For this article, we selected three aspects that seem to be of extreme importance to reflections on dwelling in exile, and in the placement of asylum seekers within their new living spaces, namely: hospitality, suffering and segregation. It is important to note that we consider these aspects, or concepts, from a spatial perspective, in light of the sensitive ambiences outlined above. In this sense, our goal is to understand how these aspects are incorporated in the collective perception of the city and how they become agents in the relationship between the individual who narrates their daily experience and living spaces, i.e. how these aspects are perceived by the asylum seeker and how she uses them to negotiate her belonging in space.

The notion of hospitality explored by Derrida (2000) brings ambiences closer to the human character of space. According to the author, the term hospitality is presented in a multifaceted way, which relativizes the roles of host and guest, and hides a sense of hostility. He thus questions the idea of unconditional hospitality and contrasts this against the hospitality of rights and a just hospitality. Although it is important to acknowledge this ambivalent aspect of hospitality, we will not go into further detail in the present study, which instead focuses on the relationship between hospitality and place/space.

"This 'giving place to place' is, it seems to me, the promise alluded to by the word (hospitality). This brings us to understand the question of place as being fundamental, underpinning the history of our culture and yet unthought within it. This would be to consent to exile, that is, to be in a native (one would say almost maternal) relationship and yet, it is in suffering with place, with the home, that thought comes to humans." (Derrida & Dufourmantelle, 2003: 16)

The above passage highlights the importance of place in understanding of hospitality, leading us to think of place as the key to an emotional conflict between the individual and their own existence, where place can represent a source of both comfort and suffering.

With respect to people seeking asylum, if the place of arrival and reception is associated with suffering, what about the place from whence they came and the places encountered during their journey? Suffering is naturally an integral part of this subject and spatial experiences. In general terms, the psychoanalyst Dunker (2017) states that suffering is something that must be fought, taking into account the "truth that infiltrates it, the truth that needs to be said". However, what we perceive, from our studies on the theme of refuge, is that suffering is central to their experience. And that, in a sense, rather than being fought, it becomes highlighted.

"At some point in modern times, being able to ask about your own happiness has become a political factor. [...] This produces a mutation in the moral status of suffering, no longer a kind of sacrifice that produces added value, necessary to justify the righteousness of a well-lived life, but an element to be politically determined in the equation of what would become a well-lived life. [...] Then, for the first time, on a mass scale, a kind of "invisible government" of our lives was formed, from which we came to expect (because it began to promise us) happiness. Expressing suffering as a certain form of fiction capable of expressing the singular experience of life appears as a great aesthetic program." (Dunker, 2018.np).

While contemporary life urges us to seek happiness incessantly in order to escape everyday suffering, asylum seekers are committed to the experience of suffering they have gone through. According to Law No. 9474, which defines the mechanisms for the implementation of the Refugee Statute in Brazil, an individual is recognized as a refugee that: "due to well-founded fears of persecution for reasons of race, religion, nationality, social group or political opinions if outside their country of nationality and cannot or does not want to accept the protection of that country" or also, "due to a serious and widespread violation of human rights, is obliged to leave their country of nationality to seek refuge in another country." Thus, the asylum seeker has, as a prerequisite for obtaining their legal status in the country, a trajectory of suffering, which must be defended in a coherent way before the State (for obvious reasons) and with society (as a result).

Finally, we must address the issue of segregation, which also carries with it notions of racism and prejudice. In his essay entitled *Necropolitica,* Mbembe argues that "To be sovereign is to exercise control over mortality and define life as the implantation of the manifestation of power." (2018: 5). We can say that this principle plays a significant role in the spatial experience of asylum seekers in Brazil. Its trajectory begins in the search for the right to exist, that is, the escape from "death". And once accepted in the country of exile, they take up residence in places known to be neglected by the government, in the poor outlying districts, outskirts and favelas (shantytowns) (Cidade, 2018), with parallel powers, where "letting live" coexists with "letting die." By focusing on the colonial heritage of Brazilian cities, we will see how the poor outlying territories took shape, together with aspects of racism and prejudice, and reaffirming the idea that "space was, therefore, the raw substance of the sovereignty and violence that sustained it. Sovereignty means occupation, and occupation means relegating the colonized to a third zone, between the status of subject and object." (Mbembe, 2018: 39) Thus, the spatial experience of refugees and asylum seekers in Brazilian cities in general, and in the city of Rio de Janeiro in particular, will be part of this perspective of segregation, deprivation of rights and lack of access to the city.

4 LISTENING TO A STORY

The theoretical frameworks presented above emerged from fieldwork involving what we call ethnotopography (Duarte, 2010), a methodology that seeks to understand the spaces experienced, inhabited and built by relating them to their individuals and socio-cultural groups. The fieldwork referred to in this article consisted of a 6-month follow-up of the Refugees at Schools program, developed and offered by PARES-Caritas-RJ. The program aims to provide education on the theme of asylum, mainly targeting children from primary and secondary education in the public and private school systems of the municipality of Rio de Janeiro and neighboring cities. The idea is that, from early childhood education, vectors of awareness are created for the theme of asylum, allowing for the possibility of forming networks of those empathetic to the cause. Children themselves embrace the role of educators within their families, thus expanding the circle of knowledge. During the period in which we followed the lectures, there was good acceptance and receptivity not only from the schools (since they invited the program organizers to come in), but from the students, who engaged in campaigns to collect donations and watch the lectures with great curiosity, posing questions that were important for our reflection.

Since our space in this article is limited, we focus our analysis on the stories told by Christina, a Venezuelan asylum seeker, during four lectures given over this period, applying ethnotopography to the narratives (Sancovschi, 2017). The analyses were performed based on the three concepts mentioned above: hospitality, suffering and segregation. However, it is worth remembering that the fieldwork also included a participatory observation of the spatial and relational experiences along the way, during transport and within the schools, yielding references of Christina's other interpersonal and spatial relationships within the space of the city. In the lectures, Christina presents a life story, telling of her trajectory in Venezuela and part of her experience after arriving in Brazil. Following her story, listeners (students and teachers) had the opportunity to ask questions, clear up misunderstandings, or make comments. With each lecture heard, the story of Christina gains nuance and depth, a natural feature of oral narratives. It is precisely from these nuances that we sought to understand the relationship between this asylum seeker and her new place of abode.

When we started our research, Christina was still an asylum seeker and throughout the period in which we accompanied her was awaiting her, and her family's, asylum status to be granted. At the end of 2019, the Brazilian government granted refugee status to 21,432 Venezuelans through prima facie recognition (UNHCR, 2019). Christina and her family were given refugee status, thus rectifying their legal existence in Brazil. It is worth remembering here that Christina is not an emblematic figure of all Venezuelan refugees, her story does not represent a pattern typical of the refugee´s journey across the Brazilian border. Unlike most Venezuelans who arrive in Brazil, Christina has a history of a more orderly arrival in the country, having entered the country documented and with a small financial reserve for the journey. However, we believe that each refugee has a unique story to tell, and that each story can contribute to the understanding of this contemporary phenomenon of mass migration and refugees.

5 TELLING A STORY AND NEGOTIATING BELONGINGS

Always accompanied by her youngest daughter, Christina starts her account with the following statement: "Good morning, I am Christina, I am from Venezuela and I speak Spanish. I am still trying to speak Brazilian. If you don't understand something, just ask." Christina speaks Portuguese mixed with Spanish, which is easy to understand. She tells her story in a very natural way, and with each story we learn a little more about her trajectory. The first time we heard her story, we had the impression that her experience as a refugee had already become naturalized. However, throughout our research, it was possible to get a sense through her story, speech and emotion how the difficulties of dealing with this label are a part of her everyday life. It is interesting to note that through this simple speech, Christina claims a place for herself. She apologizes for not being "equal", but she counts on the empathy of others to recognize her effort to learn and to try to communicate in the local language. Christina asks for hospitality while welcoming her listeners. An atmosphere of reciprocity and empathy is built here, amidst languages so similar, yet so different.

Our speaker arrived in Brazil with her family in August 2017. She was 8 months pregnant carrying her third child under conditions verging on malnutrition. The general lack of access to food and medicine, among other issues, forced them to cross the border. In her account, Christina recounts the situation her family was in before and during the period in which the degradation process affected them in Venezuela. As middle-class citizens, the family lived in their own home, had their own car, and the two oldest children studied at a private school. They had a well-structured life and were still able to help the family when needed. Upon arriving in Brazil, the family had to adapt to new living conditions quite unlike what they had been used to before the degradation process took place in Venezuela. Here in Brazil, in Rio de Janeiro, they were welcomed by a family they were acquainted with, where they were given shelter for three months. After this period, they had to move, ending up in a favela. With the help of Caritas, the family filed for asylum and Christina managed to get medical attention for the end of her pregnancy and the birth of her daughter, and the older children received a place in the public schools close to home. Christina's husband got a job and the couple's daughter was born healthy.

Christina's account has an apparently happy outcome, but suffering runs through her stories and is often emphasized. The need to be recognized in vulnerability lends Christina's speech nuances of suffering that also appear or disappear depending on the day in which she is telling her story, even though it took place in the past. We noted, for example, that the day she received a message from her family saying that the situation in Venezuela had worsened, that her mother has not been able to get the medicine she needs, this comes through in her talk. For Christina, being a refugee,

"Is important because you make the world's problems known, if you don't become a refugee, like, if I'm a Venezuelan, and every Venezuelan who leaves the country, leaves as any other migrant, nobody will know what is happening in the country, so for me the refugee is important because they have a story to tell, they have been through a situation that cannot be compared to any other. Everyone suffers, everyone goes through difficult times, but the moment you have to literally run away from a country and leave everything behind, this story takes on a great sentimental value. For me, to be a refugee is to be the voice for what is happening in the world." (transcript of Christina's speech, September 4, 2019).

Thus, the narrative of suffering opens the possibility of existence "outside of place" for Christina. She herself is committed to being here, showing her country's problems in search of help and worldwide recognition of a problem that made her flee her own home. In this sense, suffering carries the memory that she *is* here, but in reality, she *is* there.

A recurring thread in Christina's narratives is that neither she nor the family suffers prejudice in Brazil. However, she also frequently recounts situations where her daughter or husband had suffered for being foreigners and always talks about the difficulty in being included, either due to language or culture, or simply because she is Venezuelan. The need to make oneself belong is such that recognizing oneself as a victim of prejudice becomes difficult. To recognize prejudice would be to cancel out the possibility of inhabiting this place.

Even more difficult for Christina, was that she found herself in a segregated territory of the city, where powers are relativized, and spaces are dominated by social and cultural relations. After becoming refugees out of a need to stay alive, they faced new threats to life in the favela. We realized that the refugee, like the black person and the poor person, is placed in a "third zone" (Mbembe, 2018).

"Then we went to live on a *morro* (hillside shantytown) in Tijuca, then we went to live on Morro do Turano. It is also a very strong experience, because in Venezuela there are favelas just like here. And when we got there, they thought my husband was a police officer, and the next day the hill was taken overrun by police. And then everyone said you are going to be killed, you have to get out of there..." (transcript of Christina's speech, October 30, 2019).

Christina and her family managed to stay on the morro for a while longer thanks to contacts from Caritas employees who had family members living on the same morro. The church also played an important role in helping the family become recognized in the area. This illustrates the organization of segregated spaces in the city through social relations and parallel powers.

In Christina's narratives, it is still evident that adapting to this space in the city was not at all easy. She once told a story about the first shooting they heard inside the house, and how difficult it was to deal with the situation, both for her and the children. The place that should offer protection becomes a place of danger, the space that should be welcoming becomes hostile. Thus, we find together in this same space the experiences of hospitality, suffering and segregation that are sometimes evoked, and sometimes forgotten in her narratives in a constant negotiation with the space of the city.

6 CONCLUSIONS

This article describes part of a case study that sought to analyze ambiences that emerged from the narrative of a Venezuelan asylum seeker, revealing that this narrative is constructed daily in the speech and experience of the asylum applicant. This enabled us to reflect on the sensitive order of spaces, bringing the notions of hospitality, suffering and segregation to the concept of Sensitive Ambiences. Through these three aspects, otherness is revealed as essential in understanding the place that starts to be built and claimed by the relationships between different people. Construing Rio de Janeiro as a welcoming, happy and harmonious city is to see only the Rio de Janeiro of the poetic “Bossa Nova.” The daily experience of an asylum seeker in the city carries with it all the contradictions and problems of the contemporary city. We believe that this study can contribute to discussions on urban cultures, migrations and social justice, speaking to us, as they do not address solely the experience of asylum seekers, but also correspond to the experiences of a large part of the suburban population. The recognition of these nodes is essential in understanding the conditions that the city imposes on its inhabitants, and from this recognition we will be able to adapt our city projects in a more inclusive and just way.

REFERENCES

Augoyard, JF. 2008. Culture et recherche, Paris: Ministère de la Culture et de la Communication. *La construction des atmospheres quotidiennes: l'ordinaire de la culture*. 58–60.

Bachelard, G. [1957] 2008. *A poética do espaço*. Trad. Antônio de Pádua Danesi. Ed. Martins Fontes. São Paulo.

Bauman, Z. 2017. *Estranhos a nossa porta*. Zahar. São Paulo.

Brah, A. [1996] 2011. *Cartografias de la diáspora: Identidades en cuestion*. Traficantes de Sueños.

Cidade, N.C. 2018. *Refugiados Urbanos: estudo sobre a distribuição territorial de refugiados no Rio de Janeiro e seu impacto no processo de integração local*. Dissertação. UFRJ/PROURB.

Derrida, J. 2000. *Hostipitality*. Angelaki: Journal of the Theoretical Humanities, V5, n°3, 3-18, Dec.2000. Available in: https://www.tandfonline.com/doi/abs/10.1080/09697250020034706

Derrida, J. & Dufourmantelle, A. 2003. *Anne Dufourmantelle convida Jaques Derrida a falar da hospitalidade*. Tradução de Antônio Romane.Ed. Escuta. São Paulo.

Derrida, J. & Meyer, E. 2013. Uma arquitetura onde o desejo pode morar. In. Nesbitt, K. Cosac & Naify. (eds) *Uma nova Agenda para a arquitetura*. 166-173 São Paulo.

Duarte, CR. 2013 Ambience: pour a sensible approche de l'espace. In. Thibaud, JP; Duarte, CR *Ambiences urbaine en partage*. 21–30 Metis-Presses (eds) Genebra. Available in: http://lasc.fau.ufrj.br/artigos/219/ambiencia-por-uma-ciencia-do-olhar-sensivel-no-espaco

Duarte, C.R. 2010. *Olhares possíveis para o pesquisador em arquitetura*. ENANPARQ. Available in: http://lasc.fau.ufrj.br/public/upload/2019-07-15/082efbf224549714a58f9c749bc58038.pdf

Dunker, C. 2017. *A reinvenção da Intimidade: Entrevista com Christian Dunker*. Nexo Jornal. Available in: https://www.youtube.com/watch?v=0bbG5leMbK8

Dunker, C. 2018. *Reinvenção da intimidade: Politicas do sofrimento cotidiano*. Ubu (eds). Edição eletrônica, 2018.

Fischer, G.N. 1994. *Psicologia Social do Ambiente*. Piaget Institute. Lisbon.

Hiedegger, M. 1954 *Construir, Habitar, Pensar*, 1954.

Mbembe, A. 2018. *Necropolitica*.n-1 (eds). São Paulo.

Pallasmaa, J. 2012. *Os olhos da pele: a arquitetura e os sentidos*. Bookman (eds). São Paulo.

Said, E. 2003. Reflexões sobre o exílio. In: ________ *Reflexões sobre o exílio e outros ensaios*. 46-60. Companhia das letras (eds) São Paulo.

Sancovschi, I. (2017) A invenção do lugar judaico: uma vivência das ambiências nas narrativas de I.B.Singer, Moacyr Scliar e Amos Oz. Dissertação. UFRJ/PROARQ.

Lei N° 9.474, de 22 de Julho de 1997. Define mecanismos para a implementação do Estatuto dos Refugiados de 1951, e determina outras providências available in:: http://www.planalto.gov.br/ccivil_03/leis/l9474.htm

Research Tracks in Urbanism: Dynamics, Planning and Design in Contemporary Urban Territories – Allegri et al (eds)
 ISBN 978-1-032-10788-2

The (dis)connect between buildings of the tertiary sector and public space

L.M. Oliveira & M.A.J. Pisani
MACKENZIE, School of Architecture and Urbanism – Mackenzie Presbyterian University, Brazil

ABSTRACT: This theme arises from observing the recent urban transformations of an area in the southwest sector of São Paulo. The region shows a distinct pattern of urbanisation in line with the guidelines of the real-estate market of the tertiary sector, seeking to meet the demands of privileged segments of the population, to the detriment of the interests of popular classes. To reverse this pattern, the location of buildings becomes a key element to foster solutions that result in plots of land with fluidity features and permeable to urban appropriation. The study employs a survey and an analysis of quantitative and qualitative data to address the issue of the connection between buildings and public spaces for pedestrians in areas of new centralities. The results obtained demonstrate the predominance of buildings isolated on the plots and removed from their context, with the persistence of patterns that repeat the process of socio-spatial fragmentation.

1 CENTRALITIES OF SÃO PAULO CITY

The emergence and development of centralities in the city of São Paulo dates back to its historic formation. The city developed from a single centre until the late 1960s, when divided between the traditional and new centres.

"That's when, from the miracle period (1968-73), a new centre was structured along Paulista Avenue, which made the metropolitan centre develop into a Main Center and the Paulista Ave. Center: this reconfiguration would mirror the changes in the gradual passage from the Fordist regime to one of flexible accumulation" (Cordeiro, 1993 *apud* Frúgoli Jr., 2000, pg. 39).

Maricato (2001) points out that Villaça had previously identified that both the displacement and the emergence of new urban centres in Brazilian cities are related to the process of popularization of these centers. There is a devaluation effect resulting from popularization, which ultimately stimulates the emergence of another centrality targeted by companies that seek distinct spaces, not readily accessible by public transport. According to Maricato, this fact results from the elites' option to use the road transport system, which impose urban scenarios that emphasise public investment in this system. New centralities' option for the predominance of this kind of investment over public transport and public spaces, compromises the population's mobility and coexistence with elites in the public sphere.

Another issue raised by Loukaitou-Sideris & Banerjee (1998) states that the public realm of new centralities in the tertiary sector started to degenerate as commercial activities became increasingly isolated and insular, developing in closed protected environments within galleries, shopping malls, atriums, and privatized squares.

The result can be seen as a consequence of the polarization observed between traditional urban centres that display a popular public realm with social and economic diversity and, on the other hand, the new centralities that offer exclusive and excluding environments represented by privatized spaces. Hence, the polarization reinforces the differentiation in the profiles of companies, workers and residents who are interested in living and working in the traditional centres or in the new centralities.

Researcher and political scientist Lúcio Kowarick (2000) analysed the urban issue and social movements. He summarized the topic for the city of São Paulo: "São Paulo has always been a metropolis of vast contrasts, although at least until recently, one of great possibilities for social and economic ascent. But it has also and always been a city where the rich tried to get away from the poor (. . .). That is, the creation of spaces for exclusive and excluding activities has always been a prevailing dimension of political and economic groups that intervened in producing the city. This is the theme of space privatization, in the sense of making it difficult for low-income populations to access" (Kowarick, 2000, pg. 5).

Botelho (2007) also discusses the issue of popularization between the traditional centre and the desire for segregation from: "(. . .) business groups that move to the southwest sector of the city, avoiding the centres and leaving buildings in total disuse. This contrasts with the growing population living on

DOI: 10.1201/9781003220855-27

the streets and using public spaces for a network of informal relationships focused on survival" (Botelho, 2007, p. 149).

Thus, after the decline of the traditional centre of São Paulo, wide avenues emerged and co-opted public investment to produce new scenarios capable of receiving businesses from the tertiary sector, with initial interest devoted to the avenues: Paulista, then Faria Lima, Engenheiro Luís Carlos Berrini, Faria Lima, Juscelino Kubitschek, and more recently to Dr. Chucri Zaidan Avenue. This group of centralities is spatially located in the southwest sector of the city of São Paulo, which corresponds to the region identified by Villaça (2001) as an area predominantly segregated with respect to the spatial distribution of neighborhoods occupied by high-income groups.

The above-mentioned centralities shape the most vertical prominent profiles of a cityscape as an affirmation of their presence in the competition for entrepreneurial investment. In European and North-American literature, the tertiary sector centralities are called Central Business Districts (CBD) and pinpoint a concentration of office buildings in a geographically limited area that leads to the configuring of zones with a visual aspect markedly different from the larger surrounding area in which they are incorporated.

Loukaitou-Sideris & Banerjee (1998) attribute to the economic power of the corporate and business elite, expressed by market forces, the need to build an image that symbolizes the power of specific companies which, in turn, are interested in attracting the middle and upper classes to live and buy in the shopping areas where they are established. In both cases, the objective can be achieved if the built environment is attractive and offers the equipment this specific contingent of the population desires.

Hence, urban design and the qualities incorporated in the buildings become important and significant elements in promoting the diversity or segregation of the new centralities. These are factors in which public authorities can participate more strongly as agents capable of regulating and imposing urban guidelines that foster the diversification and appropriation of spaces for public use in all segments of the cities, including areas of centralities. However, the reverse quite often occurs and, according to Loukaitou-Sideris & Banerjee (1998), the alliance between political and businessmen's interests reinforces the protection of the areas of new centralities by means of zoning regulations and urban guidelines to control land use.

Public authorities' lack of interest or omission can be seen as a consequence of the transition from an industrial city to a city of services, which has led to a transformation in the role of public agencies as regulatory and management bodies, ever since the private sector started to operate openly in cities, based on the deregulation imposed by economic neoliberalism. With the private sector acting freely and extensively in cities, there was an imbalance in the performance of public authorities, who became mere providers of the infrastructure needed to make private investments viable.

According to Madanipour (2003), the nineteenth century can be identified by the development of industrial cities that housed the populations coming from the countryside to live close to workplaces then represented by industries. As a consequence, cities were organized into different sectors, with residential zones for the wealthiest classes and popular housing zones for those working in the industries. This spatial and social segregation is a characteristic of the modern city that continues today. The cities that were transformed during the industrialization process are now undergoing new shifts as a result of changes derived from the deindustrialization process and the transition to a service economy, reinforcing the growing fragmentation and polarization between different social groups in the postmodern city.

According to Ali Madanipour (2003), one of the ways to arrest and tackle the process of urban fragmentation, and the fear of the middle and upper classes of the lower-income population, may be by investing in public space and examining guaranteed access for all: "When creating areas where people mix, it is expected that different people can live together with a certain amount of tolerance. This is especially crucial at a time when the welfare state is under threat of restructuring, and social fragmentation has intensified [. . .] Recent interest in promoting urban public spaces can be interpreted as an interest in reintegrating fragmented cities" (Madanipour, 2003, p. 145).

Therefore, the subject of this study addressing the relationship between buildings and public space for pedestrians, derives from observing a recent urban transformation in the southwest region of the city of São Paulo, where Nações Unidas Avenue (Marginal Pinheiros) and Luís Carlos Berrini Avenue have already been part of this centrality axis characterized by the concentration of public investment. Today, Dr. Chucri Zaidan Avenue represents the extreme pole of this vector. The region has a distinct urbanization pattern, configured according to the guidelines of the real-estate market of the tertiary sector, which seeks to meet the demands of privileged sectors to the detriment of the interests of the more popular classes (Oliveira, 2013).

That is an example of the standard of urbanism in which processes and models can be identified that help understand the developing trends in urban centralities of São Paulo. Within this urbanization pattern, the physical and compositional features of the volumes, the choice of building cladding, and the technology available and incorporated in the list of available services become relevant, as well as the location plan for the buildings and the resulting landscape. On the subject of the connection between buildings and the public space for pedestrians, the location becomes the key element to stimulate solutions that result in plots of land with fluidity features and which are permeable to urban appropriation.

São Paulo boasts significant iconic examples that demonstrate possibilities for greater connection and integration between the ground floor of buildings and the city, namely: the Conjunto Nacional, designed by David Libeskind (1956); the Copan Building, designed by Oscar Niemeyer (opened in 1966); the Art Museum of São Paulo - MASP, designed by Lina Bo Bardi (opened on November 7, 1968); the Itaú Unibanco Business Center (1982-2005), near the Conceição metro station Line 1-blue; the Brascan Century Plaza, in the Itaim District, designed by Könisberger and Vannuchi (2003), and the Moreira Salles Institute on Paulista Avenue, designed by Andrade Morettin Architects (2017). However, this selection may be viewed as exceptions or sporadic elements in the landscape.

Each example shows one or more of the following features that encourage the connection between the building and public space, contributing to public appropriation: mixed uses provided by commercial galleries on ground floors and residences or services on the upper floors; the connection between the streets bordering the plot via public access passages through a covered gallery or square; squares for public use; living areas with seating for people without the need for consumption; and access via well-maintained and properly-dimensioned sidewalks.

The qualities found in the above-mentioned examples were interpreted in the new Master Plan of the City of São Paulo (São Paulo, Município, 2014) through the introduction of urbanistic tools that encourage the connection of buildings with the public space of the sidewalks, such as active façades to fuel mixed use and commerce on ground floors; public enjoyment to allow the passage of pedestrians through the plots; and the widening of the sidewalks.

The aim of this study was to analyse the location patterns of new office developments in the southwest region of the city, seeking to identify whether they used design guidelines that promote a connection with the surroundings. The focus on public use spaces represented by sidewalk segments and by the ground floor of corporate and office buildings results from observing the process of interiorizing commercial and services environments in new centralities.

The territorial cut-out of interest for this study is the immediate area undergoing full development around Dr. Chucri Zaidan Avenue that is organized as a linear centrality, an extension of Engenheiro Luís Carlos Berrini Avenue. This area currently features vigorous activities by private initiative, since it is included in the Urbanistic Plan of the Chucri Zaidan Sector of the Água Espraiada Urban Operation (São Paulo, Município, 2001). This urban procedure proposes a series of initiatives aimed at improving the region. Also, in January 2018, the extension of the road axis including the corridor of Dr. Chucri Zaidan Avenue with João Dias Avenue at its southern end was inaugurated. Within this territorial cut-out, the subjects of study selected for the data survey include the corporate and office buildings constructed and inaugurated in the last decade, that is, from 2010 to the present.

The study uses the case study method which, based on empirical investigation, is recommended for in-depth analysis of "a contemporary phenomenon and in its real-life context, especially when the limits between phenomena and the context are not clearly manifest" (Yin, 2010, pg. 38), as is the case with the issue explored in this article. The study procedures employed included: a survey of corporate and office developments built and installed from 2010 until the present; a survey of the locations of the selected buildings on their respective land plots to provide a cartographic base; a survey of data on the area about previously determined categories of analysis; quantitative and qualitative analysis of the data obtained in relation to the impact of the set of edifices for the connection of the building with the public spaces of the centrality.

2 THE RELATIONSHIP BETWEEN BUILDINGS AND PEDESTRIAN PUBLIC SPACES

The region established to collect data for this study is situated in a segment of Dr. Chucri Zaidan Avenue, to the south of Roque Petroni Jr. Avenue, near the Morumbi and the Market Place shopping malls. The study identified ten corporate and office developments built after the year 2010. Together these developments occupy a surface area of 207,781 square metres and nearly one million and a half square metres of constructed area, as shown in Figure 1.

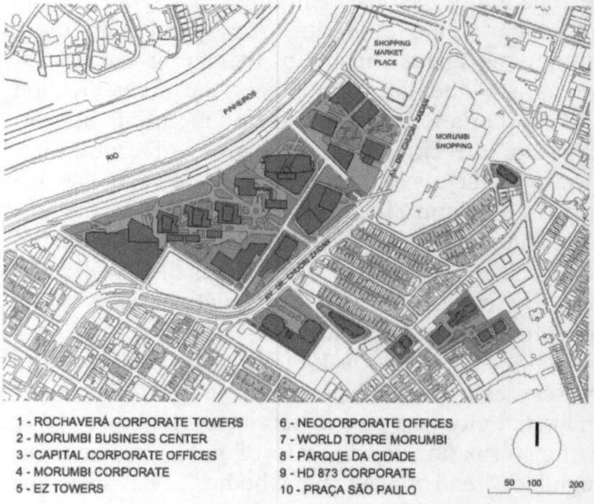

Figure 1. Map of the corporate and office developments built in the region of Dr. Chucri Zaidan Ave. after 2010. Source: Prepared by the authors from Geosampa, 2019.

After identifying the set of buildings, the research procedure applied to the ten complexes selected, included a description of the features of each, highlighting the following categories of analyses: the types of programmatic uses of the ground floors to verify the existence of commercial, cultural, or social areas for people; to verify enclosures, walls or fencing that limit direct access to the plot; and opportunities for pedestrian enjoyment within the plot.

The Rochaverá Corporate Towers (Figure 2), designed by the architectural firm Aflalo & Gasperini with landscape design by Pamela Burton, was built on a 33,000 square-metre land plot, at the confluence of Dr. Chucri Zaidan, João Dória and Nações Unidas avenues. It is a triple A commercial complex, consisting of four towers with inclined glass faces and different heights with 8, 16 and 33 floors, plus a garage building. The complex has one central square and other secondary plazas, allowing pedestrians to circulate freely on the large land area at the confluence of these three avenues. It is a well-located development in the city due to its connections: in addition to the circulations, inner squares and access to surrounding roads, the complex has connections with the Morumbi and Market Place malls, as well as the Morumbi Metropolitan Train station, which has connections to the number 4 and 5 subway lines. The architects' plan is that the entire location be connected to the city, with no fencing. The Rochaverá building received a Gold category Leadership in Energy and Environmental Design (LEED) certification (Aecweb, 2019 & Aflalo Gasperini Arquitetos, 2019).

The Morumbi Business Center (Figure 3), designed by Aflalo & Gasperini, is class A rated as a Gold LEED CS office building, with 12 typical floor plans, ground floor and three below-ground levels. Despite its plot having three sides facing streets, the ground floor access is for building users only, and the external area is intended for parking. Basically, the entire perimeter of the terrain is closed off by walls and translucent glass enclosures (Morumbi Business, 2019).

The Capital Corporate (Figure 4), designed by Itamar Berezin, is a 20-floor building, located secluded on a triangular block. The entrance faces Dr. Chucri Zaidan Avenue, with a landscaped frontal setback designed by landscape architect Benedito Abbud. The rear face, on Antônio de Oliveira Street, is formed by a blind wall around its entire perimeter, because it corresponds to the wall that surrounds the existing car park behind the ground floor lobby. There is no commerce or square for public use on the ground floor.

The Morumbi Corporate (Figure 5), designed by Aflalo & Gasperini, is located on a block surrounded by three streets, but the complex only has pedestrian access via Dr. Chucri Zaidan Ave. Near the plot boundary with Rochaverá complex there is a metal footbridge, protected by a glass enclosure which forms an elevated passage over the avenue, connecting both office complexes with the Morumbi

Figure 2. View of the Rochaverá Corporate Towers Complex and its frontal square.

Figure 3. View of the Morumbi Business Center, enclosed with walls to conceal the ground floor parking.

Figure 4. View of the Capital Corporate Offices that has no commercial base; site plan drawing.

Figure 5. View of the Morumbi Corporate Complex, with a square-made entrance.

shopping mall. The three-faced block has wide sidewalks, and a smaller width for the rear street, Rua Antônio de Oliveira. The complex consists of two towers with 25 and 17 floors, linked by the base that houses a convention area on the ground floor, and a food court on the upper floor with a skylight-topped inner garden.

The EZ Tower (Figure 6), designed by DWA and the Uruguayan architect Carlos Ott is situated on the corner of Dr. Chucri Zaidan Ave. with Enxovia Street, near the CPTM Morumbi station. It consists of two Triple A towers, with three below-ground levels, a ground floor, 26 floors in tower A and 31 in tower B, as well as 4 over-ground levels for parking. Both towers have auditoriums and a café on the ground floor, as well as helipads on the roof. A solely-contemplative square with grass and bushes was created on the corner of both streets, without seating or pedestrian routes.

The development received Leadership in Energy and Environmental Design Certification - LEED Gold and the Guia de Rodas Platinum Accessibility Seal for including more accessibility elements than Brazilian standards require (Eztec, 2014 & Estadão, 2016).

Neocorporate Offices (Figure 7) is a single office tower located on Enxovia Street, on a triangular area with Francisco Amêndola Street. DWA Arquitetura designed the building, which has six-floor basement parking and a total of 31 floors. The sidewalks around the building are wide and ample, with small landscaped areas because the building footprint occupies a large proportion of the land plot. The façade facing Dr. Francisco Amêndola Street is surrounded by metal railings and has no connection to the street. On this portion of the block, this street shows only walls and blind gables of two-floor buildings.

The World Torre WT Morumbi (Figure 8), designed by Aflalo Gasperini is located on Nações Unidas Ave., and consists of two 28-floor similar but inverted towers (from the 3rd to the 30th floor), technical areas on the first, second, 31st, 32nd and 33rd floors, a roof on one of the towers and a helipad on the other, ground floor, and five below-ground floors. The buildings are connected by five footbridges that form a single volume. The towers have individual accesses and the ground floor also houses a theater, a restaurant and the grand reception atrium. The entire volume is clad with glass in different tones (Aflalo Gasperini, 2009). The WT Morumbi was diagonally located on the land, leaving one triangular plaza facing Nações Unidas Avenue, and another facing Antônio de Oliveira Street.

The Parque da Cidade (City Park) (Figure 9) was designed by Aflalo Gasperini on a plot of land along Nações Unidas Avenue. The design consists of ten towers: a hotel tower, a shopping mall, five corporate towers, an office tower, and two residential towers with 16, 20 and 24 floors. The location concept was

Figure 6. View of the EZ Towers Complex, with a square by the entrance.

Figure 7. View of the Neocorporate Offices, with its 6-floor parking base.

Figure 8. View of the World Torre Morumbi, and the footbridge over Dr. Chucri Zaidan Ave. that connects the development and Morumbi Corporate Complex directly to the Morumbi shopping mall.

Figure 9. View of the Parque da Cidade with the Shopping Mall in front of one of the completed towers of the complex.

to integrate private external spaces with public spaces. The nearly 62,000 square-metre spaces are open, creating a linear park that follows the line of the buildings, with gardens, bikeways, restaurants and other commercial uses that are integrated into the diagonal 22,000 square-metre green area of the plot.

In addition to the accreditation of the buildings, the complex has also achieved the LEED ND Certification *(Neighborhood Development)* for bringing benefits to the surroundings by integrating and offering private areas with public uses (Aflalo Gasperini, 2010).

The HD 873 Corporate (Figure 10), designed by Israel Rewin Arquitetura and situated on Henri Dunant Street, has 22 corporate floors. The development houses a 400 square-metre plaza on the first floor, exclusively for the use of tenants, and on the ground floor a triple-high ceiling lounge, meeting rooms and an auditorium. The building does not have a wall or metal railings as enclosure on the main façade.

The Praça São Paulo Complex (Figure 11), designed by Könisberger Vannucchi and landscaped by architect Benedito Abbud on a 32,000 square-metre plot of land on Henri Dunant Street, is a development with several uses, including residential, hotel, commercial, office and corporate uses. The towers have 14, 17 and 32 floors, as well as four belowground levels zoned by use of the buildings. The site plan shows several areas of public use with

Figure 10. View of the HD 873 Building, with no commerce on the ground floor.

Figure 11. View of the Praça São Paulo Complex, with public access for pedestrians to commercial and service activities at the base.

stepped squares, gardens, a reflecting pool, seating and tables under the trees. It has an institutional area and a donated green area. Today, the complex has a metal railing that permits visual access, but inhibits pedestrians' free passage, maybe because the complex has not been completed yet. It was awarded the Aqua Certification (Bergen, 2018).

Based on the data obtained from the survey carried out on the ten developments selected for the study, it is possible to state that:

- Programmatic uses of the ground floor: none of the developments has its commercial use directly linked to the public sidewalks. Three complexes support inner commercial use: The Parque da Cidade shopping mall, the Gourmet Plaza food court on the first floor of the base block at the Morumbi Corporate Complex, and the Praça São Paulo Mall, comprising a group of 14 stores under the complex towers;
- Public enjoyment: only 3 developments allow pedestrian passage through the land plots: Rochaverá Complex, World Torre Morumbi, and Parque da Cidade (unfinished);
- Existence of enclosures around the plots: it was observed that except for two complexes (Rochaverá and the World Torre Morumbi) the others have enclosures in the form of metal railings or walls. These barriers impede direct connections between the buildings and the public areas, such as sidewalks and squares.

3 FINAL REMARKS

Lack of connection between corporate and office buildings and public space for pedestrians is a phenomenon that can be observed in the new centralities with a predominance of the tertiary sector in São Paulo. These areas have undergone a process of land valuation in the face of the investment involved to establish large tertiary sector buildings, which conflicts with the issue of spaces for public use in the valued regions of the city of São Paulo.

In view of this reality, it becomes necessary to enjoin debate and encourage other forms of production of urban space. Such forms must be based on referential repertoires that can contribute to stopping the process of socio-spatial fragmentation and promote spatial practices capable of fostering new opportunities for appropriation for public use of the spaces in between lots.

The results of the present study carried out on a stretch of Dr. Chucri Zaidan Avenue with recently launched developments revealed that the programmatic and functional decisions for the ground floors of the new buildings continue to reflect old practices of spatial segregation already condemned in the location of the buildings.

As the current Master Plan of the City of São Paulo (São Paulo, Município, 2014) proposes urbanistic instruments that stimulate connection and articulation of buildings with the surrounding areas, it is expected that the new designs incorporate these in a definitive way so that cities can be enjoyed by everyone.

REFERENCES

Aecweb. 2019. Rochaverá Corporate Towers recebe Leed Gold. São Paulo: Aecweb. Available at: https://www.aecweb.com.br/cont/m/rev/rochavera-corporate-towers-recebe-leed-gold_2018_10_20. Accessed on: Dec.07, 2019.

Aflalo Gasperini Arquitetos. 2009. WT Morumbi. São Paulo. Available at: https://aflalogasperini.com.br/blog/project/wt-morumbi/ (Accessed on: Dec.07, 2019).

Aflalo Gasperini Arquitetos. 2010. Parque da Cidade. São Paulo: Portal da internet. Available at: https://aflalogasperini.com.br/blog/project/parque-da-cidade/. Accessed on: Dec.14, 2019.

Aflalo Gasperini Arquitetos. 2019. *Rochaverá Corporate Towers*. São Paulo: Portal da internet. Available at: https://aflalogasperini.com.br/blog/project/rochavera-corporate-towers/. Accessed on: Dec.05, 2019.

Bergen Consultor de Gestão Patrimonial. 2018. *Praça São Paulo*. Presentation. Available at: https://pt.slideshare.net/bergenimoveis/praa-so-paulo. Accessed on: Dec.15, 2019.

Botelho, A. 2007. *O urbano em fragmentos: a produção do espaço e da moradia pelas práticas do setor imobiliário*. São Paulo: Annablume, Fapesp.

Estadão. 2016. Moderno e com operação mais barata. Jornal *O Estado de São Paulo*. Folha de Economia, Claudio Marques, 01 de setembro de 2016. Available at: https://economia.estadao.com.br/blogs/radar-imobiliario/moderno-e-com-operacao-mais-barata/. Accessed on: Dec. 09, 2019.

Eztec. EZ Tower. 2014. São Paulo: Portal da Eztec. Available at: https://docplayer.com.br/8385017-Ez-towers-torre-a-assinado-pelo-renomado-arquiteto-carlos-otto-o-ez-towers-e-um-projeto-de-design-unico-e-imponente-que-se-tornara-um-icone-na-cidade.html. Accessed on: Dec.03, 2019.

Fix, M. 2001. *Parceiros da exclusão: duas histórias da construção de uma 'nova cidade' em São Paulo*: Faria Lima e Água Espraiada. São Paulo: Boitempo.

Frúgoli JR, H. 2000. *Centralidade em São Paulo: trajetórias, conflitos e negociações na metrópole*. São Paulo: Cortez.

Geosampa. 2019. Mapa Digital da Cidade de São Paulo. São Paulo: Prefeitura, GEOSAMPA. Avaiable at: http://geosampa.prefeitura.sp.gov.br/PaginasPublicas/_SBC.aspx. Accessed on: Dec. 09, 2019.

Kowarick, L. 200). Prefácio In: FRÚGOLI JUNIOR, H. 2000. *Centralidade em São Paulo: trajetórias, conflitos e negociações na metrópole*. São Paulo: Cortez.

Loukaitou-Sideris, A.; Banerjee, T. 1998. *Urban Design Downtown: poetics and politics of form*. London:, University of California.

Madanipour, A. 2003. Why are the design and development of public spaces significant for cities? In Cuthbert, A. 2003. Designing cities: critical readings in urban design. Malden: Blackwell Publishers.

Maricato, E. 2001. Prefácio In: FIX, M. 2001. *Parceiros da exclusão: duas histórias da construção de uma 'nova cidade' em São Paulo*: Faria Lima e Água Espraiada. São Paulo: Boitempo.

Morumbi Business Center. 2019. São Paulo: Portal da internet. Available at: http://morumbibusinesscenter.com.br. Accessed on: Dec. 05, 2019.

Oliveira, L. M. 2013. Espaços públicos e privados das centralidades urbanas: Park Avenue, Avenida Paulista e Avenida Doutor Chucri Zaidan.Tese (Doutorado em Arquitetura e Urbanismo) – Faculdade de Arquitetura e Urbanismo, Universidade Presbiteriana Mackenzie, São Paulo.

São Paulo (Município). Law 16.050, from July 31, 2014. São Paulo, Prefeitura do Município de São Paulo. Available at: http://www.prefeitura.sp.gov.br/cidade/secretarias/desenvolvimento_urbano/legislacao/plano_diretor/index.php. Accessed on: Dec. 05, 2019.

São Paulo (Município). Law nº 13.260, from Dec. 28, 2001, mainly altered by Law nº 15.416, from July 22, 2011. São Paulo: Prefeitura do Município de São Paulo. Available at: http://www.prefeitura.sp.gov.br/cidade/secretarias/desenvolvimento_urbano/legislação. Accessed on: Dec. 05, 2019.

VILLAÇA, F. 2001. *Espaço intra-urbano no Brasil*. 2. ed. São Paulo: Studio Nobel: FAPESP: Lincoln Institute of Land Policy.

Yin, R. K. 2010. *Estudo de Caso: planejamento e método* (4. ed.). Porto Alegre: Bookman.

The Neighbourhood Units of Malvín Norte. Archaeology of an urban laboratory in the eastern periphery of Montevideo

M.C. Diotti
FADU, Faculty of Architecture, Design and Urbanism, Universidad de la República, Uruguay

ABSTRACT: The following text is part of an on-going thesis on planning and urban development within the public Faculty of Architecture of Montevideo, Uruguay (MOTDU-FADU-UDELAR) focused on the Neighbourhood Units of Montevideo´s "Director Plan" of 1956. Within this research framework, the most notable and complex unit is "Alto Malvín": a former vacant area in the eastern part of Montevideo revisited as a testing ground for social housing experiments. This exercise of urban archaeology faces the challenge of inspecting the built environment to develop the planning documents and urban projects related to the area as a guide to elucidate its spatial production and analyse the urban imaginaries deployed by the political authorities in Montevideo during the second half of the twentieth century.

Figure 1. Satellite image of the Neighbourhood Units of Malvín Norte. Source: DigitalGlobe (2015).

1 INTRODUCTION

When confronted with any contemporary satellite image of Malvín Norte –a well-connected area located in the eastern part of Montevideo- at first glance one may see a territorial mosaic consisting of multiple built fragments that coexist in an unrelated fashion. This is the material evidence of the evolution, impulses, pauses –both political and technical- that can be construed as components of an urban history which connects the ideas and imaginaries unfolded from the 1950s onwards.

Therefore, the present study aims to analyze these fragments through primary sources, bibliographic inputs and institutional records to build a historical narrative and critically review the spatial production associated with social housing in Montevideo during the second half of the 20th century.

2 BEGINNINGS AS AN INDUSTRIAL GARDEN CITY

Since the beginning of the Second World War, Uruguay –and Montevideo in particular- embarked upon a vigorous persistent process of urban expansion. The economic prosperity derived from the armed conflict, resulted in industrial development that had direct impact on the outlying urban fringes, where new factories led to the emergence of additional informal settlements (Gómez Gavazzo, 1959a).

The Regulatory Plan of Montevideo and its executive office within the Municipality was at that time incapable of conducting these new processes which triggered the appearance of new-fangled planning tools that would ultimately guide the transformations of the young metropolis.

In this context, the Municipality of Montevideo began work in the area of Malvín Norte, which had optimal conditions to tackle or even anticipate some of the issues that had appeared during the 40s. The

DOI: 10.1201/9781003220855-28

person in charge of establishing the initial diagnosis and proposing concrete solutions for this problem was the English architect Elidir Davies, addressing the commission offered by the directives of the Regulatory Plan´s office. The brief "Draft Specifications 1951-1952" of the architecture department of Montevideo´s Municipality outlines a plan to urbanize 415 hectares north of Malvín creek aiming to balance the proportion in which to build "factories, houses and open spaces" seeking "an economic equilibrium between work and daily life".

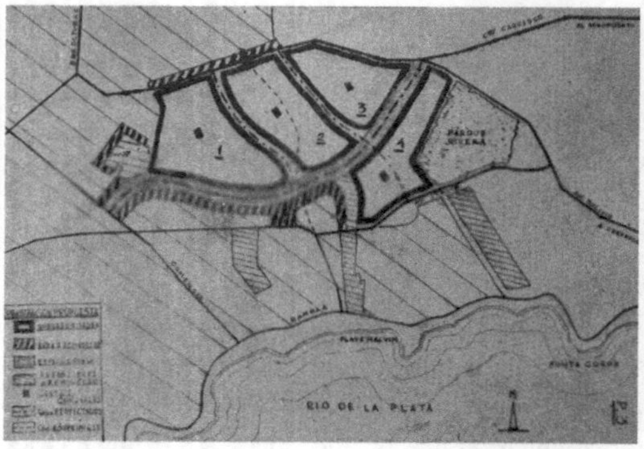

Figure 2. Elidir Davies´ proposal for the area of Malvín Norte. Source: Draft Specifications 1951-1952 – Municipality of Montevideo – Department of Architecture, 1953.

The scheme envisaged –based on economic calculus, spatial distributions, estimated densities and collective urban equipment- translates to a vast residential area adjacent to the industrial site, drawing directly on the "apartment-factories" model and taking advantage of the rapid transport links with its surrounding avenues. The whole development was intended to house almost 40,000 inhabitants in municipality-owned lots with "semi-detached houses", "detached houses" and "apartment blocks (sic) of a maximum height of 3 storeys (counting ground level)" as detailed by Davies himself in the official documents. The planning instrument and land program used –mixed with urban qualities inherited from the Garden City movement and English "New Town"- is the Neighbourhood Unit.

The Neighbourhood Unit concept, which was synthetized by the sociologist Clarence Perry during the mid-twenties in the United States of America, uses the primary school as the quantitative basis and geometric centre of composition to later deduce the population and housing capacity. Some extra parameters are then added to the overall design combo: sinuous low-speed mobility layout within the unit and major traffic routes on the perimeter. Finally, the model suggests where to locate the collective services and urban equipment, with recreational and community elements placed in the middle of the urban fabric and commercial and logistics structures on its border.

Davies´ final proposal outlined four Neighbourhood Units associated with the Carrasco road as shown in the synthetic scheme in collaboration with the Municipal Government. And it was precisely this institution –through the Regulatory Plan´s office- that was to be in charge of the final layout of the project which was finally approved.

The man in office at the time was Carlos Clémot, who had been working in Le Corbusier´s Parisian office until 1949 (Nudelman, 2013). The changes in the lexicon found in the chapter "Considerations on the project approved for the area in Malvín Norte" of the municipality document, envisions a CIAM (Congres Internationaux d'Architecture Moderne)-inspired future that would later be confirmed, along with the material transformation that took place: "Once the purpose was determined, a convenient structure was established for the dwelling function, which led to the design of large areas without rapid vehicle flow, suitable for building neighbourhood units due to its size (…) 5 islands were established to develop the dwelling projects to house 20,000 inhabitants, with an overall density of 100 persons per hectare".

These intentions were translated into specific municipal ordinances for the area which allowed further increments to the overall density through urban parameters: "The buildings to be constructed in the reference zone will be subject to a maximum height of 7 metres, with the exception of those lots with special regulations described in the specific layouts within the units. In this case, or in lots with a frontage measuring 39 metres or more, and when the building constitutes a single structure, a maximum height of 22.50 metres would be permitted".

Several other innovations were incorporated into the design, such as the organic layout of its streets, new public spaces with educational and recreational urban equipment at the centre of each unit and an incipient traffic segregation between vehicular and pedestrian circuits, as well as a new extensive Neighbourhood Centre equipped with large scale urban artefacts and services in the barycenter of the overall urban composition. The transformation of the ordinances, and the new vision for the future transformation of the area, enabled a scale-density leap for the existing scenario, considered "insufficient for the rational use of the area", paving the way for further advances carried out by "official institutes" or the "stimulation of private initiatives".

3 THE "BARRIO 16" AND EMERGENCE OF CIAM INSPIRED IDEAS

Maybe the most significant architectural contribution was made by the "Instituto de Viviendas Económicas" (INVE) with the construction of the "Barrio 16". This institute, which was within the sphere of the national government in the Ministry of Public Works (MOP), declared its future plans in 1956 through a special edition of the magazine of the local society of architects (SAU): "the housing problem must be resolved by building houses, as many as possible, directing all efforts and will to make feasible the development of broad organic plans, thus only in this way can the country accomplish the great work that is needed, on which the welfare of a large contingent of our population relies". But its goal was even more radical and ambitious: "the installation of an experimental field for prefabricated construction systems and the creation of a massive Neighbourhood Unit destined to initiate vast development plans in our country" (Revista Arquitectura 233:6). The paradigmatic case (in construction since 1955) was indeed the Barrio 16 which is mentioned as the reference for the technical aspects driven by the INVE: "regarding the Neighbourhood Unit, it has to be said that construction had already begun in the lots located in Carrasco road and Italia Avenue, which will cover an area of 50 hectares. We consider this initiative to be of the utmost national interest not only for its repercussions as a solution to the overall housing problem, but also for its significance regarding the rational use of land in the city" (Revista Arquitectura 233:6).

The architect Héctor Iglesias Chávez was in charge of the project, in which CIAM ideas were deployed for the first time in the area through 8 3-storey building blocks as an alternative to the factory worker "New Towns" that had been designed only a few years before, with a layout that still generates interest today for the solution adopted. Along with the "zeilenbau" strategy that was sponsored by the German members in the first CIAMs as the most economical layout for massive housing (Mumford, 2002) and representing a departure from the traditional building block layout and land subdivision, the building complex proposes a clear division of pedestrian and motor-vehicle flows, with alternating inter-block voids and vehicular cul-de-sacs, plus human scale leisure areas equipped with playgrounds, playing fields and sidewalks. The overall composition is strongly characterized by a diagonal line that connects the complex with the future "Neighbourhood Centre" located at the junction of the four most important Neighbourhood Units.

Regarding the architectural solution, the blocks are composed of "decimal" modules that divide the building's floor plan allowing diverse configurations according to the capacity required by each housing unit. This same diversity is observed in the design of the circulatory elements, creating specific elements for each level: on the ground floor direct access from the street, the next storey has its own stairs connecting with the street, while the upper units have an elevated street accessed by a specific vertical circulation nucleus.

By 1962, the INVE would expand the Barrio 16 by building 10 new blocks. Unfortunately, the spatial proposal and typological richness of its predecessor was not continued in this new approach, where the economic resources delivered were maximized and excessively simple urban layout and building solution adopted.

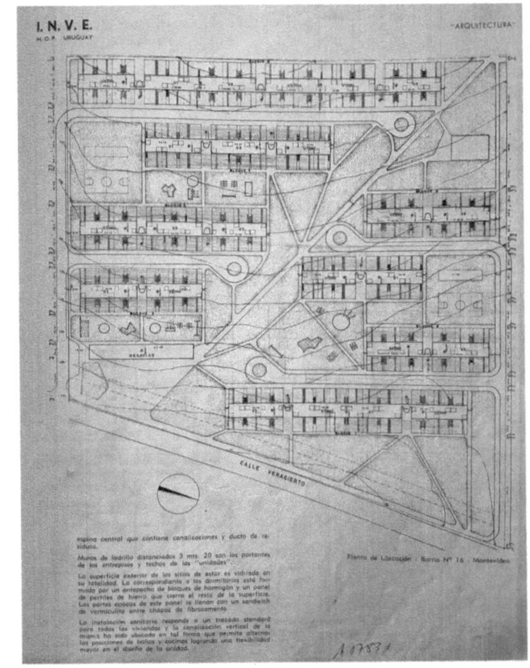

Figure 3. Floor plan of the "Barrio 16". Source: Revista Arquitectura SAU 233, 1956.

Figure 4. Street view of one of the blocks. Source: SMA, FADU-Udelar.

4 THE NEIGHBOURHOOD UNITS OF THE "DIRECTOR PLAN"

The planning document that ends up defining the global urban scheme envisaged for Malvín Norte is Montevideo's "Director Plan" of 1956. With the objective set of ordering and directing the expanding urban fabric that the "Regulatory Plan" proved incapable of controlling, the municipality implemented an organic decentralization of the city: a strategy based on territorial units ordering this urban expansion. In this sense, Montevideo is subdivided administratively and demographically, into "sectors" (areas with equipment and services on a metropolitan scale with a population of approximately 100,000 inhabitants), "districts" (housing between 20,000 and 30,000 people with mid-scale urban resources) and finally "Neighbourhood Units" (on a neighbourhood scale from 6,000 to 12,000 approx.). This model is detailed in the "Director Plan Draft" and should be understood and associated with the morphological essay proposed by the "Instituto de Teoría y Urbanismo" of the "Facultad de Arquitectura" (FADU-ITU) in its bulletin N°8; where Montevideo was "remodelled" as Neighbourhood Units and Unités d'Habitation: a radical formal approach that allows the gradual substitution of the colonial urban grid of individual lots into spatially scattered building blocks placed within green spaces.

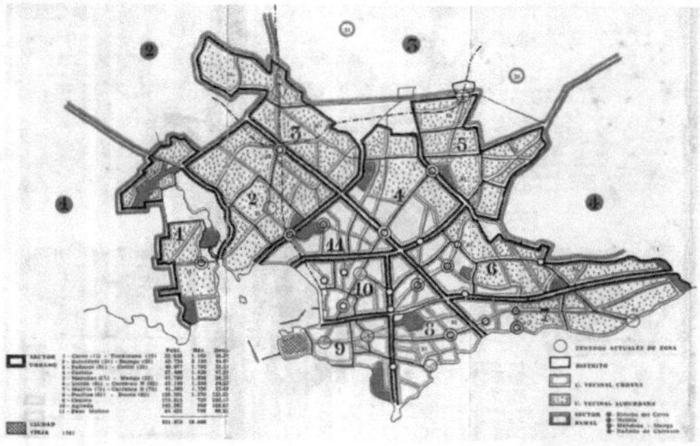

Figure 5. Administrative subdivision proposed by the Director Plan. Source: Gómez Gavazzo Archive, ITU-FADU.

The architect Carlos Gómez Gavazzo was in charge of the technical oversight in both aforementioned documents ("Director Plan Draft" and bulletin N°8), who, akin to Clemot - had been an intern in Le Corbusier's Parisian office in the thirties (Nudelman 2013). Gómez, in this double capacity as Director at the ITU-FADU and member of the "Director Plan Assessment Committee" designated by the Faculty of Architecture, was in a way the intellectual author of the formal revision of "Alto Malvín" Neighbourhood Units. In the physical model published in issue 235 of SAU's "Revista Arquitectura", Clémot's sinuous layout of individual houses in the northeast coexist with INVE's Barrio 16 - recently inaugurated and about to be expanded- and with a surprising and decisive incorporation that would mark future development in the area: various groups slabs homes that eclipsed the previous building initiatives.

The formal result was a fragment of modern city that would house 39,460 inhabitants on 170 hectares, increasing the densities proposed by the municipal office to 232 h/ha in a major scaling up in terms of utilization of available land.

Figure 6. Physical model of Malvín Alto. Source: Memoria del Concejo Departamental 1955-1959.

To accomplish these objectives, it was necessary to draw on two legal initiatives passed in 1946: the "Ley de Centros Poblados" which defined the Neighbourhood Unit as an instrument for planning and urban development, and the "Ley de Propiedad Horizontal" that promoted an increase in building height allowing the multiplication of storeys, while considering them individual property within the building.

With the new Plan approved and the legal instruments defined, the Municipality of Montevideo had delineated the future material outcome for Malvín Norte.

5 STUDENT HOUSING AND THE "MAGGIOLO PLAN"

The intentions proposed by the "Director Plan" were about to become prematurely strained by a decision related to the "Universidad de la República". An idea by the rector about to leave office in 1956, the architect Leopoldo Carlos Agorio, was to build a Student Housing complex in one of the southern Neighbourhood Units along the Malvín creek, as the foundation stone for a future University Campus.

Figure 7. Maggiolo and Clemot at the Student Housing construction site. Udelar General Archive.

Gómez Gavazzo, at that time already working with the municipality from the ITU- presented a study in 1957 where he advises against the university's initiative because it conflicted with the residential development proposed by the "Director Plan". However, in view of the advances, Gómez himself would by 1958 be in charge of creating the brief for an architecture competition for the Student Housing Complex on 9 hectares acquired by the university. The competition, finalizing the next year, voted in favour of the proposal presented by Clémot – already familiar with the area - and Justino Serralta, who also had spent time in Le Corbusier´s office recently as an intern. The project would be a massive residential slab development to house 1200 students in the spirit of the Director Plan's physical model for the area, breaking ground a few years later in the mid-sixties.

But in terms of planning, the university would follow its own path, acquiring new land and promoting a "Plan de Construcciones Universitarias" carried out by the new rector Oscar Maggiolo, who started to build a "Ciudad Universitaria" with multiple and diverse educational equipment surrounding the Student Housing Building.

6 THE "LEY DE VIVIENDA" AND THE FIRST COOPERATIVE PROJECTS

After a decade without major transformation –apart from the INVE and the university's initiative s- and only two years after enactment of law 13.728 in 1968, the complex "Vivienda Cooperativa Malvín Norte" (VICMAN) was built in the north-western part of the area, along Carrasco road: a consecrated part of the cooperative housing movement in Uruguay designed by the architect Alfredo Nebel and the engineer Eladio Dieste.

The adopted solution strongly challenges the prefiguration carried out by the "Director Plan" by proposing an urban layout that was removed from the binary and contrasting relations of the CIAM-inspired soil occupation, with 2 different housing typologies for its 1000 housing units. On the one hand, 4-storey-high "H" shaped blocks with four houses in each plan and a central circulatory nucleus, and on the other, a series of paired row-duplex-houses slightly retracted from one another. The site occupation of these two typologies is visible in the overall plan, appearing as two distinctive units, both with the individual houses in the centre and the H blocks on the perimeter, downscaling the built mass in the heart of the composition. Moreover, these two units were congruent with the time schedule for their construction, with a first building stage of 410 houses and a second of 590 along with some new buildings for daily commercial and neighbourhood services, which unfortunately never went ahead.

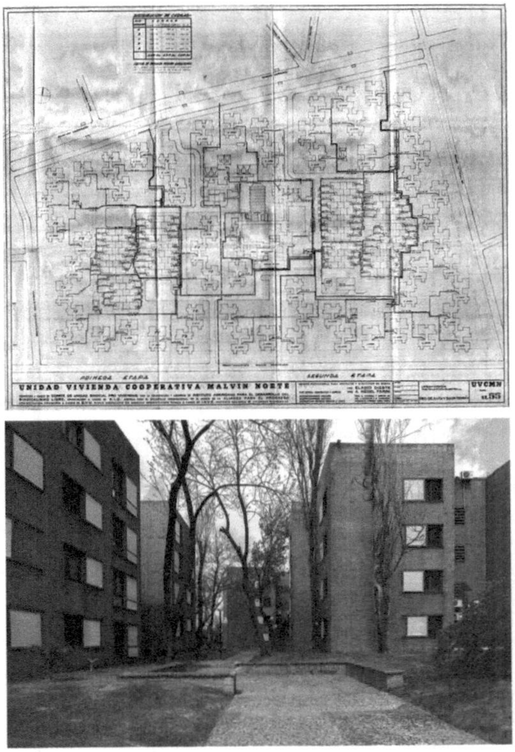

Figure 8. VICMAN floor plan and street view of one of the H blocks. Source: IHA - SMA, FADU-Udelar.

Although VICMAN is nowadays one of the most celebrated examples of the cooperative housing movement due to its excellent quality, it is important to mention that it also constitutes a closed private enclave due to its land occupation regime. This is an obstacle impossible to overcome, which in terms of urban principles and quality is contrary to those originally proposed by the Neighbourhood Units regarding the collective and bonding role of the unoccupied land.

7 HOUSING SUPER-COMPLEXES

In the years that followed, and exposing a formal dialectic of cooperative housing production system together with state-funded housing complexes, one of the most radical and polemic projects was built next to VICMAN.

During the turbulent years of the Uruguayan dictatorship, the "Banco Hipotecario del Uruguay" (BHU) pushed the design and construction of multiple housing complexes that took up the formal ideas of the "Director Plan", in a disciplinary context where the ideas of the Modern Movement in architecture were being brought into question. However, for the State-funded project, s the construction of 3 complexes named "Euskal Erría" began from 1970, and the 5 slab-housing estates that made up the Malvín Alto complex commenced during the 80s.

The Euskal Erría complexes (and its extensions, CH70 and CH71) are considered to this day the largest development built in the history of the country, with truly impressive numbers for the Uruguayan scale: 3644 housing units, distributed among 86 11-storey towers covering a total build area of 300,000 square metres (RISSO y BORONAT, 1992).

Finally, the 14-storey-slab buildings of Malvín Alto were built using the French prefabricated construction system OUTINORD, accomplishing (three decades later) both the large-scale industrialization INVE was advocating and the formal plan envisaged in Montevideo´s "Director Plan".

Figure 9. Street view of Malvín Alto slabs. Source: SMA, FADU-Udelar.

8 A CONTEXTUALIZED CRITIQUE OF THE TIUR

In a context characterized by the beginning of a new democratic government in Uruguay, the "Taller de Investigaciones Urbanas y Regionales" (TIUR) replicates the voices of the international disciplinary context in "Propuestas a la Ciudad Montevideo, 1986". After a historical narrative on the planning episodes of the city, this book develops a series of concrete proposals that serve as operative criticism of the urban development of those years. This initiative was carried out by a generation of young architects that, in the decades that followed, went on to led the most prominent academic and technical roles in the national arena.

One of the most interesting projects – by Mariella Russi and Alberto de Betolaza - directly involved the recently inaugurated housing complexes located in the Neighbourhood Units of Malvín Norte.

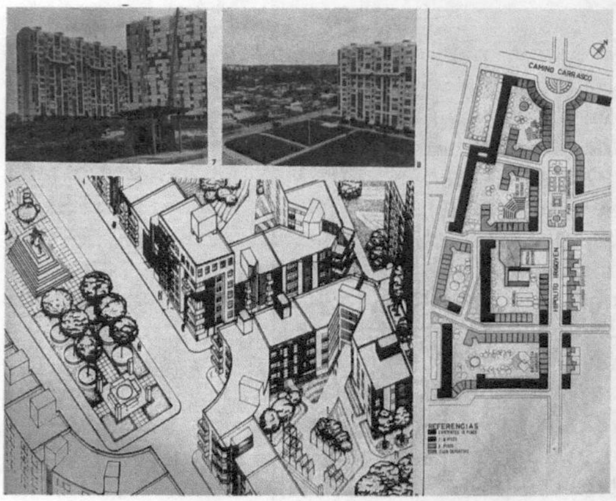

Figure 10. TIUR´s proposal for Malvín Alto area. Source: Propuestas a la ciudad Montevideo (Proposals for Montevideo city) 1986. FADU-Udelar.

The proposal was as simple and radical as the architecture it addressed. In the area where Malvín Alto and Euskal Erría complexes were located, the strategy was to recharacterize the urban block as a unit - an idea expressed throughout the book- identifying the oversized isolated slab-housing as an iconic case to deconstruct with the conceptual arsenal of the local version of New Urbanism.

In this sense, the urban Project developed by the TIUR in the Malvín Alto complex was based on a series of formal betterments with typological diversity that, once attached to the walls of the slabs, would confine the undomesticated and deteriorated exterior spaces, while providing an increase in overall density, creating a height gradient that would humanize the scale of the complex and address the spaces that were already showing signs of deterioration in the latter half of the 1980s.

Although the initiative never got off the ground and none of the actions envisaged were ever implemented, it certainly made its impact in the critical and discursive sphere of the time, amplifying those voices that were against massive housing complexes, marking an end point to the CIAM-inspired residential experiments endorsed by the "Director Plan".

9 THE ADVANCE OF INFORMAL SETTLEMENTS

But the academic criticism formulated over the TIUR projects seemed to overlook a more urgent problem, which was the main concern of the original proposals for the area: the precariousness and advance of the informal urban fabric that was already occupying the voids since the fifties.

In this context, the equipped barycenter promised by the Municipality of Montevideo in the "Director Plan" – a key aspect in this urban proposal - never came to fruition and was slowly but steadily occupied by informal houses lining the banks of the affluent of Malvín creek, constituting what would be the "Aquiles Lanza" settlement, one of the most critical areas regarding urban quality in the city. It was not only the vacant lots that became informally occupied, but also the INVE complexes, intended for low-income tenants. These became built out to a point where it was impossible to recognize the inter-block voids, generating critical issues in terms of liveability and coexistence. This concluded in an irreversible denaturalization of the collective and open nature of the Neighbourhood Units, while the housing initiatives aimed at middle and middle-lower class tenants started to fortify their perimeter to isolate the projects from the rest of the urban fabric.

Recently, the response from the local authorities has been to regularize the most critical areas where the environmental quality was unbearable. The adopted strategy was subdivision of the land into small lots of around 150 square meters under the "Ley de Ordenamiento Territorial (Land use law) 18.308" for occupation by semidetached houses. These properties, only a few years later, were already showing signs of deterioration and overcrowding, which in some cases reached a critical level, raising questions over the technical decisions and urban policies adopted by the government.

Figure 11. Boix & Merino slum (2001). Source: intgis.montevideo.gub.uy.

10 FINAL CONSIDERATIONS

The proposed exercise brought together the material and intellectual actions in the territory as a coherent narrative, where the architectural solutions and urban projects envisaged complement, overlap or even clash for five decades in the Malvín Norte area as they constitute a singular temporal and spatial sequence: a collection of fragments of the housing and urban policies that Montevideo deployed in the second half of last century.

We therefore embark on a journey that takes us from a city comprising detached factory-workers' houses, which led to an industrialized housing experiment, to later proposed CIAM-inspired slab housing questioned by the cooperative housing movement, until emerging with human scale and historic/contextual arguments ending public housing programs in a bid to halt the sprawling slums of the area.

This path has its most radical point in the CIAM-based housing complexes, where the Neighbourhood Units as a planning tool were combined with massive industrialized buildings to address the deficit in the housing stock of an area deliberately chosen as an experimental field. The gulf between these ideas and the urban conditions that ensued calls into question the solutions adopted and expose the intrinsic fragility of the experimental processes presented, seen as a territorial battle between diverse architectural typologies and urban layouts.

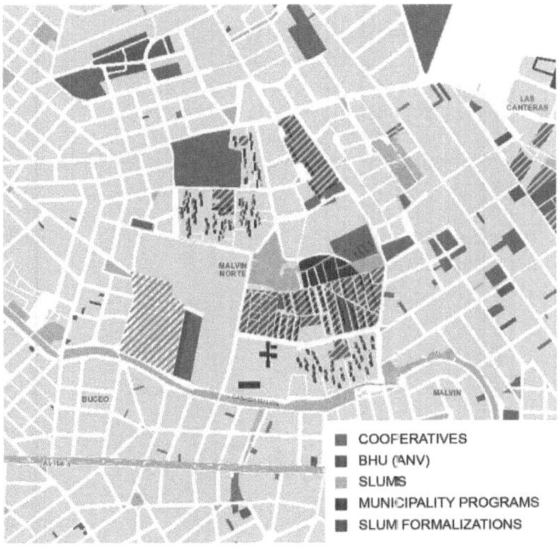

Figure 12. Map of the public housing programs in Malvín Norte by 2020. Source: SIT-MVOTMA.

The analysis of the historical evolution in these forms of occupation and land subdivision schemes is also highly relevant. The initial proposal entailed smaller urban blocks and lots with a high occupation area compatible with detached houses and private or individual solutions. But this rapidly became smaller ground floor occupations in larger lots and urban blocks, associated with an increase in density and building height. These aspects of the urban form seemed to have been taken to their limits, only to give way to smaller building associations, in smaller-scale building solutions that occupied a larger area of the lots. The contemporary real-estate initiatives both in cooperative realms and formalization of precarious settlements, curiously resembles the initial plans envisaged by Davies, far removed from the more radical transformations that took place in the area.

Coexisting with the diverse occupation formats, the planning tool permeating the whole historic process was the Neighbourhood Unit. Unfinished, denaturalized and even deprived of most of its constituent principles, it still emerges as a basic resilient territorial structure capable of accommodating the multiple schemes envisaged within the experimental urban laboratory of Malvín Norte.

REFERENCES

Alvarez Lenzi, Ricardo. 1986. *Fundación de poblados en el Uruguay.*. Montevideo: Universidad de la República, Facultad de Arquitectura Diseño y Urbanismo, Instituto de Historia de la Arquitectura.

Conti de Queiruga, Nydia. 1986. *La vivienda de interés social en el Uruguay: historia de los problemas de la arquitectura Nacional.* Montevideo: Universidad de la República, Facultad de Arquitectura Diseño y Urbanismo, Instituto de Historia de la Arquitectura.

Gómez Gavazzo, Carlos. (1959a). *Concejo Departamental De Montevideo. Memoria del Concejo Departamental 1955–1959.* Montevideo: Archivo Gómez Gavazzo Instituto de Teoría y Urbanismo, Facultad de Arquitectura Diseño y Urbanismo, Universidad de la República.

Gómez Gavazzo, Carlos. (1959b). *Informe In-Voce del Arq. Carlos Gómez Gavazzo a la Comisión de la Junta Departamental de Montevideo que estudia las condiciones de vida de la población.* Montevideo: Archivo Gómez Gavazzo Instituto de Teoría y Urbanismo, Facultad de Arquitectura Diseño y Urbanismo, Universidad de la República.

Intendencia de Montevideo. (1953). *Memoria 1951-1952. Departamento de Arquitectura.* Montevideo: Archivo Instituto de Historia de la Arquitectura, Facultad de Arquitectura Diseño y Urbanismo, Universidad de la República.

Mumford, Eric (2002). *The CIAM discourse on urbanism (1928-1960).* Cambridge, Massachusetts: Massachusetts Institute of Technology.

Nudelman, Jorge (2013). *Tres visitantes en París: los colaboradores uruguayos de Le Corbusier.* Madrid: Universidad Politécnica, Escuela Técnica Superior de Arquitectura.

Risso, Marta and Boronat, Yolanda (1992). *La Vivienda de Interés Social en el Uruguay: 1970 – 1983.* Montevideo: Fundación de Cultura Universitaria.

Sert, Luis Jos; (1947). *Can our cities survive: An ABC of urban problems, their analysis and solutions.* Cambridge: Harvard University Press.

Taller de Investigaciones Urbanas y Regionales (1986). *Propuestas a la ciudad Montevideo 1986.* Facultad de Arquitectura, Universidad de la República. 206 p.

Journals and Bulletins

ITU. *Boletín Informativo del Instituto de Teoría y Urbanismo ITU. N° 8 (1955) y 12 (1957).* Archivo Gómez Gavazzo Instituto de Teoría y Urbanismo, Facultad de Arquitectura Diseño y Urbanismo, Universidad de la República.

Sociedad de Arquitectos del Uruguay. *REVISTA ARQUITECTURA. N° 233 (1956) y 235 (1958).* Montevideo: Universidad de la República, Facultad de Arquitectura Diseño y Urbanismo, Instituto de Historia de la Arquitectura.

Ordinances

LEY N°10.723 - CENTROS POBLADOS (1946).

LEY N°10.751 - LEY DE PROPIEDAD HORIZONTAL (1946).

LEY N°13728 - PLAN NACIONAL DE VIVIENDAS (1968).

LEY N°18.308 - LEY DE ORDENAMIENTO TERRITORIAL Y DESARROLLO SOSTENIBLE (2008).

Parlamento de Uruguay

Research Tracks in Urbanism: Dynamics, Planning and Design in Contemporary Urban Territories –
Allegri et al (eds)

Urbanistic solutions in precarious settlements. The Banhado case (São José dos Campos – SP)

M. Fantin, J.C. Tavares, M.F.B. dos Santos, N.M.B. Tamanaka & R.S. Gomez
IAU, Institute of Architecture and Urbanism, Universidade de São Paulo, Brazil

ABSTRACT: The problematic land access issue of socially vulnerable populations has persisted in the Brazilian urban context, and is aggravated by the lack of urban solutions that take into account local social, productive and environmental dynamics. The complexity of populations and areas in situations of socio-environmental vulnerability addressed in proposals for urbanization and land tenure regularization warrants the attention of public authorities and academia. In this context, the present study analyzes the relationship between the socio-environmental complexity of the Jardim Nova Esperança community and the urbanistic solutions adopted in the Banhado: Urbanization and Land Regularization Popular Plan, from a qualitative-descriptive perspective. The methodology adopted envisions an inclusive city scenario and was not limited by universal models or reactive technical work.

1 INTRODUCTION

Jardim Nova Esperança, known as Banhado by its dwellers, is a singular example of urban occupation with socioenvironmental problems caused by the historic lack of access to land by socially vulnerable populations. Its singularity derives from the work, production and affective relationships established throughout its century-long history, and the site occupied. The right to the city and housing is discussed within the residents' narrative through a broader view, which sees the proximity of the neighborhood to the city center as the main factor entitling its inhabitants to enjoy the same rights and public services enjoyed by the downtown population. Residing adjacent to the central area of São José dos Campos, the community has access to its infrastructure, services and employment opportunities, despite its deregulation since the beginning of the 1930s. Therefore, the proximity to the city center - one of the reasons for settlement of the site for almost a hundred years - is considered one of the main factors that guarantee a series of rights formalized in 2016 by the Habitat III conference through the Declaration of Quito, more specifically article 13, which states cities, among other characteristics, must:

"Fulfill their social function, including the social and ecological function of land, with a view to progressively achieving the full realization of the right to adequate housing as a component of the right to an adequate standard of living, without discrimination, and universal access to safe and affordable drinking water and sanitation, as well as equal access for all to public goods and quality services in areas such as food security and nutrition, health, education, infrastructure, mobility and transportation, energy, air quality and livelihoods" (United Nations, 2019).

Global efforts have been devoted to the organization of Habitat conferences - in 1976, 1996 and 2016 – so that this theme could be consolidated and developed through an international contribution involving a host of countries. The concept of the right to the city present at the last conference must be advanced towards the inclusion of more complex elements into the study of human settlements.

The development and evolution of the term "housing right", from Habitat II, to "the right to the city", at Habitat III have stood out. This change in perspective is associated with the understanding that living with dignity requires not only adequate housing, but a series of urban services that guarantee the fulfillment of the basic functions of the urban space, such as quality and accessible mobility, health services, leisure, green areas, food, among others. (Rabelo & Fantin, 2018).

All these issues have led to a need for the development of a land tenure regularization plan for the community that assesses the feasibility of the communities' permanence, respecting the right to the city discussed in the New Urban Agenda. Consequently, the perception of the right to the city is claimed as one of the main guiding pillars of the Popular Plan of Urbanization and Land Regularization of Banhado, proposed by the PExURB group (Research, Teaching and Extension Practices in Urbanism).

DOI: 10.1201/9781003220855-29

2 SÃO JOSÉ DOS CAMPOS AND JARDIM NOVA ESPERANÇA

The city of São José dos Campos - SP is located in the eastern portion of São Paulo State, in the mesoregion of the Vale do Paraíba Paulista, approximately 90 km from the city of São Paulo. According to IBGE data (2020), the city had a population of 629,921 (97.66% urban) in 2010 and a HDI of 0.807 Its GDP per capita reached R$ 56,638.69 in 2017, and the average monthly income was 3.7 updated monthly salaries (R$2,342.00). The city, served primarily by Presidente Dutra, Tamoios, Monteiro Lobato highways and, secondly, by Governador Carvalho Pinto, lies on a privileged axis of economic investment running between São Paulo and Rio de Janeiro.

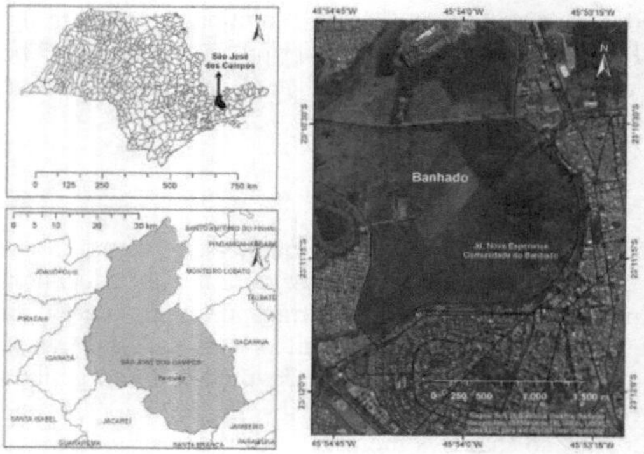

Figure 1. Location of Banhado in São Paulo. Google Earth, 2019.

Banhado's territory is close to downtown São José dos Campos, with a large portion of green area near the urban area. It is a sedimentary basin that is part of the floodplain system of the Paraíba do Sul River and characterized as a unique meandric natural amphitheatre in Brazil (Ab'Sabber apud Silva et al., 2017). On its border with the city center is a semicircular escarpment that serves as a physical barrier to floodplain fields, which extend for hundreds of meters to the northwest and up to approximately thirty meters below downtown's urban fabric to Paraíba do Sul River. Historically, this river has caused periodic flooding in the area, hence the region's name, translated as Laved. In summary, Banhado is composed of three main elements, namely the semicircular slope, the river and, intermediately, its floodplain of sedimentary soil, rich in nutrients and suitable for agricultural practice due to temporary floods.

2.1 *Jardim nova esperança community - history and work relations*

The construction of Central do Brasil's railway branch next to the terrain cavity in the early 20th century boosted the spontaneous creation of a community on the borders of Banhado´s railway tracks, comprising low-income people attracted by both work and urban dynamics from the track operations. Migrants from the south of Minas Gerais and Vale do Paraíba settled, drawn by the increased activity in the local industries (Rosa Filho, 2002). The community also has its roots in small rural farming families.

The name "Bairro Jardim Nova Esperança" was adopted for the occupied site, which expanded due to the growth of the resident families, divided into two nuclei that reflect the historical process of their formation. Nucleus I, comprised mainly of rural workers in search of space in industry between 1950 and 1980, is currently the most densely occupied area, whereas the occupation of Nucleus II by farming families started in the 1930s and has maintained its low density with agricultural production. Figure 2 depicts a characteristic scene of Nucleus I, with smaller lots and residences closer to each other, whereas Figure 3 is illustrative of Nucleus II, composed of larger lots, usually with more than one residence on the same land. The precariousness of the housing in the neighborhood is evident in both photos.

The lack of inclusive public policies in other locations in the city and the easy access to resources downtown (e.g., shops, services, infrastructure, employment and clientele for the local families' farm produce) are important historical factors for the establishment of the community in Banhado's meandric natural amphitheatre floodplain. Moreover, Banhado's privileged location provides residents with access to its entire infrastructure with no need to travel long distances.

Residing among the historical buildings that date back to the railway period and to be listed as important Heritage (Almeida, 2015) sites, are approximately 460 families, or around 2000 people, who

Figure 2. Road laid by residents in Nucleus I. Own authorship, 2019.

Figure 3. Family cohabitation in same lot as Nucleus II. Own authorship, 2019.

constitute the communities' current population, intrinsically connected with their place of settlement and rural production. According to a survey conducted by the study team, the agricultural use of the territory is second only to leisure. Around 41% of the respondents are sustained by local produce, which is also responsible for the generation of income for farming families, who sell their products in the city's central region, and 38% of the residents work within the community.

Due to the Paraíba do Sul River floodplain, an intrinsic part of Banhado's geological characteristics, its residents can farm produce adjacent to the center of São José dos Campos. This production is currently concentrated in Sector II of the settlement, but not restricted to it, since it involves cattle, pig, fish and chicken rearing, as well as growing bananas, fruit and vegetables, among other agricultural products. Figure 4 depicts the livelihood of one of the inhabitants of Banhado who, thanks to the fertility of the land, is able to harvest and sell fruit in the immediate surroundings.

However, the Banhado residents can find employment opportunities mainly in the city center and vicinity. Data from the survey revealed such opportunities for Jardim Nova Esperança are available mainly within a radius of up to 1km from the community, encompassing 56% of employed residents. Figure 5 illustrates workflows, showing that the strongest connections are close to Banhado.

The use of automobiles and public transport is therefore reduced considerably, and the residents enjoy more opportunities, time, quality of life and savings by commuting. For the city, this situation also represents cost savings through the optimization of services already consolidated in the central region and precludes the need for building new social, sanitation and transport infrastructure. Figure 6 shows that within a 2 km radius, residents of Banhado have easy access to various urban facilities.

2.2 *Eviction threats and need for plan formulation*

Over the past decade, the community has been under intense pressure for its dispersal and eviction through the establishment of local environmental issues in the form of laws, which conflict with the community's presence. In 2012, the city authorities instituted the Banhado Municipal Natural Park

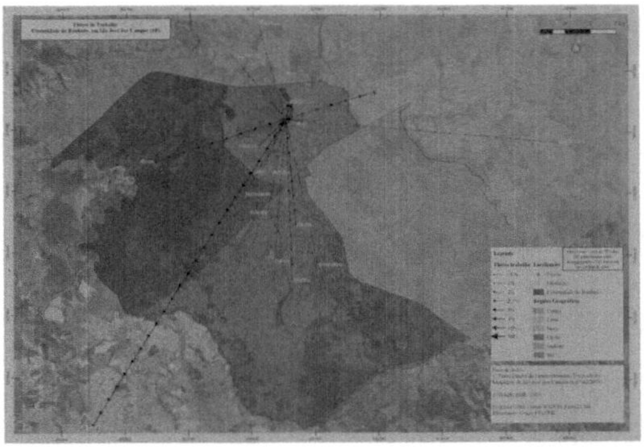

Figure 5. Map of Banhado Residents' workflow. Augusto Cesar Oyama, 2019.

Figure 4. Cart used by a Banhado resident to sell fruit. Own authorship, 2019.

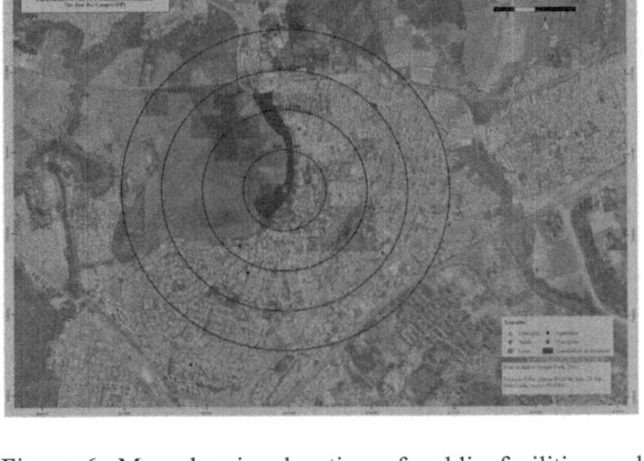

Figure 6. Map showing location of public facilities and distance radius. Augusto Cesar Oyama, 2019.

(PNMB), through Municipal Law No. 8,576/2012, in a controversial manner and without proper presentation of studies and objectives that clarified the criteria which justified the park's encroachment on the Jardim Nova Esperança community. Because it is classified as Integral Protection, this type of Conservation Unit allows only indirect use of the area.

In this context, the outlawing of occupation is a move towards the removal of the families. The change promoted in the 2018 Master Plan and the 2019 Zoning Law had the same result. Essentially, Jd. Nova Esperança's ZEIS (Social Interest Special Zone) classification, in place up until 2017, was lifted, representing another step towards the removal of the community. In addition, it should be mentioned that Banhado has layers of different legislation from different spheres (Municipal and State Environmental Protection Areas (APAs), characterizing a territory of sustainable use and subject to human occupation). Figure 7 illustrates the overlap of classifications in the area.

The community's precarious urban situation, a typical characteristic of irregular settlements, is aggravated by the lack of cleaning of drainage ditches and a gradual removal of public equipment and services. The constant presence of the municipal guard at the entrances to the community also hampers the repairing of houses, since agents actively impede the entry of vehicles loaded with construction materials, thus creating a difficult environment for dwelling.

Figure 7. Cartography of Conservation Units overlaying Banhado's cavity and Community, categorized as per São José dos Campos' Master Plan (2017). Augusto Cesar Oyama, 2019.

Figure 8 Provides a perspective for a better understanding of the existing gap between Avenida Me. Teresa, bordering Banhado, and Rua da Linha, formerly the Central Railway Line in Brazil and today Jardim Nova Esperança´s main road. The image also illustrates the contrast between the buildings in the central surrounding area and the self-built houses of the residents. Figure 9 depicts one of the water courses, which shows the high water permeability in the area due to its proximity to the Paraíba River.

Figure 8. Photo showing contrast between the houses in Jardim Nova Esperança and the residential multi-storey buildings in the city center, above. Own authorship, 2019.

Figure 9. Open water course traversing the neighborhood. Own authorship, 2019.

3 ANALYSIS OF THE URBANIZATION AND LAND REGULARIZATION POPULAR PLAN

3.1 *Methodology of the project*

The Popular Plan methodology sought to overcome the drawbacks associated with traditional models of comprehensive planning and extensive diagnosis that commonly guide plans influenced by past demands and restrictive preconceptions. The present project was guided by integrated planning that envisioned a city hypothesis and stemmed from a single question, forming the backbone of all the work: "is the safe and qualified permanence of the residents in Banhado viable?" (Tavares, 2019).

The model aimed to maintain constant learning from the social participation and dialogue with the agents involved, namely academia; residents and community leaders, public managers, and public defenders. Field visits, as well as workshops on participatory planning, training on applying socioeconomic questionnaires, and presentation and discussion of the plan with residents were conducted. The group's interdisciplinarity envisioned systemic solutions embracing the areas of law knowledge, urbanism, urban and regional planning, environmental engineering, and civil engineering.

However, the group faced other difficulties, which should be taken into consideration. Students and graduates in the areas of architecture and urbanism, law, environmental engineering and civil engineering have devoted positive efforts in building an interdisciplinary team that can work collaboratively on different fronts regarding the issues involving the area of the project. However, the team's range of skills should have been expanded further, since it lacked members from areas to act directly on some of the farming issues in the community (e.g., agronomists) and in social relations (e.g., social scientists). Among the possible consequences of the lack of social workers was a perceived lack of dialogue with the social groups of Jardim Nova Esperança. The team was restricted to relations with a small group of community leaders linked to the Residents' Association, which hampered assessments of their real social situation and possible internal disputes.

The initiative was developed outside governmental centralization, but consulting its instruments for regulating urban space and current public policies. Dialogues with the City Hall, however, were limited, and the work process took place independently of government agents. Due to the impossibility of negotiation, the final result presented to the City Hall was fundamentally external to its own activities, dynamics and interests, and hence was flatly rejected by the municipal management - mainly local executive and legislative powers. Negotiation was accepted only with the judiciary.

In view of the lack of political will of the City Hall for holding a dialogue with the Residents' Association, together with the development of the Plan with no direct dialogue, its effectiveness is jeopardized.

Figure 10. Workshop held with members of the Community. Own authorship, 2019.

Social participation took place through workshops, seminars and field visits as a process of decision-making events that combined technical and political aspects. The active participation of the community - approximately 10% of the total population - in devising the plan led to proposals that directly addressed issues pertinent to Jardim Nova Esperança. However, insecurity, constant threats of eviction, and lack of dialogue with the City Hall meant residents agreed with most solutions proposed. They disagreed, however, with some decisions and strongly claimed that "staying in Banhado" would be the best choice. The public neglect and long lack of maintenance in the area also contributed to the residents being enthralled at even the simplest of proposals that hinted at improvement.

During the development of the Plan, alternative techniques of cartographic production from aerial surveys conducted by Remote Piloted Aircraft (RPA) were also used. Combined with geotechnologies, these techniques enabled the consolidation of a detailed physical base of the territory, essential for the final outcome, i.e., lending autonomy to the work and decision-making without investing major resources in the generation and pooling of the minimum data necessary for developing the project.

The methodology associated with productive relationships explained the community's bond with water - conflicting in times of flood, but present in the population's imagery due to the Paraíba do Sul River and the numerous spouts in the area – thus constituting one of the main relationships learned and used in the scope of the Plan, as addressed in the ensuing section.

3.2 *Components of the plan*

The Plan proposed a set of urban solutions towards forming a network of services and layout of roads through branch infrastructures that limit and order the density via the design of plots and blocks, helping to restore the hydrological cycle and integrate the community with the city centre through public squares, called "fair squares".

The solution starts out from the land-work and the community-water relationships, which are directly related. The abundant presence of water and the soil formation directly benefit the farming practiced by a large part of the community. The squares support the correlations between these two factors, enabling the sale of the produce and establishing a relationship of local production and trade associated with the city centre, thus completing the cycle. The community has a genuine interest in preserving the environmental protection area, and not only to meet legal obligations. A clear relationship between residents and nature permeates Jardim Nova Esperança, mainly through its productive, water, and ecological cycles. After delimitation of the area, its urban design was developed and the branched network model adopted following the current logic of the blocks, lots and implementation of buildings. Unlike a mesh composed of articulated rings (grid-shaped), the infrastructural system based on the trunk (branched) logic provides services to currently occupied areas and inhibits further settlement. The infrastructure preserves the existing form of urbanization without expanding it. The road layout was fundamental for supporting the planned infrastructure network, guaranteeing public and private services through urban (mobility, sanitation, electricity and communication), social (public and community facilities, urban furniture, and landscaping), and housing (improvements and construction) infrastructure for the entire community.

Another issue addressed in the Plan concerns water drainage. Currently, the community has no microdrain elements, but only trunk collectors, which collect rainwater runoff from the center and release it within the community. These collectors must be buried and extended to master ditches outside the community. Today, ditches dug by residents help the removal of water resulting from the constant flooding of the water-saturated soil.

The community's relationship with water, however, is not always one of conflict, as a technical view might point out. The workshops conducted revealed a relationship of affectionate memory with the hydrological cycle present in Jardim Nova Esperança, related to both the Paraíba do Sul River - the great excavator of the meandric natural amphitheater where the community is located – and the several bodies of water predominantlyon the main slope that marks the physical separation from the city centre. Therefore, while drawing up the proposal, care was taken towards the preservation of the hydrological cycle and addressing this with the population, while meeting the necessary technical standards.

The proposal aimed at guaranteeing the maximum possible water permeability along its path through perforated drainage networks and surface channels so that no downstream flooding or accumulation of standing water are generated. The community is expected to develop a caring and affectionate relationship with water, contributing to the preservation of the hydrological cycle that permeates its territory. However, slope studies must be conducted and waterpoints identified, as well as borehole and pluviometric studies for the development of a detailed project for the local context. Figure 11 shows a scheme that overlays the layers covered in the Plan.

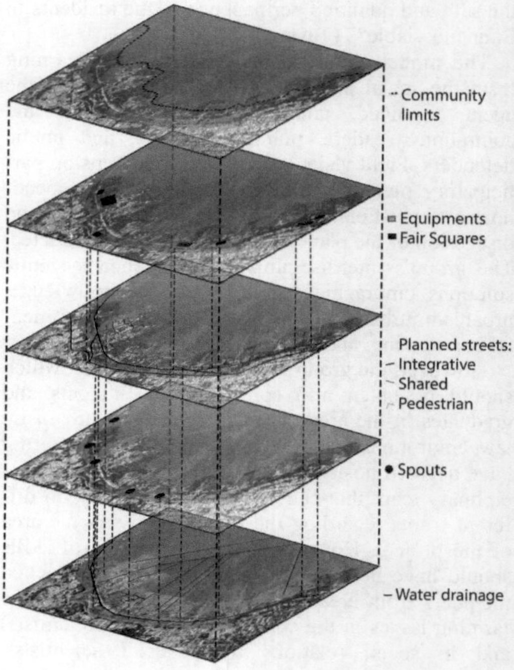

Figure 11. Overlaying of different layers of the plan. Produced by the authors, 2019.

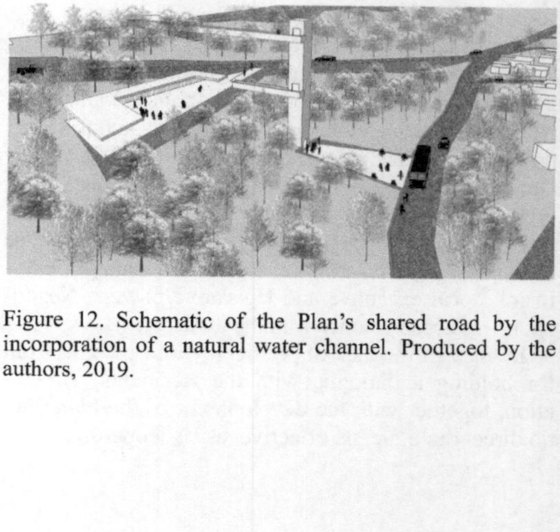

Figure 12. Schematic of the Plan's shared road by the incorporation of a natural water channel. Produced by the authors, 2019.

Some residents attribute the presence of insects in the area to water. However, water mixes with sanitary sewage in some open ditches. Towards solving this issue, the plan devotes special attention to water and sewage treatment, proposing specific solutions for each area of Banhado, and not limited to the traditional model of removal and collective treatment in conventional stations. Through interdisciplinary work, the team proposed two individual solutions for local treatment and a collective solution, always consulting the community on the present hydrological cycle.

Regarding the more structural elements of the plan, the blocks and plots were the urban units adopted for the subdivision in this stage, in view of the proposal for land regularization. They are delimited and ordered by the trunk infrastructure, and always served by their respective basic services, such as electricity, water, telephone, etc. The blocks were adopted in Nucleus I, where the individualization of land lots predominates, whereas the plots were included in Nucleus II, where family farms prevail and buildings are not defined by land lots. The division, therefore, respected the historical occupation of the community and avoided major physical rearrangements that might compromise the Plan´s feasibility. The parceling in land lots was detailed in the next stage of the project, through a survey and planimetric demarcation that considers the minimum size regulated by the specific municipal, state and federal laws.

While still taking into account the history of the occupation and the current land-work relationship in close association with the local hydrological cycle and the subsistence agricultural production of families, the Plan encompasses four fair squares at strategic points for enhancing its integration with the city center. These squares create dialogue directly with the characteristics of their surroundings, becoming spaces for the sale of local produce and for socializing. The need for their creation became evident from a relationship established between the technical team and the community during the preparation of the plan.

Figure 13. View of one of the fair squares proposed under the Plan. Produced by the authors, 2019.

This social interaction permeated by a critical eye enabled the team to assess the need for spaces that would enhance the relationships between Jardim Nova Esperança and the city center, combining their productive relations with the territory and the social interaction of the population with the city.

4 CONCLUSIONS

Within the topic addressed, the mapping of issues in proposals for urbanization and land regularization must be highlighted in situations of socio-environmental vulnerability, common in many Brazilian cities, and which warrant attention from both government and academia.

Based on the potentialities of the social and economic dynamics of Jardim Nova Esperança, the project devised a plan aimed at promoting local productive activities, guided by the community´s land and work relationship. According to the work methodology, the project was developed through the interplay on several urban scales, beginning with the functioning of the urban complex that forms the region of São José dos Campos, its metropolitan region, and, finally, on municipal and neighborhood scales.

This study focused on urban solutions for precarious settlements based on the case of the Banhado Popular Urbanization and Land Regularization Plan - Jardim Nova Esperança, in São José dos Campos-SP, threatened by urban renewal and road infrastructure projects by the city government which overlooked the presence of the community and its dynamics. Moreover, it aimed at understanding the way the complexity present in the community - its particular situation within its rural-urban connection, the dynamics established between residents and the environment, working relationships within the space, etc. - could support the urban solutions adopted during the preparation of the Popular Plan.

The urban solutions proposed and developed jointly with the various actors involved and the relationships identified within the local socio-environmental specificities revealed the possibility of adopting a work methodology that envisions an inclusive city scenario and was not limited by ready universal models or reactive technical work dictated by local conditions.

The local, regional and working conditions must be appropriated as tools for project prospecting, guiding the urban solutions adopted for land regularization plans.

From the analysis of the elements outlined above - despite criticisms - and the suitability of the solutions adopted by the Popular Plan of Banhado for their socio-environmental context, the alternative and critical urban planning developed with the actors involved acquires qualities that, in addition to remedying the issues detected, highlights the qualities of the local reality towards productive arrangements and a sustainable, integrated, safe and inclusive city.

REFERENCES

Ibge (Instituto Brasileiro De Geografia E Estatística). Ibge Cidades: São José dos Campos. Available from: <https://cidades.ibge.gov.br/brasil/sp/sao-jose-dos-campos/panorama>. [Accessed 3 February 2020].

Jaramillo, S.G. 2008. Reflexiones sobre la "informalidad" fundiaria como peculiaridad de los mercados del suelo en las ciudades de América Latina. *Territorio*: 18–19, Bogotá, Colombia.

Moore, E. & Garzón, C. 2010. Social Cartography: The Art of Using Maps to Build Community Power. *Race, Poverty & the Environment* 17(2): 66–67.

UN. 2016. New Urban Agend. Quito 1–29.

Rabelo, E. & Fantin, M. 2018. Prioridades e diretrizes das conferências Habitat: entre o neoliberalismo e o bem comum. In Natacha Rena; Daniel Freitas; Ana Isabel Sá; Marcela Brandão (eds), *I Seminário Internacional Urbanismo Biopolítico (v. 1)*: 466–476. Belo Horizonte: Fluxos.

Rosa Filho, A. 2002. *As políticas públicas do poder executivo municipal na remoção e/ou reurbanização das favelas do município de São José dos Campos.* Masther Thesis, Universidade do Vale do Paraíba, São José dos Campos.

Silva, D. V. 2017. Banhado Resiste: estratégias discursivas de uma comunidade protegida. São Paulo. In *Anais do Encontro Nacional da Associação Nacional de Pós Graduação e Pesquisa em Planejamento Urbano e Regional.*

Silva, D. V. & Vianna, P. V. C. Zanetti, V. R. 2017. Planejamento urbano, agentes e representações: criação do Banhado, cartão postal de São José dos Campos. *Ambiente & Sociedade*, 20(2): 163–184.

Tavares, J. & Fantin, M. 2019. *Banhado: plano popular de urbanização e regularização fundiária.* São Carlos: IAU/USP.

Research Tracks in Urbanism: Dynamics, Planning and Design in Contemporary Urban Territories – Allegri et al (eds)

Sant Adrià Viaduct and João Goulart elevated highway: Urbanity in areas under viaducts

M.I. Camañes Guillén & E.H. Abascal
FAU-MACK, Post-graduate Program in Architecture and Urban Planning; Mackenzie Presbyterian University, São Paulo, Brazil

ABSTRACT: This paper aims to identify similarities in the ways of experiencing the spaces under the Sant Adrià Viaduct, located in the homonymous city, on the outskirts of Barcelona, and the João Goulart elevated highway, in the central region of São Paulo, appreciating that the barriers resulting from the viaducts reveal different dynamics and that the spaces under them can act as connections between the surrounding urban meshes. The study of the social, material and production dimensions comprises the methodological approach employed through the study of urban morphology and the presence of quality attributes of public spaces. The article demonstrates, conclusively, that the viaducts both fragment and structure the urbanization in these cities and that the spaces under these viaducts preserve an important urbanity, expressed by the daily dynamics of the place and urban form materiality, acting as urban bridges to segregated regions.

1 INTRODUCTION

Within the Architecture and Urban Planning fields, the importance of studying the structuring of the metropolitan area and the urban dynamics resulting from the implementation of mobility networks is clear, paying closer attention to the user on a local scale. The purpose of the investigation reported in this article was to claim the areas under viaducts, the residual spaces resulting from the road system network, as project areas for architecture, highlighting the opportunity to connect and reactivate the local level of these places with the city. The interest in understanding these dynamics in the areas under viaducts as city-structuring elements is established here as an urban scenario.

Currently, it is possible to identify the impact of urban infrastructure in the city by its relationship with the architecture, perceiving that dynamics increasingly determine the environment we live in. In many cases, the infrastructure splits or integrates entire neighborhoods, causing a high social and spatial impact. This problem has emerged in recent years, going beyond the city area to include the outskirts and conflict management in consolidated areas.

The daily routine observed in areas under elevated highways of metropolis is our focus, a topic derived from a doctoral study in progress. The two structures observed in this investigation are in the cities of São Paulo (Brazil) and Barcelona (Spain). The João Goulart elevated highway, popularly referred to as "Minhocão" and previously "Via Elevada Presidente Costa e Silva", in São Paulo, and the Sant Adrià Viaduct, part of the C-31 highway, which connects the cities of Sant Adrià, Sant Roc and Badalona, in the outlying region of Barcelona. The history of the implementation of these structures is similar: both were built to promote mobility and connection between different areas of the cities.

The C-31 highway, in Sant Adrià de Besòs, and the east-west diametral route, of which the João Goulart elevated highway is part, delineate distinct areas with different spatiality, revealing the barrier factor. However, in both cases, the spaces under them, expressed by the duplicity of the higher and the lower grounds, are squatted, leading in both cases to conflicts among the inhabitants of the two areas. The methodological approach comprised empirical observation of these places and record of the activities engaged in by users, in conjunction with a mapping assessment of the different territories. The line of questioning of the investigation starts from the observation of the occupation and urban forms present in the surrounding area, analyzing the perceptions about these spaces, and understanding what can imprint urbanity in these places.

2 THE SANT ADRIÀ VIADUCT: GROWTH OF THE CITY FOSTERED BY THE VIADUCT

The implementation and situation of the city around the C-31 highway, which crosses the cities of Sant

DOI: 10.1201/9781003220855-30

Adrià and Badalona, have been questioned since the road´s inauguration. The question is whether the highway implementation has invaded the city, or the city has adapted to it. Several readings confirm that both situations have occurred, and, after the implementation of the road, there followed a growth in the urban mesh in spaces that were previously empty. (Figure 1) The Barcelona-Mataró Project[1] was developed in two stages, in 1964 and 1965, by 5ª Regional de Jefatura de Obras Públicas de la Provincia de Barcelona[2], starting at Les Glories Catalanes square, in the city of Barcelona, and finishing at the city of Montgat. Figure 1 shows the extension of the road and the built areas (darker spots) which emerged after its implementation.

The design of the elevated viaduct was defined in 1965, which starts with the Besòs river transposition, in the form of a bridge, then continues as an elevated road for 1,646m, until meeting the Alfons-XIII Highway (Figures 2 and 3). In practice, both the city and the viaduct are contemporary, for they serve the functionality of the physical spaces of the city, which operates by means of differentiated elements (Solà Morales, 2008). The viaduct and the urban mesh of the city of Sant Roc have grown at the same time: the road network construction started in 1962 and its buildings were erected between 1965 and 1968 (Jordi, 2012).

The Sant Adrià viaduct and the space beneath it have promoted claims from the neighborhood about the occupation and conservation of the underneath area. This space accommodates, besides the Encants, the tramvia station, a huge number of parked cars and a street market called "Encants de Sant Adrià'[3]. (Figure 4). Besides claiming the space under the viaduct for other activities, the inhabitants of the neighborhood currently request acoustic pollution control due to the loud noise level. The large area under the viaduct is occupied by a structural sequence of pillars

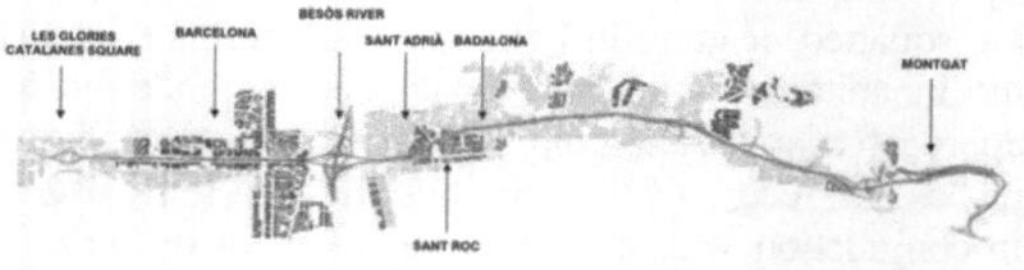

Figure 1. The Barcelona-Mataró highway, from Plaça das Glóries Catalanes to Montgat. Source: Map of Jordi (2012), edited by author.

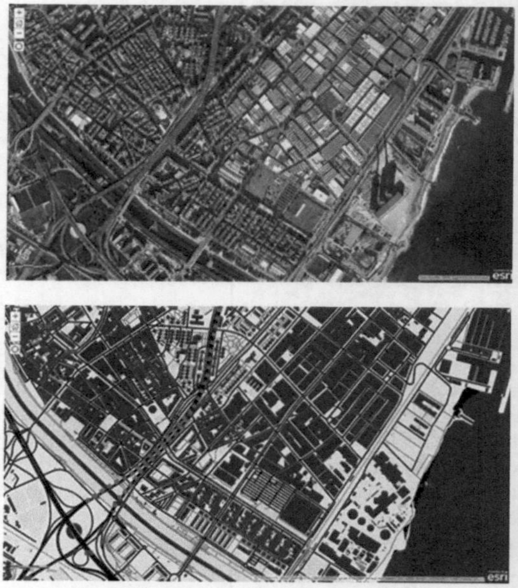

Figures 2 and 3. The Sant Adrià viaduct and urban mesh of surrounding areas, articulated in relation to the viaduct, and on the right the relation of full and empty spaces in the urban mesh, highlighting the viaduct. Source: Google Earth and SIG (Geoportal Barcelona), edited by the author. Available at: https://geoportalplanejament.amb.cat. Access in: January 2020.

to support the viaduct lanes (Figure 5). Among the huge number of pillars, there has always been, since its inauguration, many parked cars, as well as drug users, rendering it a danger zone, with high rates of car theft and violence. Among the neighborhood inhabitants' achievements is the use of the space under the viaduct for urban art projects and events, such as concerts, fairs, and local parties, contributing to the vitality and safety of the place. (Jordi, 2012)

2.1 *Title, author and affiliation frame*

Immediately after crossing the Besòs river, the highway passes through a built-up area of Sant Adrià de Besòs, crossing the city of Sant Roc and entering the city of Badalona, after a slight change in direction, adjacent to a 50m large area[4] reserved for the urbanization of social housing. At the end of the viaduct,

1 The Barcelona-Mataró road was the first section of the highway built in Spain, partially inaugurated on March 14th, 1968 (Badalona - Montgat), and officially inaugurated on July 3rd, 1969, when the whole section Pl. de les Glories Catalanas de Barcelona – Montgat was ready.

2 The 5th Provincial Branch of the General Roads Directorate and neighborhoods of the Ministry of Public Works of the Franco regime was the department that approved the commission and monitoring of the control and implementation projects of the first sections of the C-31 highway.

3 The Encants de Sant Adrià is a weekly street market like the informal street markets in São Paulo; however, permission has been granted by the municipal authorities for use of the public space in Sant Adrià.

4 7-B-307 Blue prints Archive, "Nova Construção da rodovia Barcelona-Mataró, Praça Glórias – Trecho Badalona – New Project for building of the Barcelona-Mataró highway, Praça Glórias – Badalona Section", drafted by EPTISA, SA, in August, 1965.

Figure 4. Sequence of images of viaduct, its surrounding area and Encants de Sant Adrià street market. Source: Author's collection.

Figure 5. Sequence of images of The Sant Adrià viaduct, the porches supporting the highway platform, and the pillars with the urban art project. Source: Images provided by Badalona's central photo archive - Figures 5 and 6 (Jordi, 2012) and Figure 7: El Periódico. Available at: https://www.elperiodico.com. Access on: January 2020.

there has been a gradual implementation of blocks, to the north of the viaduct, built after the road network planning in the 70s, and sufficiently distant from the viaduct structure (Figure 5).

Expressions of urban art adorn the sequence of supporting columns of the viaduct. A large proportion of the Sant Adrià supporting pillars have been recently painted under a project led by the Sant Adrià municipality called 'Bosc de Colunas'. The graffiti artist Zosen, in conjunction with a Japanese artist, Mina Hamada, have painted around 200 columns, with the support of the municipal authorities (Figure 5).

3 THE JOÃO GOULART ELEVATED HIGHWAY: AN URBAN SCAR

Previously named Presidente Costa e Silva Elevated Highway, popularly referred to as "Minhocão" (Big Earthworm), and recently renamed as João Goulart Elevated Highway, this is a road connection conceived in the early 70s as an alternative for vehicle flow between the East and West zones of São Paulo. The long viaduct measures 3.6 km in length and has been controversial since its inauguration for favoring private vehicle circulation over public transportation. The purpose of "Minhocão" was to relieve the traffic while avoiding intersections and traffic lights. The viaduct was built within meters of the windows of existing buildings in a consolidated region of the city center, soon becoming a synonym for degradation of the region. The elevated highway was inaugurated in 1971, when the city had 8 million inhabitants and an authoritarian government. The implementation of "Minhocão" was decided under President Juscelino Kubitschek's technocrat government (1956-1961), when the expression "Brazilian economic miracle" was coined to designate the magnitude of the country´s expansion.

The large, covered area created by the elevated highway gave rise to shaded, segregated, and squatted areas in the vicinity, mostly over existing avenues, with some land expropriations (Figure 8). Shortly after its inauguration, the viaduct served as shelter for the homeless, drug users and prostitutes. At the time, several actions arose aimed at attenuating the impact generated by the construction, such as art works in the space under the viaduct and its closure to car traffic during the night. The surrounding area has undergone a period of transformation involving the region's commerce and services, as well as the inhabitants, fueled by the decrease in housing prices. Car traffic is now also banned on weekends, when the huge platform is used for sports, leisure and relaxing activities.

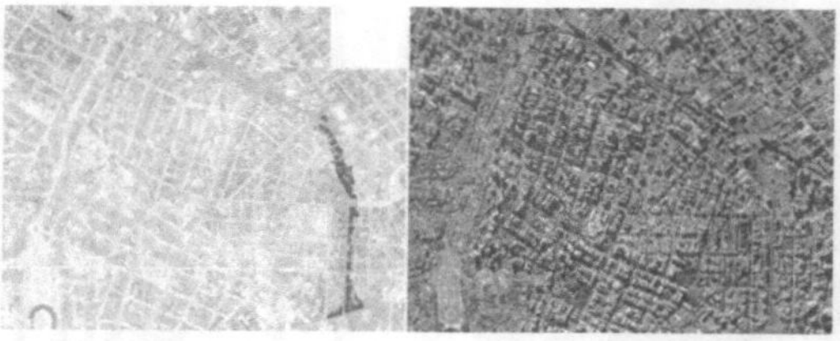

Figure 6. Montage on left: the João Goulart elevated highway projection over VASP/Cruzeiro do Sul basis (1954), highlighting the land expropriation; on right, projection over aerial photograph. Source: Hereñu (2016).

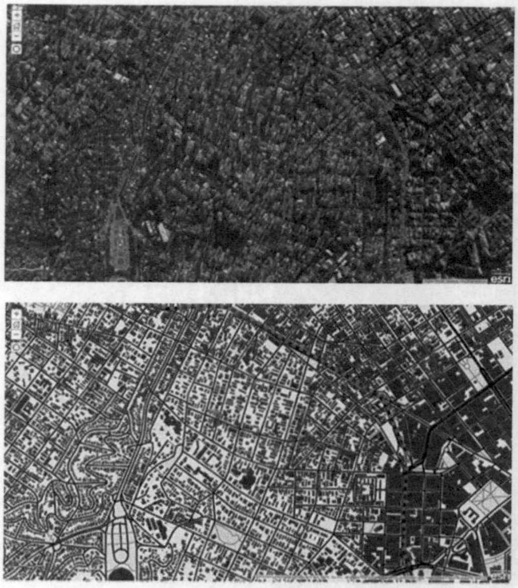

Figures 7 and 8. Map showing occupied and empty spaces in relation to the João Goulart elevated highway and, on the right, aerial photograph (2020). Source: SIG, edited by the author.

Due to the change in the usage profile and its gradual deactivation provided for in the current Urban Master Plan, the region has experienced real estate appreciation, with the creation of several new real-estate ventures exploring the idea of transforming the elevated highway into a huge elevated urban park, like the famous High Line[5], in New York. There has also been a noticeable change in the structure of the socio-economic profile of the region´s residents, as well as the emergence of new establishments catering to the profile of these users.

The viaduct structure was erected in a consolidated, highly verticalized and dense region of the city (Figures 9 and 10). Different layouts follow the avenues overpassed by the elevated highway, starting at Roosevelt square and ending at Francisco Matarazzo Avenue, from downtown to the West zone of São Paulo. One of the biggest issues of all the controversy generated in São Paulo over the João Goulart elevated highway and Minhocão Park is its proximity to existing buildings. With an average distance of six meters from apartment windows, at some points the buildings are as close as 2 meters from the highway as it follows the overpassed avenues. The decision on building the park, demolition or retention of the elevated highway for car traffic divides the population.

Urban art has been introduced since project inauguration, as an expression that breaks with the rigidity of the concrete structure along the avenues. Different projects have been developed, using the pillar as the support. Ranging from the first intervention by Flávio Motta and Marcello Nitsche, who applied kinetic art concepts into their art intervention, followed by unauthorized graffiti, which is constantly found on the pillars in layers of overlapping paintings, through to official interventions, such as "Máscaras Afro-Brasileiras" – *Afro-Brazilian Masks* (2014) and a recently promoted project by SESC (Social Service of Commerce) named Projeto Giganto (Figure 9), by Raquel Brust, carried out in 2013 and again in 2019-20. The purpose of this intervention was to promote the use of the city architecture as a platform for the hyper dimensional photographs of regular people living in the region, inviting the population to reflect on the perfection of the conventional faces seen in ads.

The art exhibited in the spaces under the João Goulart elevated highway is undergoing constant construction and renewal, often expressing the social conflicts of the city of São Paulo. The alternative urban space use is questioned by the different people occupying the areas under the João Goulart elevated highway: the homeless share the space with people who live in the region or passersby, graffiti artists and vandals (Figure 10). "The populations incorporated in the informal dynamics transcend the traditional spatial frontiers [...] They slip into the cracks of the urban fabric, the built areas [...] live in the folds and cracks of the city." (Peixoto, 2004, pg. 426).

Figure 9. Sequence of pictures part of the Giganto Project under the "Minhocão". Source: Giganto. Available at: http://projetogiganto.com.br/. Access in January, 2020.

5 The park is located in an elevated highway in the East part of the city, in the extinguished New York Central Railroad.

Figure 10. Sequence of pictures under the "Minhocão". Source: Author's Collection.

4 AREAS UNDER VIADUCTS: A POSSIBLE URBANITY BETWEEN FRONTIERS AND CONNECTIONS

In several areas, the contemporary urban space presents a fragmented fabric with missing elements that could promote articulation and aggregation among different sectors and morphological units. This situation exacerbates segregation and social rupture leading to the loss of essential guidelines for the population and the smooth flow of the urban organism (Portas, 2011). These frontiers serve as examples of this fragmentation for imposing different contours; however, in the same way as the contours of lakes and rivers form frontiers, they also interact and nurture themselves through the exchange of organisms. These boundaries and frontiers can be compared to territories of exchange and interaction, presenting, therefore, great opportunities. Following this argument, the areas under viaducts have the potential to create bonds, since they are public spaces belonging to the city and are formed according to the use and permanence of people.

The edges of the highways tend to become stunted spaces outlining invisible walls (Sennet, 2014). One can affirm that urbanism in the 20th century has been highly responsible for defining limits instead of thresholds, leading to a loss in the new urbanization diversity (Jacobs 1960), and understood according to the relationship between massive and "naturally grown" urbanization.

The potential urbanity in the areas under viaducts is expressed in the flexible capacity of these places, which can change according to the needs, thus imprinting an urban character on the space (Krafta, 2012) According to the author, this urbanity is not related to the pleasantness or liveliness of the places; it is related to their diverse aspect, as well as the activities, forms and inclusion of people who can access these places. Foucault (1984) uses the term real spaces to dissolve the idea of a place with a single identity and an exclusive function; this space-place projected and constructed according to strict functional approaches bears multiple overlaid and overlapping strata and changing temporalities, redefining the boundaries, sections, and the usual temporalities of urban spaces (GUATELLI, 2012). These multiple diverse strata manifest in the form of different architectural dimensions, building typology, public spaces, as well as activities expressed by different social presences, both economically and ethnically (AGUIAR, apud NETTO, 2012).

These spaces devoid of formality yet with strong identity have great potential to respond to and follow changes in a major part of social relations in urban life. To confirm this statement, we have established here a set of concrete attributes in the spaces under viaducts based on categories derived from the theory of Carmona et al. (2003), where baselines are discussed to promote understanding of the most relevant urban parameters and criteria in the urban project. Thus, we can highlight three traditions of thought, in practical and theoretical terms: the art and visual tradition, which emphasizes the visual and aesthetic aspects in the built environment; the social use tradition, which focuses on the inhabitants' urban experience; and, finally, the place production tradition, which can be considered a synthesis of the other two, with emphasis on the idea of the public space serving as the stage for urban life, thus turning it into a place. The proposed diagram (Table 1) depicts a set of attributes that can be applied on a neighborhood scale. These attributes are visuality, identity, legibility, contextuality, connectivity, accessibility, inclusiveness, safety, density, diversity, and conservation. Clearly, not all attributes mentioned are applicable in the examples given; however, we intend to demonstrate that there is a significant presence of attributes in the areas under these viaducts that can promote new use possibilities at these places.

Table 1. Urbanity: References, Elements and Narratives. Source: Produced by the author, based on Carmona (2010) and Rezende (2019).

URBANITY: ELEMENTS, REFERENCES AND NARRATIVES											
Theoretical references of the attributes: main theoretical traditions that include the urbanity attributes detailed in the thesis:	VISUALITY	IDENTITY	LEGIBILITY	CONTEXTUALITY	CONNECTIVITY	ACCESSIBILITY	INCLUSIVENESS	SAFETY	DENSITY	DIVERSITY	CONSERVATION
AESTHETIC NARRATIVES											
COEXISTENCE TACTICS											
SUSTAINABLE PLACES											

Drawing on the study by Rezende (2019), highlighting the most relevant urbanity attributes for the assessment of public space appropriation in precarious settlements and proposes a critical review of the main conceptual trends, we note a variety of attributes that can be structured to generate information on the appropriation of public spaces by means of urban perception, reading and interpretation studies. In Table 2, we have identified elements in the areas under these viaducts that are responsible for assigning these characteristics which, based on the theoretical basis reported by Rezende (2019), imprint urbanity onto these places. According to the author, both normative and experiential urbanity trends exist, derived from theoretical and practical discussions on urban transformations that occurred in the 20th century, when critics on the concepts of urbanity fostered consensus and dissensus regarding patterns of contemporary public space appropriation (Rezende, 2019).

The premises presented by the author stem from Solà-Morales's argumentation about the city, proposing three essential qualities: coexistence, memory, and alterity associated with homology, agents that absorb the urban structure, objects, people, and activities, respectively. Therefore, the approach involves the coexistence of objects, memory - collective or otherwise – people and alterity of different activities. According to Rezende (2019), these qualities are not unidimensional in concrete reality, especially because there are different levels of objects, people, and activity abstraction. Thus, they are expressed in space representation and practice or in representing space. These three types of urban experience can be understood based on Lefebvre's space theory (1991); whose main premise is that space is a social product.

The social product of public spaces, however, is related to aspects of urban form. The distance from buildings, presence, or absence of natural elements in the areas under the viaducts assessed will determine different perceptions. Analysis of the different sections of the two viaducts along their courses (Figure 12) reveals the Sant Adrià viaduct area has broader dimensions in its transversal profile, as well as a distance ranging from 15 to 30m away from buildings, while this spacing for the João Goulart

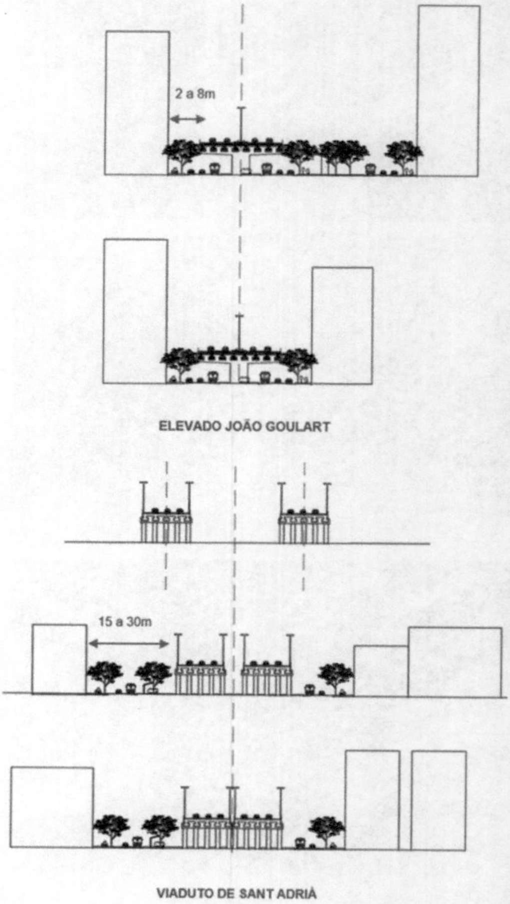

Figure 11. Comparison of viaduct profiles: João Goulart elevated highway on the left and Sant Adrià viaduct on the right. Different profiles presented along their courses. Source: Drawn by the author.

elevated highway ranges from 2 to 8m, based on the overhang relative to the avenue below, which is more distal only at one point over Marechal Deodoro da Fonseca square, an area that existed before the implementation of the elevated highway.

There are certain similarities between the territories of the viaducts analyzed, as well as distinct perceptions from the set of materiality of each of the structures transformed according to the use and permanence of people; however, the way the viaducts were implemented and the changes which occurred in the urban mesh of the surrounding areas after their construction are very important factors for the perception of the areas under them (Figure 14). The perspectives of the areas under the viaducts are directly related to the architecture in the surrounding area, as well as the presence or absence of vegetation and leisure areas, directly impacting on the light intensity that permeates the areas shaded by the viaduct.

Table 2. Urbanity: References, Elements and Narratives. Source: Produced by the author, based on Carmona (2010) and Rezende (2019).

	URBANITY: ELEMENTS AND TENDENCES	
	PUBLIC SPACE	
URBAN ELEMENTS (main components, functions and forms)	TENDENCY TO NORMATIVE URBANITY	TENDENCY TO EXPERIENTIAL URBANITY
VISUALITY	INVENTED	SIGNIFICANT
Façades. Signs. Artistic and decorative elements. Composition. Sculptures. Colors. Lighting. Style. Sceneries. Parties. Artistic exhibits.	Intense propagation of symbolic public spaces of touristic, artistic or environmental interest conceived with no connection to the local history and culture.	Elements of the local history can contribute for the formation of significant public spaces due to their importance, thus enabling the urban experience.
LEGIBILITY	GENERIC	SINGLE
Landscape. Uniqueness. Boundaries. Marks. Façades. Multifunctional. Familiarity.	Decontextualized project solutions and public space maintenance derived from functional approaches, resulting in no self-identity places.	More efficient strategies and tactics for the sustainability of the places are developed based on reflective actions promoting the preexistence.
IDENTITY	CONSUMPTION	HABITABLE
Culture. Regionalism. Urban tribes. Music. Gender. Age. Collective issues. History. Parties.	Spaces with access conditioned to income and consumption-oriented, usually excluding the less prosperous segments of the society.	Consumption spaces give people from different social classes the opportunity for meetings, appropriations and exchange of experience.
DIVERSITY	FRAGMENTED	PLURAL
Diversified commerce/service offers	Impediment of the appropriation of public spaces by great part of the population due to concessions to private institutions.	Promote social organization and institutional transparency aiming to guarantee equal rights to access and remain in all public spaces for all people.
DENSITY	INVADED	SHARED
Urban fabric full and empty relations. People. Vehicles. Bicycles. Motorcycles. Turmoil. Children. Elderly people. Parties. Markets. Pedestrian streets. Tables on the sidewalk.	Exclusive public spaces occupation by mobility structures, markets, illegal street vendors, parking lots, among others, obstructing the way and impeding leisure and relaxing activities.	Space shared by pedestrians, bikers, cars, market vendors, promoting the urban interaction among users and allowing the optimization of the urban traffic.
ACESSIBILITY	ISOLATED/PRIVATIZED	SOCIALIZED/ EQUITABLE
Topography. Ramps. Human scale. Ergonomics. Public phone. Public restrooms. Adequate paving. Banks. Playgrounds. Squares. Protection fences.	Spaces with access conditioned to income and consumption-oriented, usually excluding the less prosperous segments of the society.	It is possible to provide the city with public spaces with different privacy scales, adapted to the sociocultural conditions of each area of the city, since the question involving walled sectors is more complex. Public spaces have still been the highest collective expression despite of technology and globalization. com a tecnologia e a globalização, o espaço público continua desempenhando a expressão máxima coletiva. Mesmo apropriados de maneira específica por alguns grupos sociais e atividades transgressoras, os espaços devem ser inclusivos aos mais vulneráveis.

(*Continued*)

Table 2. (Cont.)

	URBANITY: ELEMENTS AND TENDENCES	
	PUBLIC SPACE	
URBAN ELEMENTS (main components, functions and forms)	TENDENCY TO NORMATIVE URBANITY	TENDENCY TO EXPERIENTIAL URBANITY
CONTROL	INTERDICTORY/ UNSAFE	CONFORTABLE/ SAFE
Care. Maintenance. Garbage. Several broken objects. Dirt. Rubble. Control. Damaged elements on the streets, sidewalks, façades. Lack of public illumination. Environmental comfort. Cleaning. Materials. Safety. Surveillance. Monitoring. Cameras. Private security services. City guard. "Eyes on the Street". Walls. Bars. Public illumination. Illumination. Neighborhood. Control.	Public spaces abandonment, resulting in lack of maintenance and planning. The high criminality level and the perception of significant violence stimulate often oppressing strategies of prevention and fight against crime, preventing the free appropriation of the public space. frequentemente repressivas, impedindo a livre apropriação do espaço público	Reversal of several public spaces abandonment conditions by means of restoration. Public spaces can have different surveillance and protection levels to ensure the same safety and comfort levels without preventing their appropriation by people. níveis de segurança e conforto, sem, no entanto serem interditórios à apropriação pelas pessoas.

According to Solà Morales (2008), the urban project is developed in a sequence of concrete elements and episodes that connect people and city, as well as having a relationship with materiality, or "city skin", which is the surface where social and physical connections are developed. The author states that "urbanity is made of touch and vision, sensations and suggestions "[…]" It is different from the idea of "urban structure" or the concept of "urban system", mostly applicable to the interdependence of activities and positions, or in "functional areas" used to classify and allocate spaces (Solà Morales, 2008, pg. 23). The author argues that simultaneity, temporality, and diversity are the main attributes of the city, without which its concrete existence is not possible. We have defined these attribute indicators: coexistence, memory, and difference, respectively. Urbanity cannot be based solely on a set of attributes or the result of certain interactions in specific urban contexts. On the other hand, Choay (2005) argues that transformations in the urbanity concept are strictly connected to the transformations of cities since the 19th century. Solà-Morales is assertive in defending the existence of urbanity in unplanned or poorly planned cities. Based on this, we can affirm that there is urbanity in the areas under the viaduct.

Krafta, 2012, holds that urbanity is historically produced as a sequence of "amalgamated urbanities" in time, by the urban space and examining the characteristics of each object in the light of their preexistence. The pre-existing objects, associated with historical timepoints, possibly induce specific urban forms, found in every object of study.

For Rezende (2019), the condition of urbanity emerges from the interaction among local agents considering the pre-existent; it emerges in the presence or absence of ephemeral or daily events. According to the author, everyday life is a condition

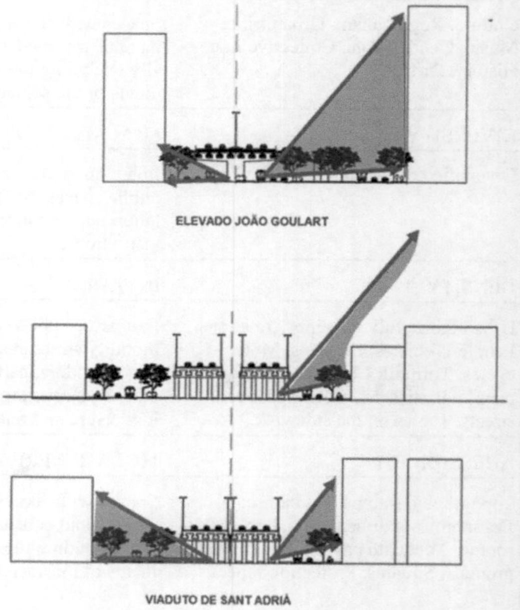

Figure 12. Comparison of viaduct profiles, from the perspective of the areas beneath them. Source: Drawn by the author.

of urbanity, especially in the informal city, the theme of his research. The diagram below (Figure 13) depicts a qualitative assessment of the spaces under the viaducts analyzed, showing that the experiential urbanity, based on the everyday life condition, prevails in both cases; and all attributes, according to Table 2, are present in these areas.

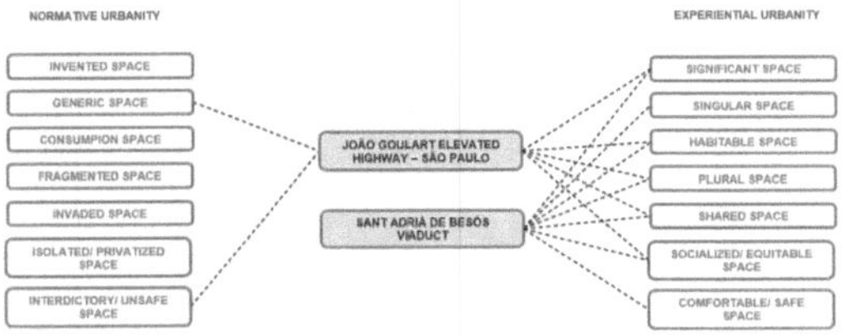

Figure 13. Qualitative analysis of the spaces under the viaducts analyzed. Source: Produced by the author, based on Carmona (2010) and Rezende (2019).

5 FINAL CONSIDERATIONS

The unexpected activities occurring in the areas under the viaducts analyzed confer meaning to these places, with an emphasis on actions that contribute toward their identity. Besides the commercial, service and transportation activities in the spaces under the Sant Adrià viaduct, the cars, and pedestrian flow through a generous open area, permeated by vegetation and large public spaces in the areas bordering the buildings and viaduct structure. For the João Goulart elevated highway, however, the activities overlap in a limited area in the interstices of the viaduct pillars and sidewalks of the avenues. The only open perspective, resembling that of Sant Adrià's, is when the elevated highway meets the connections from Marechal Deodoro square and Largo do Arouche.

The activities within the empty spaces found under these large structures promote an interchange between the two sides of the city. There is a lack of situations promoting more connectivity between the two parts segregated by the viaducts. However, by identifying attributes in these places, the potential of these spaces is elucidated.

By exploring the density of the city, introducing empty spaces and varying distances between buildings and the viaduct, the Sant Adrià viaduct seems to be able to breathe between the two sections of the city. However, given the characteristics of the city's growth, this was possible because of the adaptation of the urban mesh to the viaduct, which occurred concurrently. Likewise, by means of everyday life expression, the João Goulart elevated highway exhibits connection and permeability with the surrounding area; the everyday lives of inhabitants carry on amid the chaos of overlapping activities. The adequate space through daily practices goes beyond the concrete space: it is noticed, experienced, subjective, reframed. It is a space not only for consumption, but an affective and symbolic place related to the anthropological, poetical, and mythical space. Certeau (2011).

REFERENCES

Augé, M. (2012). Não lugares: Introdução a uma antropologia da supermodernidade. 9° ed. Campinas: Papirus.

Brenner, N. (2018). Espaços da urbanização: o urbano a partir da teoria crítica. Rio de Janeiro, Letra Capital: Observatório das Metrópoles.

Campos, C.M. (2008). Eixo da Ambiguidade: a região da Avenida São João nas inversões do tempo. In: Artigas, Rosa; Mello, J.; Castro, A.C. Caminhos do Elevado: Memórias e Projetos. (19-45) São Paulo: Imprensa Oficial do Estado de São Paulo.

Carmona, M. (2003). et al. Public Places Urban Spaces: The Dimensions of Urban Design. Oxford: Architectural Press.

Carmona, M. (2010). Contemporary Public Space: Critique and Classification, Part One: Critique Journal of Urban Design, v.15. n.1, 123–148.

Contemporary Public Space, Part Two: Classification. Journal of Urban Design, v.15. n. 2, 157–173.

Certeau. M. (2011). The invention of everyday life: performing Arts. Petrópolis: Vozes

Choay, F. (1961). El reino de lo urbano y la muerte de la Ciudad. In: RAMOS, A. M. Lo urbano em 20 autores contemporáneos (61-72). Barcelona: Edicions UPC, 2004.

Cullen, G. (1961). Townscape. London: Architectural Press.

Foucault, M. (1984). O corpo utópico, as heterotopias. São Paulo: n-1 Edições.

Guatelli, I. (2012). Arquitetura dos Entre Lugares: Sobre A Importância Do Trabalho Conceitual. São Paulo: Editora Senac.

Jacobs, J. (1960). Morte e vida de grandes cidades. 3ª edição. São Paulo: Martins.

Jordi, A. V. (2012). O Viaduto de Sant Adrià e Sant Roc. Barcelona: MTPPA _ Mestre da Teoria e prática do projeto na Universidade Politécnica da Catalunha. [Web page]. Available at: https://upcommons.upc.edu/handle/2099.1/17176. (Access in: 20/ 01/2020).

Krafta, R. (2012). Impressões digitais da urbanidade. In: Aguiar, D., Netto, V. M. (Ed.), Urbanidades (15-134). Rio de Janeiro: Folio Digital: Letra e Imagem.

Netto, V. M. (2012). A urbanidade como devir do urbano. In: Aguiar, D., Netto, V. M. (Ed.), Urbanidades (33-60). Rio de Janeiro: Folio Digital: Letra e Imagem.

Peixoto, N.B. (2004). Paisagens Urbanas. 3ª ed. São Paulo: Senac.

Portas, N. (2011). Políticas Urbanas II: Transformações, regulação e projetos. Lisboa: EFA/FCG.

Rezende, W.S. (2019) Urbanidade experiencial: situações e interações no cotidiano de Paraisópolis, São Paulo. Tese de doutorado em Arquitetura e Urbanismo. Universidade Presbiteriana Mackenzie. [Web page]. Available at: http://tede.mackenzie.br/jspui/handle/tede/4051 (Access in: 20/ 01/2020).

Sennett, R. (2014). O declínio do homem público. Rio de Janeiro: Editora Record.

Shaftoe, H. (2008). Convivial urban spaces: creating effective public places. London: Earthscan.

Sitte, C. (1965). City planning according to artistic principles. New York: Random House.

Solà-Morales, M. (2008). De Cosas Urbanas, 81. Barcelona: Gustavo Gili.

Whyte, W. H. (1980). The Social Life of Small Urban Spaces. Washington, DC: The Conservation Foundation.

Cultural Policies and Governance in the conversion of industrial areas: The case of the LX Factory, Lisbon

N. Somekh & T. Cardoso
PPGFAU, Faculdade de Arquitetura e Urbanismo; Universidade Presbiteriana Mackenzie, São Paulo, SP, Brazil

ABSTRACT: This article aims to study the case of the LX Factory, in the city of Lisbon, Portugal, and seeks to understand whether contemporary transformations, encouraged by cultural policies, are capable of promoting and guaranteeing social inclusion. Based on multiple reviews of the LX Factory project and two field trips (in 2013 and 2019), the circumstances in which this initiative was consolidated and developed were identified, as well as how the issue of social inclusion was addressed, in addition to the possible process of gentrification in the area and its urban surroundings. Thus, this article seeks to understand whether this iconic case can serve as a reference to be replicated in other cities undergoing transformation. A description of the LX Factory is provided, outlining its origin, development, and impact on the Alcantara neighbourhood and city of Lisbon.

1 THE LX FACTORY AND THE TRANSFORMATIONS IN THE CONTEMPORARY CITY

Based on cultural policies embraced in the European Union and adopted in Portugal since 2000, the LX Factory can be considered a possibility for contemporary transformation. Established in an old industrial complex in the zone of Alcântara, has increasingly housed new alternatives for work and income since its foundation. In general, a series of companies and professionals from different areas, but especially those of technological, cultural, and artistic propensities have set up at the site, which is also used for hosting events and festivals. It was created in 2007 by a group of intellectuals and the company Catumbel, part of the MainSide Investments group which, at the time, bought a space at Mirandela Printshop (Gráfica Mirandela) with the intention of investing in initiatives within cultural and artistic fields to return the space to the city, as it was identified as an opportunity for investment and contemporary activities (Paulino, 2015).

We learned that the creation of this new hub was the result of a cultural policy which provided conditions for it to grow and thrive within its urban context. The Municipal Master Plan (2012) and the Strategies for Culture of the City of Lisbon (2017) are important examples of tools which promote the process in this case - as in the face of transformations taking place in contemporary society and the fact that the Portuguese capital is a node in a global web of flow of people, ideas and capital - Portugal sought to draft a project for Lisbon considering the main trends in human evolution, such as: growth in the world's population, the continuous process of urbanization and globalization, the economic interaction, and the digital and informational revolutions (Seixas, 2018). We also understand the importance of strengthening the webs of cooperation as opposed to competitiveness, in such a way as to solve common problems based on a model of horizontal non-hierarchical management.

In Portugal's case, the city of Lisbon – hit hard by the financial crisis of 2008 - invested in strategies involving cultural policies and application of the concept of the creative economy as a way of reinventing the city's degraded spaces and tackling the consequences brought on by the crisis, in both economic and social spheres. Characterized as a city which in recent years has reinvented itself as being increasingly cosmopolitan and integrated with the global vanguard trends, the Portuguese capital has sought to structure new perspectives in regard to issues related to quality of life and entrepreneurship, allying these to its policies of interventions in urban spaces (both public, such as equipment, and ecological areas). Thus, a new perspective for transformation of the city was drafted and thereby a new projection in which culture is seen as a means of promoting well-being and quality of life for the people (Seixas, 2018).

Initiatives were proposed by Lisbon's City Council to counter the effects of the financial crisis. Notable among these, in 2009, the institution drafted a document entitled Strategies for Culture in the City

DOI: 10.1201/9781003220855-31

of Lisbon - reviewed and updated in 2017 - and invested in the concept of a creative economy as a way to reinvent the city and its social and economic relations. The incentive for the development of start-ups and public incubators for small businesses was one of the main actions carried out by the local authorities, generating a series of small businesses which sought to meet a vast variety of market demands (Markun, 2019). An example of this action is the programme Startup Lisbon, which acts as a public incubator for digital and creative businesses, seeking to attract mainly young and foreign investors.

During this period, the city also adhered to Agenda 21 for Culture, whose objective is to integrate culture into local policies of sustainable development, encompassing 9 commitments: cultural rights; cultural heritage, diversity, and creativity; culture and education; culture and environment; culture and economy; culture, equity, and social inclusion; culture, urban planning and public space; culture, information and knowledge; and culture governance (Seixas, 2018). The first document (Strategies for Culture of the City of Lisbon) sought to identify the relation existing between the cultural field and the development of cities, i.e., how quality of life and the well-being of people who live in and benefit from the city can be promoted through culture.

The fact that Lisbon adopted cultural guidelines as a strategy to counter the effects of the crisis resulted in a scenario conducive for the emergence and consolidation of small businesses, startups, and co-working spaces especially those connected to the field of culture and art. João Seixas, in his book Projections of Lisbon: utopias and strategies for a city in perpetual movement (2018), p.323, states that:

> On the other hand, in Lisbon there are clear rising trends of development in shared work activities, by means of co-working spaces and new clusters of proto creative, technological, and cultural activities. In these areas, policies of promotion of a network of startups and of support to creative economic dynamics may be shaping a new pattern of essentially young work activity, with high rates of corporate dynamism (including high rates of corporate turnover) and with involving training of knowledge bases, technology, and innovation.

Based on this initiative, a series of creative clusters emerged in the Portuguese capital, such as the case of the LX Factory in Alcântara, which consolidated over time and remained, giving rise to a major flow of people in the region to this day. Thus, in identifying this scenario, the following question emerges: is it possible to promote social inclusion in a sustainable way, through a project of urban intervention based on cultural policies?

Firstly, it is important to highlight that the LX Factory project was not carried out by public bodies, but via private initiative through the company Catumbel of the MainSide Investments Group which, in 2007, as aforementioned, identified an opportunity in the field to invest in initiatives related to cultural and art areas and return the space, which was formerly not in use by the city. The LX Factory is located in the civil parish of Alcântara, an old industrial area of Lisbon situated in a region neighbouring the Tejo River, far from the historical centre, but near important points of the city, such as: Alcântara's Sea Station (Gare Marítima de Alcântara), April 25th Bridge (Ponte 25 de Abril), the National Museum of Coaches (Museu Nacional dos Coches) and the Museum of Arts, Architecture and Technology (MAAT). The following photograph depicts its implementation:

Figure 1. Aerial photo of LX Factory. The image was produced based on images and data available on Google Earth in May 2019.

2 THE LX FACTORY AND INDUSTRIAL BUILDING

The industrial complex which currently houses the LX Factory project was implemented with the arrival of the Lisbonese Company of Wiring and Fabrics in the area, in 1846, when it set up in the riverside area of Santo Amaro. The design for the first building was carried out by the architect José Pires da Fonte, who sought inspiration in the mills of Manchester, whose model was based on large, tall, imposing buildings, which had wide internal spaces, given they were designed according to an almost open plan. This first building had four floors, which took advantage of the natural lighting of the environment, besides housing operations such as: weaving, carding, and spinning of warps and wefts. There was also a second building (smaller and separate from the workshop areas) where the steam machines and boiler house were located. Between 1851 and 1855, the plant expanded, with the construction of five more buildings next to the main one, which were called the Small Plant (Paulino, 2015).

In 1873, due to the large number of operators who worked there, the first worker village was built by the company, situated on the street behind the plant, now called First of May Street (Primeiro de Maio). In 1900, the industrial complex grew once more, with the construction of a new workshop. The plant thrived until the crisis of 1917, which resulted in its liquidation and sale to a new proprietor, the Industrial Company of Portugal and Colonies. There it remained until 1961, when the Annual Commercial Typography occupied the site, which added a floor to the plant's main building. Thus, the ground floor was used for typographic printing, the first floor for offices, the second one for photography and offset, the third was used for bookbinding and the fourth as a cafeteria and convenience services. In the 80s, the site received a new proprietor: the Mirandela Printing Shop.

In 1997, the Portuguese Association of Industrial Archaeology proposed the building be listed. However, a few years later, the Institute of Management of Cultural Heritage dropped the idea, deeming that the building lacked the required status for protection. Ultimately, the space was bought by the company Catumbel, of MainSide Investments in 2005, which set up the LX Factory there in 2008, with the objective to only use the plant's main building. During this period, the industrial complex of 10 buildings was generally in reasonable condition. On the land plot there was a small, grassed area with few trees, but no gardened areas, given that the space used to be an old industrial site where these types of spaces were not considered. The surrounding area was characterized as enjoying ample access to road infrastructure and public transport (Ribeiro, 2012).

To implement the LX Factory, the project's creators chose to carry out a simple and minor intervention on the buildings. The architects Ana Duarte Pinto and João Manoel Alves were in charge of this project, whose objective was to only carry out cleaning, painting, and replacement of broken windows. This project of minimal intervention was:

> "[...] motivated, on the one hand, by the containment of costs in terms of initial investment, in face of the presumably uncertainty of the project, and on the other hand, by the desire to preserve the industrial image of the space. The original volumes, fenestration and materiality were maintained. The space's compartmentalisation, which essentially consisted of big empty naves, was achieved using gypsum board walls which guaranteed the intervention's reversibility. Exposed in an untreated way without using ornaments or superfluous artificialism, the setting was revealed in its authentic materiality. The identity and contemporaneity are invoked by the artificial illumination and by the materiality of new elements." (Paulino, 2015, p. 86).

The new tenants initially occupied the first floor of the main building. Givens this used to house the administration of the Mirandela Printing Shop (Gráfica Mirandela), it was already adapted for use as offices. The other floors were occupied according to the need to accommodate new companies in the area. The administrators of the LX Factory gave each tenant freedom to transform their space according to their needs, however, with the commitment of always maintaining the industrial characteristics of the building. This adaptability of the space was one of the factors that most attracted companies from the creative field to the LX Factory, because they valued the air of informality which the environment offered, configuring its identity and singularity, besides the proposal of reusing and recycling an old space (Paulino, 2015).

Figure 2. Photo in the top-left corner shows the industrial buildings in June 2013. The bottom-left image shows flyers for shows and cultural events held there, also in June 2013. The image to the right, depicts the ground floor and industrial buildings of the LX Factory in March 2019. Photos: Nadia Somekh.

3 LX FACTORY AND ITS MANAGEMENT MODEL

The proposal that LX Factory offered, renting large floor spaces at a low price, called the attention of small businesses and entrepreneurs, who were attracted by the viability of the space, the excellent location and by its innovative concept. Most of these companies were involved in the area of arts and culture, such as photography, plastic arts, theatre, fashion, architecture, design, among others, and even though this was not the project's initial intention, the site soon began to be associated with these business activities. However, from the very outset, MainSide Investments had to deal with legal issues to be able to approve the LX Factory project, because the Alcântara Plan XXI defined the space as an area of mixed urbanistic reconversion, where some businesses were allowed, but not connected with commerce and/or restoration. As they awaited the license approval for commercial use on the ground floor, one of the solutions to bypass the problem was to present the cafeteria as a canteen supporting factories and not restoration (Paulino, 2015).

Another preliminary solution was to offer space to companies free of charge, and then contractually it would be able to evict the tenants at any time if necessary. The strategy to establish anchor companies which lend the space prestige, thus generating a large flow of people, was also used. In LX Factory's case, the main companies that initially held this role and contributed to its development were the bookshop *Ler Devagar* and the school *Fórum de Dança* which attracted a great number of people to the place as they were known by the public. A relationship of exchange between them was thus existent: for example, the LX Factory offered the bookshop a prime location on the site, as well as a symbolic income and, in exchange *Ler Devagar* revitalized the space through its products and/or by holding cultural events in the space.

In 2009, as described in the Master's dissertation *The recycling of industrial uses and new typologies of activities and cultural spaces. Case Study: LX Factory* (2009), by Gonçalo José Veloso Queirós de Carvalho, there were 98 companies in the building, predominantly connected with the fields of culture, design and advertising. During this period, the LX Factory attracted a public comprising professionals from the area of culture: people who went there to carry out castings, use the bookshop, the restaurant or take part in workshop/courses; tenants who rented spaces within the LX Factory to hold events and those who visited the space to go to a specific event, without necessarily having ties with the space. This public embraced the concept proposed by the company of maintaining the industrial aspect of the space, which gave it an authenticity and enabled not only that the main building be preserved, but also 95% of the whole industrial complex, which had great historical value (Carvalho, 2009).

As the space consolidated itself, it increasingly became a creative cluster. Besides being a space that valued activities connected to art and culture, the LX Factory frequently stimulated initiatives which was of a lucid, commercial and cultural nature, always stimulating dynamics to generate the flow of people in the area (entry and exit), as this dynamic of people is of great importance to such projects. Events were planned, such as: the LX Market, a fair held on Sundays for vendors of second-hand items, urban craftsmanship, plants, food, art and design; the Open Days, which took place twice a year during which the factory complex could be visited, while several activities were run by some of the resident companies. Besides attracting anchor companies and holding these types of fairs, the LX Factory supported, since its first year of operation, the promotion of several informal cultural e artistic initiatives, also serving as an exhibition space.

Among these initiatives, there are two events which had immense importance for the consolidation of the space and publicizing its name: the OFFF Festival (f08-10 May, 2008) and the Peter Zumthor exhibition: Buildings and Projects (07 September to 02 November, 2018). This is due to the fact that both were able to attract a flow of people to the space, especially groups involved in the fields of culture and art. In both these cases, the LX Factory granted free use of the space and in exchange gained improvements and the local publicity of the company which organized the event. After hosting these two major events, the LX Factory continued offering its space for cultural and art activities, although none had this same magnitude. This may have been due to the growth of the LX Factory which, following an increase in resident companies, made less free space available for such large initiatives (Paulino, 2015).

Besides the OFFF Festival and the Exhibition Peter Zumthor, other events were held over the ensuing years, such as: the Red Bull House of Art (2010 to 2012), Brazil Space (Espaço Brasil) (2012), the Kairos Pavilion (o Pavilhão Kairos) (2013) and Wool on Tour (2012, 2013 and 2014). It is important to highlight that, in the many cases in which the LX Factory provided its space for events, one of the conditions of these agreements was that the occupying organization had to invest in the restoration of the architectural heritage of the space in question. An example of this practice was for the Space Brazil (Espaço Brasil) event in 2012 which occupied the space of Plant L, which was completely empty. Thereby, "[...] the space was offered free of charge to serve as a cultural center of Brazil in Lisbon during the celebrations of the Year of Brazil in Portugal. In exchange, the LX Factory benefitted from a completely remodelled space" (Paulino, 2015, p.101). The installation of the Kairos Pavilion (Pavilhão Kairos) in 2013 also helped to redevelop the gardened area where it was located.

Over the years, the LX Factory, which began as a temporary project, established itself in Lisbon, becoming a model of local urban dynamics, making it a permanent project, preserving a large part of the existing industrial complex. In 2013, there were 115 companies in the LX Factory, over half of which were related to the fields of culture and art (Gabriel; Vale; Silva; Azevedo, 2013). However, as the project matured, the number of cultural and art initiatives declined, especially those related to fine arts which were prevalent in the initial phase of the project.

Thus, the LX Factory's exponential growth, as well as the increase in rental of the spaces, became a factor hindering the establishment of small cultural businesses and entrepreneurs in place. Currently (2019), as shown in Figure 3, the LX Factory houses 169 companies, namely: 20 restaurants, 35 shops, 92 offices and 22 other units, and boasts an events calendar that includes workshops, meetings and conversations, festivals and exhibitions. Despite the fact that these events occur on a smaller scale, due to the more dense occupation of the space by the resident companies, they continue to promote activities connected to art and culture.

4 FINAL CONSIDERATIONS

By investing in cultural and art initiatives in an abandoned space in the city where there was an opportunity for investment in urban rehabilitation, with the objective of returning the space to the city and promoting it, the LX Factory has the physical capacity to recuperate old buildings with historical value, as well as open spaces, while incorporating innovation into the model of occupation and management of the site. For example, the appropriation of the space occurred in unison with its restoration (which demonstrates the importance of the proposal of use during the recuperation of an asset of historical value); the new use proposed for the space fits into the contemporary city's dynamics by housing spaces supporting coworking, small and medium start-ups, valuing those engaged in cultural and art activities; by investing in the recovery of historical buildings, on some occasions, an exchange system was applied in which use of an area was granted to a given company, which then financed the restoration of the space occupied, in lieu of paying rent. Besides that, a strategy was used to bring anchor companies onto the site, as a way to attract people and help strengthen the project's financial side, especially in its first years of activity.

Regarding social scope, the LX Factory has promoted the opening of a redeveloped inclusive space for the local population, with activities open to the public, such as debates, workshops, exhibits and festivals. Another important point to highlight is related to the fact that, initially, the space enabled the development of small startups and entrepreneurs (especially artists) that saw it as an opportunity to set up in the space, given the more affordable rents. However, it is important to point out that as the LX Factory grew and stabilized in the city's dynamics, this first intake generally had to give up the rented space due to its valuation on the back of the space's success and appreciation by the population and private sector, this same success led to the valuation of the surrounding areas. Despite that, it is important to highlight that the space generated a series of jobs by bringing a new use and always investing in attracting people from different income brackets, tastes, and cultures, thus demonstrating an open and welcoming space for diverse groups. The LX Factory project also gave rise to initiatives which mobilized society in a shared innovative way, besides valuing work alternatives within the contemporary city's characteristics.

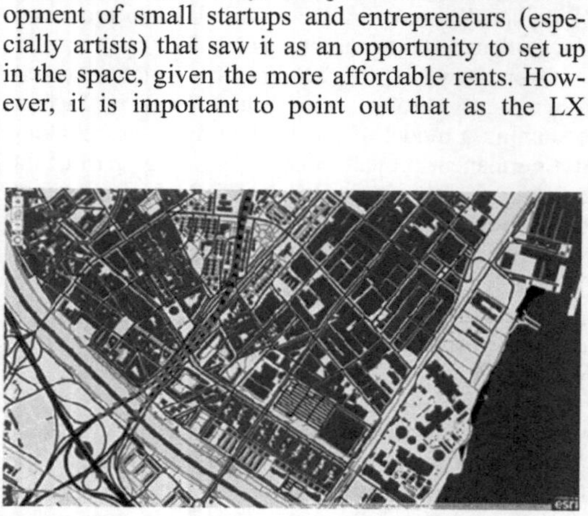

Figure 3. Floor plan of the LX Factory complex indicating its areas in use (light grey), free spaces (grey) and garden (dark grey). Elaborated by the authors based on figures and data from the site: https://lxfactory.com/ Last access: February 2020.

As the LX Factory project established a presence and recognition in the city, it not only became more values, but so did the vicinity. Nevertheless, the transformations in the neighbourhood due to the occurrence of this cluster warrant further evaluation. However, it is understood that an urban intervention, albeit public or private, is likely to trigger the process of the area's gentrification, a scenario which can be circumvented by offering space to tenants with diverse income brackets, including the poorer population. Thus, the LX Factory is an example of urban intervention in an industrial area of Lisbon previously abandoned and devalued, which was able to maintain a large part of the built heritage, highlighting it within the urban fabric and attracting people to the space.

The LX Factory was able to grow and consolidate during a period of uncertainty by offering the financially struggling public the chance to invest in their own companies at an affordable price. This initiative was sustained by the discourse of Lisbon's City Council of promoting activities in the fields of culture and art within the framework of the Municipal Master Plan with Strategies for Culture in Lisbon (2009 and 2017) and by applying the concept of a creative economy as a way of combating the financial crisis the country was also facing. Although the process of the space's valuation over the years led to the exit of some small entrepreneurs who did not keep up with its growth, ushering in medium-to-large companies, the project can be considered inclusive in as far as it drove a public space relevant to urbanity, ceasing to be a private space for the city and fostered new alternatives of working and income generation and social mobilization by disseminating local memory and culture. Could this be a possible way forward?

REFERENCES, SYMBOLS AND UNITS

Câmara Municipal de Lisboa. Estratégias para a cultura da cidade de Lisboa. Lisboa, 2017. Disponível em <http://www.cm-lisboa.pt/fileadmin/VIVER/Cultura_Lazer/ESTRA_CULTURA_LISBOA_2017_01.pdf> Último acesso em: 29 de abril de 2019.

CARVALHO, Gonçalo José Veloso Queirós de. A Reciclagem dos usos industriais e as novas tipologias de Actividades e espaços de cultura. Caso de Estudo: *LX Factory*. Dissertação de mestrado em Arquitetura e Urbanismo. 189p. Universidade Técnica de Lisboa, 2009.

GABRIEL, Leandro et al.Formação de espaços criativos: o caso da LX Factory em Lisboa. Artigo apresentado ao IX Congresso da Geografia Portuguesa. Universidade de Évora, 2013.

MARKUN, Paulo. Texto inédito de Paulo Markun para a *Folha de São Paulo. Folha de São Paulo*. Texto sem data de publicação. São Paulo, 2019.

PAULINO, Diogo Miguel Mendes. Adaptação de património industrial a espaços expositivos informais. Dissertação de Mestrado em Arquitetura. 205p. Universidade Técnica de Lisboa, 2015.

RIBEIRO, Claudio André dos Santos. Reabilitação Urbana e Sustentabilidade. LX Factory: um exemplo de Reabilitação Sustentável na Cidade de Lisboa. Dissertação de Mestrado em Arquitectura com Especialização em Planeamento Urbano e Territorial. 91p. Universidade de Lisboa, 2012.

SEIXAS, João. Projeções de Lisboa: utopias e estratégias para uma cidade em movimento perpétuo. Lisboa: Caleidoscópio, 2018.

The tentacles of the cathedral city: A reflection on architecture and urbanity

P.C. Gonsales
São Paulo, Faculty of Philosophy, Languages and Humane Sciences, University of São Paulo, Brazil

ABSTRACT: This aim of this paper was to present Roger Bastide´s reflections and analysis on the city of São Paulo and the records he left during his time in the city from 1938 to 1954. When commenting on the verticalization process of the city, Roger Bastide called it a cathedral city. He glimpsed in the curves of some modern buildings a possibility for Baroque modernism and analysed, from a sociological point of view, the so-called "machines for living". When addressing urbanity, the French sociologist imagined the city of São Paulo as a tentacular city, with its tentacles moving outwards in all directions. In these movements, Bastide saw the movements of aggregation and segregation of the black population in the city. However, Bastide's metaphor can be extended to a series of other inclusions and exclusions produced by the São Paulo metropolis both then and now

"The São Paulo metropolis, after having eliminated its past, was born *incognito*. Its only signs of identification were not stable elements, but vertiginous ongoing processes: fusion, growth, acceleration, speculation. Its official symbol in the coat of arms was two crossed branches of a coffee bush, but for the population in general, it was a locomotive at maximum acceleration" (SEVCENKO, 2000: 85)[i]. (Translated by the author)

1 INTRODUCTION

Roger Bastide was a French sociologist who joined the French mission to form the first classes of students at the School of Philosophy, Sciences and Languages of the University of São Paulo. While in Brazil (from 1938 to 1954), Bastide held the chair of Sociology II at the Faculty and, while on Brazilian soil, sought to integrate with the environment by establishing bonds of friendship with the intellectuals of São Paulo and dialoguing with many others, especially in Northeastern Brazil.

Bastide's intellectual output is vast, but his prominence in the French academic system was achieved through his studies on Bahian Candomblé. However, as he with Brazilian intellectuals of São Paulo in particular, Bastide wrote articles in newspapers commenting not only on topics in vogue in the intellectual and academic milieu, but also recording in these texts the journey he took to understand the national culture. These texts reveal him as an observer who was unswayed by Brazilian groups or intellectual currents, which gave him greater license to comment on major themes or the finer details of cultural or daily aspects of our society. He published his texts in several newspapers of the day, in São Paulo and Rio de Janeiro, and in many other publications throughout Brazil. In some of these texts, Bastide was outspoken on the urban landscape of São Paulo. At other times he commented on modern Brazilian architecture through a sociological analysis on new ways of living. He reflected on the inclusion and exclusion movements of the metropolis of São Paulo.

2 CONTEXTUALIZING AND DISCUSSING BASTIDE´S COMMENTS ON SÃO PAULO ARCHITETURE AND URBANISM WITH INTELECTUALS AND ARCHITECTS OF THE DAY

A few years after this definitive return to France, Roger Bastide wrote a book for the French people describing the Brazil he had experienced. In depicting the city of São Paulo, Bastide called it a cathedral city, alerting readers to the nature of São Paulo´s landscape: "The hand of the architect here replaced the hand of God" (BASTIDE, 1979: 14)[ii]. Bastide alluded to the

---

i In original: "A São Paulo metrópole, portanto, tendo eliminado seu passado, nasceu como uma incógnita. Seus únicos signos de identificação não eram elementos estáveis, mas processos em curso vertiginoso: fusão, crescimento, aceleração, especulação. Seu símbolo oficial eram dois ramos cruzados de café, mas para a população em geral, era uma locomotiva em aceleração máxima"

ii In original: "A mão do arquiteto, aqui, substituiu a mão de Deus"

DOI: 10.1201/9781003220855-32

intense verticalization of the city centre of São Paulo, building a metaphor, where the skyscrapers that stood out on the horizon were like the cathedral towers characterizing the skyline of old European cities.

"... its centre rises in tower buildings, arrow-like buildings, campanile buildings-, and when the sun lays to rest in all its glory, thousands of glazed windows and terraces, form a single stained-glass window of gold, pink and blue" (BASTIDE, 1979: 145)[iii]. (Translated by the author, as are all other Bastide's citations).

Roger Bastide attributes this accelerated process of São Paulo's verticalization to the high price of properties, the result of real estate speculation, and also to mobility difficulties commuting from the peripheral areas, where the wealthy classes used to live, into the city centre workplace. However, this was not all. There was something in the air which the people of São Paulo breathed at that time impelling them to search incessantly for novelty, a need to be modern. This modernness was materialized in the form of "symbolic objects" such as the automobile, the cinema, the skyscraper. The incessant search for modernization was an important impulse for the verticalization of downtown São Paulo. Bastide observed that the past did not feature in the daily life of São Paulo, "hence the desire to always create newthings, not to be satisfied with old environments (...), but to modernize, to seek everything that is new" (1979: 146)[iv] and then the French sociologist concluded: "The people from São Paulo build the scenario in the image of their century, and then demolish it when the world taste begins to transform." (1979: 147)[v].

The first skyscrapers of São Paulo date back to the 1920s, when international winds brought to Brazil the novelties of modern architecture that made, in the ensuing decades, the spread of skyscrapers in the São Paulo scenario possible. The development of new materials such as reinforced concrete made technical innovations possible and that, in turn, gave architects greater freedom to promote the conciliation of technique with art. Lending themselves to mass production, these buildings would be more economical, making them more affordable to the general public. There was also the search for the autonomy of architects who sought to distance themselves from the field of engineering. These were, in very general terms, some of the arguments defended by Lúcio Costa, one of the central figures of modern Brazilian architecture, to legitimize the new architecture among us (COSTA, 2003: 39 – 52).

This new architecture, which proposed the reconciliation of technique and art plus the rational use of spaces and materials, was inspired by the theories of the architect and urban planner Le Corbusier, who influenced the circle of Rio de Janeiro's architects, led by Lúcio Costa. Travelling through Brazil in the late 1920s, Le Corbusier gave lectures at which he disseminated his theories on architecture and urbanism. He returned the following decade by invitation from the Brazilian government, leaving a draft design for the building that would house the Ministry of Education and Health, inspiring the final project of the group of architects led by Lúcio Costa, the first push toward making modern Brazilian architecture known on the world stage.

The most general context of this surge in modern architecture in Brazil was analysed by the art critic Mário Pedrosa (2003). In his text, Pedrosa recalls that in the 1930s, Brazil was experiencing difficulties due the world economic crisis of 1929. Moreover, Brazil was in the grips of the dictatorship of Getúlio Vargas. Having risen to power on the promise of economic and social advances, Vargas's dictatorship made intensive use of propaganda to convey the idea that Brazil was moving in great strides towards progress and prosperity. Pedrosa recalls that the dictatorship "sought in its totalitarian tendency to draw in young architects, whose ideas and conceptions, however, were of the complete opposite leaning" (2003: 100)[vi]. The builders, in turn, "made use of the dictator's power of action to put their ideas into practice" (2003: 100)[vii]. Pedrosa then concludes:

"The dictatorship offered them this possibility, but resulted in a contradiction, not yet totally overcome, between the democratic and social ideals implicit in the new architecture, between its rational and functionalist principles and the concerns of self-propaganda, displaying strength, and the taste of sumptuousness and wealth to impress those responsible for the dictatorship, symbolized perhaps then, by the sometimes excessive nerve and the free forms that have become fashionable" (PEDROSA, 2003: 101)[viii]. (All of Pedrosa's quotes translated by the author).

---

iii In original: ... seu centro eriça-se em edifícios-torres, edifícios-flechas, edifícios-campanários, e quando o Sol se deita com toda a sua glória, milhares de terraços, de vidraças, de janelas formam como que um só vitral de ouro, rosa e azul.

iv In original: daí a vontade de criar sempre coisas novas, de não se satisfazer com ambientes antigos (...), mas de se modernizar, de buscar tudo que é novidade.

v In original: O paulista constrói o cenário à imagem de seu século, para depois demolir quando o gosto mundial começa a se transformar.

vi In original: buscou em sua tendência totalitária atrair a si os jovens arquitetos cujas ideias e concepções era, entretanto, de inspiração completamente oposta

vii In original: utilizaram-se do poder de ação dos ditadores para pôr em prática as suas ideias

viii In original: A ditadura lhes ofereceu esta possibilidade, mas resultou daí uma contradição, ainda não totalmente superada, entre os ideais democráticos e sociais implícitos na nova arquitetura, entre seus princípios racionais e funcionalistas e as preocupações da autopropaganda, de exibição de força, o gosto do suntuoso e da riqueza para impressionar os responsáveis pela ditadura, simbolizada, talvez, então, pelo 'brio' às vezes excessivo e as formas gratuitas que se tornaram moda.

From Pedrosa's analysis of self-promotion that the builders engaged in with the State, it must be pointed out what propaganda served both sides. While, on the one hand, the architects could use the co-optation of the State to put their ideas into practice, the State, as well as the elite, here also represented by the great Brazilian industrialists, hired these builders and architects to build their residences and commercial buildings as a way of conveying a message of power, reinforcing what the sociologist, anthropologist and playwright Jean Duvignaud said when analysing American industrial cities:

"The 'bourgeois-conqueror' in his universal exhibitions and in American cities decorates and modifies the world he dominates. The parties he conceives are of security and power because he wants to impose the symbol of an unalterable force" (DUVIGNAUD, 1983: 166-167)[ix]. (Translated by the author as are all of Duvignaud's citations).

Looking at the American cities of Chicago and New York, Duvignaud regards the architecture that he calls industrial because it is mass produced, as a symbol of the power of capital, which materialized in the form of reinforced concrete, steel and glass, materials from which the skyscrapers rise. This symbol of economic power still carries the sense of permanence; the symbol is the very place where man works and lives (DUVIGNAUD 1983). But Duvignaud looks at the other question also commented by Pedrosa, namely, the social and democratic ideals of modern architecture. The French intellectual recalls the projects of Wright, Gropius and Le Corbusier who thought of spaces of reconciliation of men, regardless of social status, but who were relegated to a certain utopia, because the harmony of the people does not dialogue with industrial society. Even here in Brazil, perhaps more than in many other places in the world, this social dimension of modern architecture has remained in the background, at least in the early years, an issue that will be better explored later.

Another important aspect that can be gleaned from Pedrosa's analysis, mentioned above, is when he criticizes the excessive use of gratuitous forms as a way of seducing those in power. It is true that although modern architecture in Brazil was marked by internationalization, Brazilian architects sought to develop original solutions that would address problems inherent to Brazil, such as the solution found in *brise-soleil*. This constitutes an element that would provide thermal comfort, while protecting the interior of the building from the heat of the tropical sun, yet at the same time was an element that conferred elegance and lightness to the façades.

There was also the search for originality fueled by a strong nationalist feeling that implied differentiating Brazilian culture from other cultures. The Brazilian intellectuals – also co-opted for the propaganda of the State, which wished to build the image of a prosperous, united, and singular country – were immersed in the task of delineating Brazilianness and seeking to attenuate the ambiguity translated by tradition, represented by rural and archaic Brazil, and the modern, represented by the metropolises, seeking to build a line of continuity between these two opposites.

In an analysis of the context of São Paulo architecture, Luís Saia shows the proximity of architects to intellectuals (2003). At this point, modern Brazilian architecture, which until then had expressed itself much more by means of internationalization, rational modernization, dedicated to catering to the needs of modern man, should also become part of Brazilianness. The proposal was to imagine a line that would link the originality of the Brazilian colonial baroque to modern Brazilian architecture. Hence, this ambiguity between the archaic and the modern would be resolved through architecture, as this was still an impossible issue to resolve in the economic and social fields.

This link between modern architecture in evidence on the world scene and colonial baroque has enlivened some of the most influential intellectuals in the cultural scene of São Paulo and Rio de Janeiro, among them Mário Pedrosa, who comments on the curved forms, often adopted in Oscar Niemeyer's projects as a take on Baroque, attributing this approach to the Brazilian cultural trait:

"… Oscar Niemeyer, undoubtedly obeying the demands of his temperament, indulges more and more in the Baroque taste through large irregular shapes and wide curves. (…) There is a danger in confusing the articulated volumes with the winding profile of the curves, but, in any case, this trend corresponds perhaps to a cultural, if not racial, constant. Brazil, let us not forget, was born under the sign of Portuguese and, in part, Spanish baroque" (PEDROSA, 2003: 103)[x].

However, this proximity between modern Brazilian architecture and the Baroque style was not well-accepted initially by Oscar Niemeyer, who, still overly adhered to the ideals of rationality of modern architecture, following the internationalising tendency of architecture, declared in a 1944 text his respect for the lessons of the past, but stated that the

---

ix In original: O 'burguês-conquistador' em suas exposições universais e nas cidades americanas decora e modifica o mundo que ele domina. As festas que concebe são de segurança e poderio porque quer impor o símbolo de uma força inalterável.

x In original: (…) Oscar Niemeyer, obedecendo sem dúvida às exigências de seu temperamento, se entrega cada vez mais a um gosto barroco pelas grandes formas irregulares e amplas curvas. (…) Há perigo em confundir os volumes articulados com o perfil sinuoso das curvas, mas, de todo modo, esta tendência corresponde talvez a uma constante cultural, se não for racial. O Brasil, não esqueçamos, nasceu sob o signo do barroco português e, parcialmente, do espanhol.

old architectural forms no longer had meaning in the face of new technical possibilities (NIEMEYER, 2003a). "We believe only in architecture made without compromises, based on new construction processes and new materials used in all ways possible" (NIEMEYER, 2003a: 244)[xi] (Translated by the author). However, years later, in 1958, Niemeyer recalled the need he and other modern architects felt to follow the teachings of purist architecture to the letter; but after some time, when modern architecture had established itself, there came the need for architects to imagine new forms and new solutions that reflected dreams and fantasies, ones which cold and rational architecture would be unable to express. At the end of the text, Niemeyer recalls the meeting with Le Corbusier that took place in 1947, when the French architect recognized that Niemeyer made the baroque with concrete (NIEMENYER, 2003 b).

The appearance of curved lines on the façades of buildings in São Paulo was noted and commented on by Roger Bastide, who glimpsed a type of modern Baroque that would at the same time mark the uniqueness of modern Brazilian architecture and break with the rationality of the cement cubes of modern skyscrapers, bringing harmony to the urban whole.

"This emerging Baroque that seems to me to be portraying architectural and pictorial modernism in the same way as the Baroque of the 18th century sought to express simplicity of the Jesuit style of the Renaissance. There's evolution, no reaction, going back.

But this evolution (...) seems to be of enormous interest to Brazil. Brazilian cities, today, juxtapose different and contradictory arts instead of synthesising them in a harmonious fraternity of lines. (...) It is not a matter of waging war against these modern buildings, for they meet their needs, the new needs created by men in modern civilization and by the birth of countless new occupations that did not exist at the time of the mansions and shanties. But the transformations I mention above, ranging from classical modernism to baroque modernism, have enabled the cities of Brazil to take on a more harmonious aspect. No more skyscrapers squeezed between islands of the past are perceived, but organic assemblages in a unified aesthetic climate" (BASTIDE, 21/09/1947)[xii].

Bastide identified in the flowing lines the shape of the Brazilian colonial baroque. "Now, countless modern façades have already become curved, sinuous; in the alignment of the streets, they form motionless waves, like waves breaking on an asphalted river, onto the dark pavements" (BASTIDE 09/21/1947)[xiii]. For Bastide, the appearance of the curved line in the urban landscape of São Paulo introduced a little sensitivity to the cubic rationality of the skyscrapers of the time. Bastide believed that "Baroque is not only an architectural style, capable of offering formal and/or ornamental solutions, but also a lifestyle and sensitivity that mobilises the domains of the unconscious, stoking the oneiric imagination and the Dionysian paths of creative freedom" (PEIXOTO 2015: 42)[xiv].(Translated by the author).

Lourival Gomes Machado, who also carried out systematic studies on the Brazilian colonial baroque, also highlighted this oneiric dimension that the baroque inspirations gave to modern Brazilian architecture and, above all, emphasized this line of continuity established between colonial baroque and modern Brazilian architecture:

"... to us, the certainty that the new Brazilian architecture faithfully matches the needs, conditions and mentality of the country and, between that and the authentic architecture of the colonial time, the only difference imposed by the temporal dimension is that society has changed, transforming in terms of living conditions and the scope of the technique. So perfect is the match that, at times, the observer feels depicted in the lines of new buildings, not the European cubism of the master, but the gratuity and illusions of perspective that were, in their time, the blessings of Baroque, a historical element of

---

xi In original: Acreditamos somente na arquitetura feita sem compromissos, baseada nos novos processos construtivos e nos novos materiais aproveitados em todas as suas possibilidades.

xii In original: O barroco que me parece estar nascendo se pretende ao modernismo arquitetural e pictural do mesmo modo como o barroco do século XVIII se pretendia à simplicidade de estilo jesuítico da Renascença. Há evolução, não reação, volta atrás.

Mas essa evolução (...) parece-me de um enorme interesse para o Brasil. As cidades brasileiras, atualmente, justapõem artes diferentes e contraditórias em lugar de sintetizá-las numa harmoniosa fraternidade de linhas. (...) Não se trata de iniciar uma guerra contra esses imóveis modernos, eles correspondem a necessidades, às novas necessidades criadas entre os homens pela civilização moderna e ao nascimento de uma multidão de novas ocupações que não existiam na época dos sobrados e mucambos. Mas, as transformações a que me referi acima, que vão do modernismo clássico, ao modernismo barroco permitiram às cidades do Brasil tomar um aspecto mais harmonioso. Não mais arranha-céus espremidos entre os quais se percebem ilhas do passado, mas conjuntos orgânicos em clima estético unificado.

xiii In original: Ora, inúmeras fachadas modernas já se curvam, tornam-se sinuosas, no alinhamento das ruas formam como vagas imóveis, como ondas quebrando-se num rio asfaltado, sobre as calçadas escuras.

xiv In original: ...o barroco não é somente um estilo arquitetônico, capaz de oferecer soluções formais e/ou ornamentais, mas é também estilo de vida e sensibilidade que mobiliza os domínios do inconsciente, atiçando a imaginação onírica e os percursos dionisíacos da liberdade criadora.

Brazilian art" (MACHADO, 2003: 77 – 78)[xv] (Translated by the author).

In the excerpt above, Lourival Gomes Machado, a student of Roger Bastide, reinforces the impressions of his former master by highlighting the breaking down of cubism´s rationality by introducing baroque-inspired elements into modern Brazilian architecture. Elements that would not only break with the rationality of modern buildings but would also provide continuation of the Brazilian architectural tradition that had begun with the colonial baroque.

The rationality of modern architecture in Le Corbusier´s terms used to be widely criticized, not only by Lourival Gomes Machado, as mentioned above, or Roger Bastide, also cited earlier, but by other influential intellectuals of the time, such as Sérgio Milliet and Mário de Andrade. The so-called cement cubes, minimalist decoration, and the stripping down of details on façades – purism preached not only by Le Corbusier, but also by Mies Van der Rohe and others – was the object of criticism from these and other intellectuals, including Europeans, who saw the rationalization of the modern architectural landscape as a kind of negation of human sensitivity, natural and necessary to all men. "… this utilitarian art leaves a void in the imagination, a kind of gap in sensitivity. Something is missing and, suffering somewhat, we miss it" (BASTIDE, 31/09/1947)[xvi].

There was a consensus that modern art was better able to cater to the needs of modern life. These needs were defined by Flávio de Carvalho (a modernist artist and architect) in a 1938 text where he was pondering what the house of man of the 20th Century should be like. In the text, Flávio de Carvalho observes that the 20th century man would use the house as a point of passage and no longer as a place of protection. The widening of the notion of collectivity would make the city the centre of man's life and the consequent reduction in the importance of the home as the only centre of man´s activity (2003: 52 – 53). At the end of the text, Carvalho concludes that:

"The patriarchal and matriarchal family, as almost divine heads, gradually loses its importance as a social and religious centre and is replaced by a general civic centre, the city.

This diminished importance of the family in man´s life is one of the most powerful determinants of the forms of the city and the home of the 20th century man" (CARVALHO 2003: 55)[xvii]. (All citations by Carvalho are translated by the author).

The arguments defended by Flávio de Carvalho that reinforce the so-called "machines for living" (as defined in Le Corbusier's theories) contrast with the sociological analysis that Roger Bastide published on this issue of modern housing. In a text written in 1951, Bastide argues that "the house is not made for an abstract man, reduced to needs and functions – but for a man of such and such community" (BASTIDE, 27/06/1951)[xviii] rejecting homogenizing internationalism, but respecting the particularities of each community. In the Brazilian case, society was formed out of a strong patriarchal mentality that remained in society at that time and still persists to this day. As an example, Bastide borrowed from Gilberto Freyre's analysis, which observed that the apartments reproduced this patriarchal society by situating the employees′ quarters close to the kitchen area.

Bastide continues his sociological analysis of the "machine for inhabiting" by observing the replacement of stone, material from which ancient houses were made, by glass. He notes that "The window that often occupies the entire façade places the inhabitant virtually on the street" (27/06/1951)[xix]. This integration of the interior with the exterior was also a trend of modern international architecture, but Bastide noted that the residents of the first floors ended up dissolving their privacy in the collective. While for Flávio de Carvalho the natural way of man was to dissolve in the collective, Bastide defended the idea that Brazilian society maintained the patriarchal mentality, that the consequences of these changes in living habits could lead to the dissolution of the family nucleus, the basic unit of a society. As a suggestion, Bastide proposed to the architects, to think of a way of reconciling the "centripetal house" whose centre is the intimacy of the family with the "centrifugal house" where intimacy dissolves in the collective, vertical city with its skyscrapers and horizontal structure of the family (27/06/1951), building a new aesthetic of the house, "that would be at the

xv In original: … a nós, nos basta a certeza de que a nova arquitetura brasileira corresponde fielmente às necessidades, às condições e à mentalidade do país e que, entre ela e a autêntica arquitetura dos tempos da colônia, só vai a diferença imposta pela dimensão temporal em cuja função variou a sociedade transformando-se no tocante às condições de vida e ao alcance da técnica. Tão perfeita é a correspondência que, por vezes, o observador sente revelar-se no risco dos edifícios novos, não o cubismo europeu do mestre, mas a gratuidade e as ilusões de perspectiva que foram, em seu tempo, o apanário do barroco, elemento histórico da arte do Brasil.

xvi In original: … essa arte utilitária deixe[a] um vazio na imaginação, uma espécie de lacuna na sensibilidade. Alguma coisa falta-nos e, com um pouco de sofrimento, sentimos-lhe a falta.

xvii In original: A família patriarcal e matriarcal com os chefes quase divinizados perde gradualmente a sua importância como centro social e religioso e é substituída por um centro cívico geral, a cidade.

Essa diminuição da importância da família na vida do homem é dos fatores determinantes mais poderosos das formas da cidade e da casa do homem do século XX

xviii In original: a casa não é feita para um homem abstrato, reduzido a necessidades e funções – mas para um homem de tal ou qual comunidade.

xix In original: A janela que frequentemente ocupa toda a fachada põe o habitante virtualmente na rua.

same time a machine of living and a family focus – where the elements are placed and where you can relive the intimacy, but now around beautiful objects" (27/06/1951)[xx].

Commenting on the architecture, Bastide analysed the skyscrapers, celebrated the originality that the baroque inspirations brought to modern Brazilian architecture, and viewed the modern ways of living with a sociological eye. He also looked at the city's morphology and noted its sociological dimension. It is true that modern Brazilian architecture reflected a very general characteristic of Brazilian society and preserved itself as an elitist architecture in its early years, somewhat removed from the socializing proposals of becoming accessible to the general public. This subject was widely discussed among architects and thinkers of the day. The democratization of modern architecture took time to materialize in housing developments, schools, and hospitals – the Pedregulho housing complex, designed by Affonso Reidy in Rio de Janeiro was completed in 1952, while one of the country's first housing developments was inaugurated in 1966 in Campinas city, located 100km from São Paulo. This occurred because the industrialization of the country had yet to offer the infrastructure needed to support businesses focused on the production of prefabricated materials for use in mass construction projects. Moreover, the construction of housing developments, hospitals and schools depended on state funding, which did little about these issues. Therefore, the social benefits of modern architecture were still inaccessible to the Brazilian population.

Also at the urban level, there were debates related to college education teaching curricula that should incorporate urbanism with more emphasis on development courses in the area of city planning. Some of these debates about the socialising nature of architecture and urbanism can be followed in the annals of congresses of architects and specialised magazines of the time such us *Habitat*, *Revista Módulo*, among others.

The accelerated growth of cities implied the need for an in-depth morphological study, which would make it possible to develop plans for adapting and regulating the uses of urban space which should be compiled into Master Plans that also provide guidelines for future urban interventions. The regulations that would govern the uses of urban space were the central issues of modern architecture and urbanism. After all, the innovations of this new architecture emerged in response to the need for rebuilding European cities destroyed in World War I. The recovery of these cities was to be rapid; hence, the use of prefabricated materials and techniques that would allow mass construction. The reconstruction of these cities should be guided by urban plans that would entail rational organization of space, preferably divided into zones (residential, commercial, leisure, etc.) that should be interconnected through wide avenues, surrounded by large landscaped areas. Le Corbusier had developed several urban plans to achieve his ideal city. Among these plans was *Ville Radieuse*, first presented in 1924. Plans, however, that never left the drawing board. The two fully planned cities that did emerge were Brasilia, a Brazilian federal capital whose urban plan was designed by Lúcio Costa, and the city of Chandigarh in India, whose project had the contribution of Le Corbusier. Furthermore, those plans where the ideal city was designed were only in the utopian plan and were the plans that Duvignaud suggested in a study cited above.

While, on the one hand, the plans of modern urbanists envisaged a more democratic urban space, the reality, at least in São Paulo, was far from accessible to people of all classes. Roger Bastide published a text in which he borrowed part of the title of the famous work by the poet Verhaeren to refer to São Paulo as *La Cité Tentaculaire*. The tentacular city for Bastide was not only that which advanced to the lands down the valley, a city which extended north, south, east, and west, without any urban discipline. The tentacles of the city of São Paulo moved in all directions and, in its dynamics, aggregated and segregated. Bastide noted that World War II had interrupted the flow of immigrants from Europe and the East to Brazil at the same time as Brazilian industrialisation was accelerating. With the lack of Europeans, blacks were absorbed into the labour market and enjoyed an equal working relationship with whites. This, for Bastide, was the movement of aggregation of the city's tentacles. However, these blacks, especially those who improved their lives and eventually ascended in social position, were rejected by the whites in social relations, forcing them to build interracial relations, founding their own athletic associations, clubs, etc. Consequently, this began causing divides to appear in these relations, where the rising black population rejected the poorer black population, reproducing the prejudice that they had suffered by whites – these were the segregation movements.

However, these were not the only aggregation and segregation movements of the city's tentacles. São Paulo was a cosmopolitan metropolis home to an immense immigrant population of various nationalities and ethnic groups. These immigrants from different origins organized in sometimes rather closed communities and built their own social centres as a way of preserving the culture of their homeland, at least partially. Each of these colonies was concentrated in neighbourhoods such as in Braz with the Italians, Vila Mariana with the Germans, Penha with

---

xx In original: que seria ao mesmo tempo máquina de morar e foco familiar – onde os elementos daquela sejam colocados e onde possa reviver a intimidade, mas agora em torno de objetos belos.

This work was carried out with the support of the Coordination for the Improvement of Higher Education Personnel - Brazil (CAPES) - Financing Code 001.

the Spanish, Liberdade with the Japanese, the Lebanese Rua 25 de Março trade and so on. As with blacks, the city's tentacles aggregated everyone inviting them to produce riches in industry and commerce. However, the coexistence of so many nationalities produced tensions and exclusions, even on the part of Brazilians in relation to foreigners who were generally higher educated and more skilled in certain crafts. There was some fear that the labour market would privilege these better prepared immigrants, placing Brazilians in an inferior position. The Brazilian acculturation of these immigrants was the object of public policies in the area of education in São Paulo throughout the 1930s and 1940s. It is true that, as Bastide observed, the Brazilian was traditionally friendly (BASTIDE, 1952), the tensions generated by the contrasts between different cultures were settled, supported by a relatively peaceful relationship of coexistence in a multiracial and still young country, whose culture was in a fluid state and whose crystallizing would occur over time (BASTIDE, 1979).

However, it is possible to imagine that these movements of inclusions and exclusions were exacerbated by aspects such as lack of spaces for interaction - like parks - in a clear discontinuity in the initiatives of Councilor Antônio Prado, mayor of the city of São Paulo in the 1910s, which encouraged the construction of parks as spaces for mixing, promoting ideals of civility and citizenship among the general population (SEVCENKO, 2000). In addition, the concentration of entertainment in the city centre made it difficult for residents of the surrounding districts - far from the downtown area - to access it, not to mention difficulties in mobility and other problems such as housing deficit and lack of basic infrastructure. These issues could be mitigated through planned urban interventions. However, these were not problems to be solved through the sole initiative of architects, but in the 1960s the group *Nova Arquitetura* emerged in São Paulo and proposed to discuss in more depth the architect's role articulated with a cultural and political action. These were, however, social, economic, and political issues. Since the beginning of the 20th century, São Paulo has been strongly influenced by real-estate speculation, which, with the help of public authorities, has led to new neighbourhoods, verticalized the city centre and orchestrated city sprawl. It is noteworthy that São Paulo´s first Master Plan was approved only in 1972.

3 A BRIEF FINAL ASSESSMENT

The tentacles of the city of São Paulo noted by Bastide expanded in a disorderly fashion. The French sociologist observed and commented on the city and its fast pace, with its landscape built by the hands of architects, predominantly in cubic form, but with some curves serving as a reference to the distant colonial past. He also noted that these tentacles promoted more exclusions than inclusions. Amid this disorderly dynamic, the city increased inequalities and reinforced the decline into the chaos that disturbed the souls of São Paulo in those (and these) times.

As can be seen in this text, Bastide's analysis always had the mark of a sociologist and was often endowed with certain poetics. For the French sociologist, society was built by men. In his understanding, there were no pre-established formulas which reduced phenomena to static things. Bastide's sociology was endowed with sensitivity that made him more susceptible to perceiving movements. For this reason, his texts offered the possibility of dialogue with other thinkers who were his contemporaries or even with thinkers of the present day.

REFERENCES

Bastide, R. 1979. *Brasil, terra de contrastes*. São Paulo: DIFEL.

Bastide, R. 1952. São Paulo *La cité tentaculaire*. Le Courrier, ago – set, p. 9.

Bastide, R. 1951. Estética de S. Paulo II – A cidade vertical. Jornal *O Estado de S. Paulo* (27 jun.).

Bastide, R. 1947. A volta ao barroco ou a lição do Brasil. *Diário de Notícias* (21 set.).

Bastide, R. 1943. A França noturna. *Diário de S. Paulo* (26jun.).

Costa, L. 2003. Razões da nova arquitetura. In: Xavier, A. *Depoimento de uma geração*. São Paulo: Cosac & Naif, p. 39–52.

Carvalho, F. 2003. A casa do homem do século XX. In: Xavier, A. *Depoimento de uma geração*. São Paulo: Cosac & Naif, p. 52–55.

Duvignaud, J. 1983. *Festas e Civilizações*. Fortaleza: Ed. Edições Universidade Federal do Ceará, p. 149–174.

Machado, L. G. 2003. A renovação da arquitetura brasileira. In: Xavier, A. *Depoimento de uma geração*. São Paulo: Cosac & Naif, p. 75–78.

Niemeyer, O. 2003a. Pampulha: arquitetura. In: Xavier, A. *Depoimento de uma geração*. São Paulo: Cosac & Naif, p. 244–245

Niemeyer, O. 2003 b. Contradição na arquitetura. In: Xavier, A. *Depoimento de uma geração*. São Paulo: Cosac & Naif, p. 245–248.

Pedrosa, M. 2003. A arquitetura moderna no Brasil. In: Xavier, A. *Depoimento de uma geração*. São Paulo: Cosac & Naif, p. 98–106.

Peixoto, F A. 2015. *A viagem como vocação. Itinerários parcerias e formas de conhecimento*. São Paulo: EDUSP.

Queiroz, M. I. P. 1983. Nostalgia do outro e do alhures: a obra sociológica de Roger Bastide. In: Fernandes F. & Queiroz, M. I. P. (orgs). *Roger Bastide*. São Paulo: Ed. Ática, p. 7–75.

Saia, L. 2003. Arquitetura paulista. In: Xavier, A. *Depoimento de uma geração*. São Paulo: Cosac & Naif, p. 106–119.

Sevcenko, N. 2000. *Pindorama Revisitada cultura e sociedade em tempos de virada*. São Paulo: Fundação Peirópolis, p. 68–112.

Xavier, A. (org.). 2003. *Depoimento de uma geração*. São Paulo: Cosac & Naif.

http://www.cdhu.sp.gov.br/institucional/quem-somos (CONSULT: 12/02/2020).

The Afro-Brazilian cultural experience on the streets of the city of Alfredo Chaves-ES

R.C. Bettcher
PGAU-UFES, Postgraduate studies in Architecture and Urbanism of the Federal University of Espírito Santo, Vitória- ES, Brazil

ABSTRACT: The Macrina neighborhood has been insurgent since its conception: a marginal territory donated to freed slaves, consisting of a hill without urban planning, where alleys and narrow streets snake through its undulating terrain. In the same historical period, the city of Alfredo Chaves proper was planned for Italian immigrants and affinity with European culture prevails. However, the ghetto inhabited by a black minority gains prominence through cultural dynamics in the streets, which territorialize African culture references on Alfredo Chaves soil. The empirical analysis transits between sacred and profane territories, represented in the form of three festivals: St. Benedict, Our Lady of Aparecida and Carnival. Recording through photographs and experience with the festive activities through attentive and active observation of and on the street are the main strategies for perceiving these festive collectives.

1 INTRODUCTION

Popular culture manifests in different ways in the Brazilian territory, showing singularities of memories passed on from generation to generation through orality and which are consolidated in cities, repeated annually in rites and celebrations. In the words of Sodré (2016), it is a problematic term, as "popular" denotes a culture that is different from the official and hegemonic one, defining customs, festive events and life styles, from culturally peripheral and economically subordinate regions. Another issue pointed out by the author (SODRÉ, 2019) is that architecture and urbanism contributed to the consolidation of European cultural centrality by linking culture to the rational ordering of spaces, as a unique truth. The rustic, the peasant, the urbanism that deviates from the European standard are still subjugated in the Brazilian cultural sphere.

An urban context that exemplifies this cultural inequality is that of the city of Alfredo Chaves-ES, which has 15,000 inhabitants (IBGE, 2010 Census), comprised mostly of descendants of Italian immigrants. It also has a black minority descended from blacks freed from slavery that is concentrated in the Macrina neighbourhood. This neighbourhood is the object of an urban analysis adopted in this study, understood as a cut-out on a small scale where its inhabitants exercise intimacy in their narrow and irregular paths, where the use of public spaces is predisposed both in daily life, due to the restriction of private spaces, and during their festive events.

The inclusion in this territory is based on my affective experience, since childhood, as a resident of the neighbourhood. Familiarity gained first by the landscape painting seen from the window of my house (Picture 1) which depicts Macrina hill, with its diversity of customs and architectures, distinguishing it from other neighbourhoods in the city.

Figure 1. The Macrina neighbourhood – gouache on canvas.

Source: Roberta Bettcher (2003).

DOI: 10.1201/9781003220855-33

The desire for freedom was achieved early by the few residents, the owners of the street. Without fear of subversion, the walls are skillfully jumped by them, the pick-up soccer game takes place over a clay field, they are the natural explorers of public and private spaces. The cemetery, at the top of the hill, is the place to look at the sky, a lookout point, ideal for flying kites, as free space, without the obstacle of power lines from telegraph poles. They seem to be more courageous than the other inhabitants of Alfredo Chaves. They live in a feared place, a kind of medieval fortress, circumscribed around the hill, but in this case without physical walls, the wall built is prejudice: racial, economic, and social. With ancient features of a territory formed in 1888 after the end of slavery in Brazil, the Macrina neighbourhood was Alfredo Chaves' first refuge for blacks to exercise their freedom. At the same time, European immigrants arrived in the region, most of them Italians, and started building the urban centre, planning what is now the city centre. At that time, resources were scarce for everyone, but Macrina was left out of the plan and, with no urban rules to be followed, the density of houses increased around the hill, with the low parts flooding on rainy days.

This segregation was maintained with physical and invisible artifices. Symbolic equality of all citizens. Neighbourhood vulnerability to prostitution, violence, and marginality, which extends to the present day. But this irreverence in the ways of life is also reflected as joy, the parties are a great asset of the neighbourhood. The concentration of festivities linked to blacks is quite striking, ranging from religious events, with devotion to the so-called black saints, to the libertine carnival block. Macrina is the name of the neighbourhood, but also of the street around the entire hill itself. This street is the main focus of the parties in this cultural territory. Ethnography is the methodological basis for the empirical description of the festive details of this insurgent urbanism. The photographic record and attentive presence during the events are strategies for perceiving the relevant aspects of each of the festivities. The three parties selected for this analysis are: the feast of Saint Benedict, the feast of Our Lady of Aparecida and Carnival. And what do they have in common? All of them have the street as the protagonist of the parties, and although temporary cycles, the territory is permanently marked by these cultural interventions.

2 FESTIVE SUBVERSIONS IN THE CITY

Afro-Brazilian culture has remained in Brazil through adaptations, according to the Brazilian region, exhibiting singularities in its expression. The street is part of the ritual of passing through this culture, it is necessary to parade its belief and art beyond the territory of origin, to deterritorialize the limits imposed, in this case, by racial prejudice. It is on the street that new symbols for urban living are conceived, a space for everyday meeting or one which, on festive days, allows for greater interaction among people.

According to Guattari, each group or individual leads their own model of subjectivity, "[...] cognitive demarcations, but also mythical, ritual, symptomatological ones, from which he/she positions themselves in relation to their affections, anxieties and tries to manage their inhibitions and drives" (GUATTARI, 1992, 21, our translation)[1]. He uses the term "collective" as a generator of multiplicity, in a derivation more linked to the logic of affects. The cyclical repetition of the festive rites introduces an impregnated symbolic content and allows the appreciation of the meaning of life, through a clear representation, going in the opposite direction to the formal serialization produced by technological machines, which smooth out the roughness of singularity.

In times when individualism is so questioned in the contemporary city, collective initiatives that remain in urban territory gain value. According to Raffestin (1993), the action of these organizations formulates an abstract, symbolic, relational space, which is closely related with real space, but has another time scale, a time "invented" by men that happens in parallel to real time. Social actors confer meaning to space through their representations, while power, according to Raffestin (1993), is part of every relational space. He divides these collective actors into two categories: syntagmatic and paradigmatic. The first category is part of a programmed process linked to an organization, such as the family, the State, churches, political parties and companies. In the latter there is a need for control and domination, such that the actor is integrated into the process. In the second type, the individual is not integrated into a defined programme, so it becomes an asset in relation to syntagmatics, since organizations want their control to integrate it into the various forms of processes.

Blacks were abruptly forced into these programmed processes, while racism encouraged them to adhere to their own religious practices, which were considered witchcraft or dark magic, as did the great supremacy of the Catholic Church and slave masters during the years of slavery:

"The coexistence of blacks with the Portuguese changed their customs and, in this respect, the

---

1 "[...] demarcações cognitivas, mas também míticas, rituais, sintomatológicas, a partir da qual ele se posiciona em relação aos seus afetos, suas angústias e tenta gerir suas inibições e suas pulsões".

influence of the Catholic Church was great, because it prevented them from worshiping their Orixás, fearful of repression by their masters, blacks also transferred their religious devotion to some Catholic saints, revering, for example, Our Lady of the Rosary, Aparecida, Conceição and Saint Benedict, whose worship in Espírito Santo is one of the most fervent to date" (MACIEL, 2016, 124, our translation)[2].

Sodré (2019) stated that blacks used religious sentiment as a strategy of resistance to oppression: "Freedom and independence are states sought with great creativity by the faith of the black community" (SODRÉ, 2019, 99, our translation)[3]. Creativity is present at parties in the Macrina neighbourhood, which follow the religious calendar and take over the streets of the city.

Figure 2. Saint Benedict Shack.
Source: Roberta Bettcher (2019).

2.1 *Saint Benedict's Feast*

The homage to the black saint, who was a cook, son of African slaves and protector of the poor, starts from an old narrative. Legend has it[4] that a slave ship coming to Brazil and on the verge of a shipwreck, carried a flag with the image of Saint Benedict hanging over the mast. The drowning blacks look at the black saint and ask for his intercession to be saved. At that moment, the mast broke and fell to the sea, where blacks could support themselves and reach the beach. From then on, they promised the saint to hold an annual feast in his honour, in which the ritual that re-enacts the saving mast is exalted.

The day of the saint in the State of Espírito Santo is December 26, does not follow the Catholic calendar, and coincides with the day off given to slaves after the celebration of Christmas. A small chapel was built in his honour in the Macrina neighbourhood and next to it is the so-called "Saint Benedict shack" (Picture 2), where the festive and devotional expression is free, the territoriality of the Congo and the Jongo, which are representative manifestations of culture of black people from Espírito Santo.

The Congo is very widespread in the State of Espírito Santo and has a characteristic identity, it is widely exploited in the cultural sphere, as it represents a symbol of the Espírito Santo culture. The *jongo* is also recognized as a part of Brazilian cultural heritage, as its invention occurred in Brazil, the result of the poetry and art of the Africans and of their descendants. In Espírito Santo, it refers to the songs of "jongo circles", which are also called "cax ambu circles". The presentation in a circle, suggests horizontal social relations of communication, without the hierarchy established between slaves and masters of the colonial period, and communicating the rituals to the new generations is a task that the Masters encourage, leaving children free to use instruments or to "start the jongo", when they are participating in the circle. The volunteers are random, children and adults, deprogrammed of rules they take turns on the drums, as well as riding horses. The rhythmic music of the drums is remarkable in this cultural manifestation.

In this celebration, the dogmas propagated by the Christian faith, contrast with the freedom and singularity of the faith of the black community. Chapel and Shack indicate through the form of appropriation of their spaces, what Lefebvre (2008) defines as "sociological segregation", where the dispute for the territory occurs between different ethnicities, cultures and subcultures, which fuel the formation of ghettos in the city. The ghetto of the Christian faith is concentrated in front of the chapel, while the ghetto of African ancestry is the shack.

On the 27th of December is the eagerly awaited procession of the "Pull of the mast". The lack of a set program lends a subversive tone to the event, the spontaneous meeting of people, the silent and relaxed wait, the attention turned to the drums that are handled by the children, the respect for the ritual that is closer to the profane world than to the sacred one.

The shack is the temple of this rite, a place providing for dance and music, which welcomes people who participate in the event. The Alfredo Chaves parade has characteristics that are different from

2 "A convivência dos negros com os portugueses modificou seus costumes e, nesse sentido, a influência da igreja católica foi grande, pois, impedidos de cultuar os Orixás, temerosos da repressão dos senhores, os negros também transferiram sua devoção religiosa para alguns santos católicos, reverenciando, por exemplo, Nossa Senhora do Rosário, Aparecida, da Conceição e São Benedito, cujo culto no Espírito Santo é um dos mais fervorosos até os dias atuais"

3 "Alforria e independência são estados buscados com muita criatividade pela fé da comunidade negra"

4 Bicalho, 2000.

other municipalities in Espírito Santo, since the mast appears in stages, and the population moves through the city, only to return to the neighbourhood with the completed article. Seven homes were visited on the route in 2019, and a cycle of actions is repeated at each stop: Mestre Afonso, an elder in the community, greets the owners of the house. Hot dogs, soft drinks and wine are given to participants. Captain Tito's whistle signals the band's performance: the stop and the continuation of the music. An improvised jongo in honour of the host is sung, to thank for the reception. The part of the mast that was hidden in the residence is removed. Sweets are placed on the painting of the saint, on the flag, which is carried by children.

Residents called up annually celebrate in their own way, extolling its festive essence. Because it is similar to the carnival in its dynamics through the streets and the consumption of alcoholic drinks, the "Pull of the mast" has implicit rules of respect for others, where harassment and excess, so common in carnival celebrations, are forbidden.

This game of hiding the mast of the saint, in different parts of the city, and often outside the neighbourhood, creates an interesting dynamic between public and private spaces. It is as if by expanding this cultural manifestation beyond the territory of origin, it temporarily brings down the walls and obstacles erected by prejudice and event participants can move freely. In one of the residences, in the Ipanema neighbourhood, something unusual happens: the residents welcome visitors in the yard, then the crowd enters the private space, huddling in the cramped environment. The right to the city, conceptualized by Lefebvre, takes place in this urban breakthrough, which moves away from the massive culture of contemporary festive events: "Commercialized, industrialized, institutionally organized leisure activities destroy this "naturalness" from which people take possession of in order to walk through them" (LEFEBVRE, 2008, 117, our translation)[5]. The last residence visited is where the large mast is held, decorated with flowers, leaves and ribbons, it is carried by men to the chapel of Saint Benedict, where a large audience awaits its arrival.

The festive atmosphere is contagious for people. The mast is taken from the chapel to assemble its parts. Mestre Afonso is the one who performs the task, he fits the parts together, the flag with the sweets, the rooster at the top. When the mast rises, the sweets are scattered on the street and the children strategically positioned, collect as many as possible, fun always abounds. The imposing mast next to the shack (Picture.3), is received by fireworks and applause and much music and dance in celebration of Saint Benedict.

Figure 3. Feast of the Mast with the Shack on the left and the Chapel on the right.

Source: Roberta Bettcher (2019).

2.2 *Our Lady of Aparecida feast*

The protective figure of Saint Benedict alludes to the female protective side, the black and welcoming mother, represented by Our Lady of Aparecida. The feast takes place on the national holiday of October 12 and is a religious custom to be preceded by the *novena*, that is, nine days of prayers to the saint. But, also customary in the community of Macrina, is decorating the streets around the church with the colours that symbolize the image of Aparecida, and thus, it is common for the group to concentrate on the street to make and distribute the ornaments.

As the space of the church is restricted, the production of the ornaments for the feast is extended to the street. The interesting thing is that the relaxed and fun atmosphere of this collective production attracts other people from the community who are not regular churchgoers, but on the eve of the feast, get involved. It is as if, by demanding time and voluntary work from the community, the feeling of belonging spreads among the participants and the public spaces have the value of a dwelling, in which the esteem of preparing the street to receive the guests for the feast, gains great importance. Each individual sees themselves represented at the celebration, thus showing a creative and organized community while also displaying the value of everyone participating in the process.

Gradually the street becomes coloured with blue and white ornaments (Picture 4) and the movement of the wind brings a new rhythm to this territory of the city, a calm and festive rhythm, which makes us walk while gazing upwards: "Leaving the representation, ornament and decoration aside, art can become *praxis* and *poiésis* on a social scale: the art of living in the city as a work of art. Returning to the style, the work, that is, the sense of the monument

5 "Os lazeres comercializados, industrializados, organizados institucionalmente, destroem essa "naturalidade" da qual as pessoas se ocupam a fim de traficá-las e trafegar por elas".

Figure 4. Mass on the Dona Macrina Street.
Source: Roberta Bettcher (2019)

and the appropriate space at the Feast, art can pre prepare structures of enchantment" (LEFEBVRE, 2008, 134, our translation)[6].

On a *novena* day, the mass is extended to the street, people explore the possibilities of accommodation, steps, walls, curbs, and the pavement itself are used for rest. Offerings and wafers, which are rituals of the celebration, go beyond the limits of the sacred, go out into the street making it temporarily sacred. The value of art and community belief permeates the territory. The black saint is in the microcosm of the houses, in private oratories, on the wooden rosary in front of the church, collective pieces of art that remain throughout the year, adorning the proud Dona Macrina street, with all the sacred content that the devotion to the image of Aparecida encourages in its residents.

2.3 *Carnival*

The street as a sacred territory via religious processions is the same in which it is allowed to transcend during libertine carnival celebrations. This celebration has the "Unidos da Macrina" carnival parade group representing the neighbourhood and it is within this that local inventions can emerge. The "bichos", for example, are monstrous dolls whose outer structure conceals a person and instill fear during the carnival days.

Another famous figure at this festive event are the so-called "Zé Pereiras" (picture 5), who are children and teenagers who cover all parts of their bodies with clothes, gloves and accessories and use monster masks to scare people, weeks before carnival celebrations begin. They leave in groups, most of them

Figure 5. Zé Pereiras in public spaces.
Source: Roberta Bettcher (2020)

departing from the Macrina hill, at dusk, and roam the city spaces, inciting curiosity and fear in those who catch sight of them.

The fright in the form of play also mirrors the narratives of Exu, the cheating god, who has fun spreading havoc. Cited in African tales, he is part of a frightened providence and "[...] contradictions stem from it: good and evil, life and death, pain and joy, prosperity and deprivation" (CAMPBELL, 1949, 78, our translation)[7]. The mystery of identity, the masking, ceasing to be who you are, and becoming something else, replicates an artifact widely used in African culture, which are masks, representative in rituals and personalized in their aesthetics and meaning, according to the tribe from which they come. Their function is to communicate, therefore they can play a cultural, religious or political role, not only in primitive peoples, but in contemporary movements: the mask maintains anonymity in public manifestations, protects the identity of the heroes of fiction. However, it can also be an instrument of resistance: the "Careta de Acupe, in Bahia's Recôncavo region, was used by slaves to frighten their master allowing escapes and

6 "Deixando a representação, o ornamento, a decoração, a arte pode se tornar *práxis* e *poiésis* em escala social: a arte de viver na cidade como obra de arte. Voltando ao estilo, à obra, isto é, ao sentido do monumento e do espaço apropriado na Festa, a arte pode preparar estruturas de encantamento"

7 "[...] dela procedem as contradições: o bem e o mal, a vida e a morte, a dor e o gozo, a prosperidade e a privação".

fueling the imagination that supernatural demons inhabited those lands." (ETNOS, 2019, 14, our translation)[8].

Death is present in the daily life of a territory that grows around the cemetery and fosters the emergence of urban legends related to fear, such as the Werewolf and ghosts. The collective imagination creates fears about the mysteries of the after-life, as well as resisting cultural differences that dialogue in the same urban context. Thus, the Bichos and Zé Pereiras, materialize this fear ingrained in the Macrina neighborhood, for housing the cemetery and for racial prejudice, and spread fear and mystery through the city streets, before and during the carnival.

3 CONCLUSIONS

Amid traditions and adaptations, Afro-Brazilian culture resists in the city, as a fundamental process of the presence of collectivity in Brazilian public spaces. Communication through its practices and orality facilitate its spread in a familiar language and without the repression of the body.

The rituals of the Saint Benedict feast connect public and private spaces in an irreverent way. A type of festive devotion, where the disciplinary rigidity of the church is in the background, the drums and the simple verses of the songs of *jongo* and *congo*, in the midst of permitted drunkenness, promote a street party with a devotional content which is highly focused on the religious culture of black ancestors, in which religiosity is incorporated into the feelings and values of the group. Saint Benedict is an accessible deity, hiding in the houses, re-enacting the myth of the mast that initiates its approach with blacks and promotes playful devotion through the city streets.

The Our Lady of Aparecida Feast uses the street as an extension of the sacred temple and thus, improvisation in the accommodation of churchgoers, encompasses curbs, walls, stairs and even the pavement of the street. With this adornment, Dona Macrina street receives a party dress, with ornaments hanging from above, dancing on the blue and white ribbons, moving with its shadows on the flagstones. And each year, collective art appropriates the street for its temporary expression, but gradually permeating itself, permanently, singularizing its composition.

At carnival, fear and joy go together. The masked, who incite fear, join the fray before the event. The "Bichos" portray the contradictions existing in African culture, since fear can be represented in masks, as a way of defending the oppression suffered by the black culture itself, but in the Brazilian way, that is, mixed with carnival joy. Free, the "Zé Pereiras", walk aimlessly through the city streets, covered from head to toe and with a monstrous mask on their face. Child and youth play make-believe in the Macrina neighbourhood, which has remained a tradition in the city of Alfredo Chaves. Theatrical performances, with improvised dynamics in urban spaces, all actors without a predefined script or prior organization, as Raffestin would say: strengths in the interaction with the city. Each year, following the cyclical movement of the parties, the creative Macrina neighbourhood sets out on a new journey, in a territory that accumulates collective art in its city expression.

REFERENCES

Bicalho, L. 2000. *O Congo de São Benedito na ilha de Nossa Senhora da Vitória*. Vitória City Hall.

Campbell, J. 10 Ed. 1997. *O herói de mil faces*. São Paulo: Cultrix/Pensamento.

Etnos: faces da diversidade. São Paulo: Farol Santander, 2019, p.9. Exhibition catalog.

Guatari, F. 1992. *Caosmose: Um novo paradigma estético*. Rio de Janeiro: Ed. 34.

Instituto Brasileiro de Geografia e Estatística (IBGE). Availableon:<https://cidades.ibge.gov.br/brasil/es/alfredo-chaves/panorama>. Accessed on February 14, 2020.

Lefebvre, H. 5 Ed. 2008. *O direito à cidade*. São Paulo: Centauro.

Maciel, C. 2ª Ed. 2016. *Negros no Espírito Santo*. Vitória: Public Archive of the State of Espírito Santo.

Raffestin, C. 1993. *Por uma geografia do poder*. São Paulo: Editora Ática S.A.

Sodré, M. 2 Ed. 2016. *As estratégias sensíveis: afeto, mídia e política*. Rio de Janeiro: Mauad X.

Sodré, M. 3 Ed. 2019. *O terreiro e a cidade: A forma social negro-brasileira*. Rio de Janeiro: Mauad X.

8 "A Careta de Acupe, no Recôncavo baiano, era utilizada por escravos para assustar os senhores dos engenhos e fazendas, permitindo fugas e alimentando o imaginário de que demônios sobrenaturais habitaram aquelas terras".

Environmental policy: To the state, institutionalisation, to civil society, action. Comments on the Rio Cocó State Park

V.H.G. Sampaio, J.A. Farias, C.F.S. Freitas & D.R. Cardoso
PPGAU+D, Department of Architecture and Urbanism, Universidade Federal do Ceará, Brazil

ABSTRACT: Should it be up to the State to assume all the duties in an environmental policy, or should it involve the action of civil society in fostering sustainable local solutions? This paper aims to discuss this issue based on an investigation of the historical process of delimitation and regulation of the Cocó State Park, located in the city of Fortaleza and a biophysical component of the largest hydrographic basin in the region. The methodology entails bibliographic and desk research, in addition to examining cartographic databases. The results of the study revealed that urban green areas, a constant target of the fragmentation imposed by real-estate speculation, suffer directly from the slowness of government legal responses and point to the decisive role of socio-environmental movements in the process of recognizing and safeguarding protected areas.

1 INTRODUCTION

Governments change and public policies change with them. The most obvious example of this fact comes from the current management of the Federal Government (2019-2022), which has systematically dismantled the most basic principles of Brazilian environmental policy hard fought for over the last several decades. For many experts, the State must remain a central element for the institutionalisation of environmental issues for the success of the formulation, implementation, and management of sustainability (Ferreira, 2019). We must accept that the State is not a monolithic body, endowed with an inflexible homogeneity, where public policies, as an expression of the action of State, result from the interaction between collective and individual actors, who liaise to assert their unique interests.

However, given the clout of a State that has been historically patrimonialist and fragmented into social extracts, we know that environmental challenges can only be faced and overcome through better articulation and collaboration between the several sectors of society. Thus, in a network of interdependent actions, social movements and other civil organisations can push the State towards restructuring the environmental agenda, by taking effective measures capable of reversing the numerous devastation processes underway, especially those related to an urban ecosystem that is essential for the quality of life of its residents.

This study aims to discuss this issue through an investigation of the lessons learned from the historical process of delimitation and regulation of the Cocó State Park, located in the city of Fortaleza, and a biophysical component of the largest watershed of a metropolitan region with over 4 million inhabitants (IBGE/2018 estimate). As Cocó State Park is the main green area in the city setting and a focus of constant political demonstrations, we seek to carry out a historical survey of the process of delimitation and regulation of the region from the 1970s, through the definitive polygonal, in 2017, to the actions of maintenance and improvement of its infrastructure undertaken by the government of the state of Ceará.

We consider the hypothesis that State action, fundamental in the processes of regulation and legal control, loses prominence over time because a well-managed environment, with good results, depends essentially on what people do or fail to do. In this sense, a consistent environmental policy can only be achieved under political regimes of participatory democracy that guarantee creativity and the autonomous management of society. The State, naturally remains central to the implementation of this set of transformations of public management, especially concerning institutionalisation, regulation, and supervision. But civil society action is essential to control and correct directions according to popular desires (Freitas, 2019).

The methodology adopted was a case study approach. For the understanding and analysis of the historical process of the social mobilisation in defence of Cocó State Park, we researched bibliography (scientific articles, dissertations, journalistic articles) and documents (Master Plan and other legal instruments), and examined official cartographic

DOI: 10.1201/9781003220855-34

databases. For the analysis of the evolution of the vegetation over time, we used images from LANDSAT 5 MSS C1 Level-1 (1985) and Sentinel-2 (2019), made available by the Earth Explorer website (https://earthexplorer.usgs.gov/) of the United States Geological Survey (USGS). The selection of images was made according to the quality of the satellite capture and percentage of the block obscured by cloud cover, and months of the year with approximate rainfall characteristics (July to September, since these represent the end of the rainy season in the state of Ceará). Image processing was performed using the QGIS 3.6.3 software version Noosa, a free and open Geographic Information System (GIS).

2 GREEN AREAS AND ECOSYSTEM SERVICES

Three simultaneous shifts generally characterise the transformation processes of urban green areas. The first is a more immediate change, consisting of a decrease in the total green area and natural habitats. As a result, there is an increase in fragmentation that also decreases continuous habitat patches, and greater distance between habitats, one of the main threats to urban biodiversity. There is also a functional homogenisation of green areas, due to the need to 'design' and 'recreate' nature in the city, emphasising its 'domestication' (Madureira et al. 2011).

This state of movement can only be interrupted if society understands the importance and scope of services and benefits associated with urban green infrastructure. The green infrastructure is composed of multifunctional meshes of permeable and vegetation covered fragments, preferably wooded and interconnected, allowing the recomposition of the several elements that integrate the landscape mosaic. This is why green infrastructure poses a challenge for urban planners and public policy formulation beyond the sole management of the urban landscape (Pulighe et al. 2016).

The Millennium Ecosystem Assessment (MEA) report considers urban green spaces as contributors to a society's regulatory and cultural ecosystem services (MEA, 2005). Access to ecosystem services, available through urban green spaces, is directly related to the quality of life in cities and is also considered one of the main aspects of sustainability in a highly anthropised environment. Urban green areas provide recreational facilities that perform playful and educational roles, promote the physical and psychic health of humans, offer a place of refuge for everyday occupations, and are the primary source of contact with nature (Bennett et al. 2016; Sandifer et al. 2015).

The benefits of urban green areas are relevant. They contribute toward mitigating floods, acting as collectors of part of the rainfall in cities due to their minimal impermeability. Green areas also help to reduce risk of landslides, acid rain and heat islands, because they are responsible for increasing local humidity and forming islands of freshness (Moura, 2008). They reduce urban discomfort and help improve atmospheric composition, soil balance, weather, climate, while also diminishing pollution through CO2 absorption, bacterial purification, and fixation of dust and particulate matter (Rossatto et al. 2018).

As the main link between society and the environment, the discussion on the importance of urban green spaces is gaining visibility in debates on sustainability, well-being, and quality of life in the built environment (Stessens et al. 2017). However, the real mastery of the correlation between green infrastructure and its respective ecosystem services for urban planning and management processes is still limited (Baró et al. 2015). If the study of this infrastructure is done in isolation, without dialogue and cooperation between the stakeholders - whether public or private -, or if policymakers cannot access public consultation, the benefits to the population may not be satisfactory (Kabisch et al. 2016).

3 FORTALEZA AND THE COCÓ RIVER BASIN

The Cocó river basin is part of the metropolitan setting of Fortaleza. The total drainage area of the basin of approximately 513.84 km² covers part of the municipalities of Pacatuba, Maranguape, Itaitinga, Maracanaú, Eusébio, Aquiraz, and Fortaleza. Its headwaters are located on the eastern slope of Serra da Aratanha, Pacatuba city, draining into the Atlantic Ocean (Figure 1). It is the most important basin of Fortaleza city, today the fifth most populous in the country, with an estimated 2,643,247 inhabitants (IBGE, 2018) in an area of 312.407 km², giving it a population density of 7,786.44 inhabitants/km² as the 9th densest city in the country (IBGE, 2010).

On the metropolitan scale, the inadequacy of urban planning and management policies and

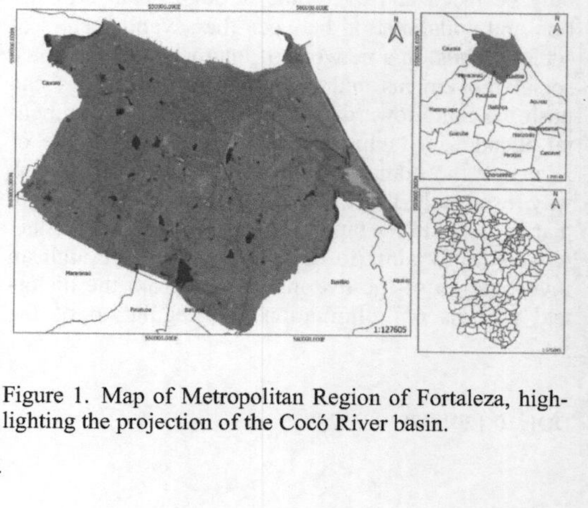

Figure 1. Map of Metropolitan Region of Fortaleza, highlighting the projection of the Cocó River basin.

instruments is due, among other factors, to the characteristics of the Brazilian Federal pact, which does not incorporate metropolitan regions nor the use of the territorial scale of watersheds for urban development purposes as a sphere of effective articulation of public policies. Most cities limited institutional capacity is chiefly reflected in the formulation and implementation of environmental policy instruments.

Fortaleza, akin to many other Brazilian capitals, is undergoing major population expansion and a significant process of transformation and verticalization of the urban space (Alves, 2013). As the urban area of the capital of Ceará expands, vegetation coverage is shrinking significantly (Figure 2).

According to the Environmental Inventory of Fortaleza (2003), its municipal territory underwent an intense process of urbanisation and environmental degradation. The city, which originally was almost totally covered with native forest vegetation, in 1968 had only 65.79% of its 312,407 km² area preserved. In the ensuing years, the loss of vegetation coverage was significant. In 1990, the percentage remaining was 16.64%, while in 2002 this number was even more significant, decreasing to an alarming 7.06%. In other words, within the space of 34 years, there had been a loss of green area equivalent to over half of its entire territory, with irreparable damage to the region's fauna and flora.

For several years, the Cocó River basin was a region of low population density, considered a natural barrier to the expansion of the eastern sector of Fortaleza. It was only from 1950, with the subdivision of the "Sítio Cocó" estate, that the city traversed the River - then described as an "obstacle" to development. The construction of the shopping mall Iguatemi, the administrative centre of the state of Ceará, the university of Fortaleza, the courthouse Clóvis Beviláqua, the beltway, and the landfilling of mangrove areas intensified the process of occupation of this region and brought rural areas closer to the urban nucleus (Costa, 2005; Rufino, 2012). The construction of the Cidade 2000 housing complex, in 1972, and the extension of Santos Dumont Avenue, the main access to the new infrastructure, represented another milestone in the process of expansion and subdivision of the region, strengthened by the Integrated Master Plan of Greater Fortaleza (PLANDIRF), drafted in 1972 (Soares, 2005)

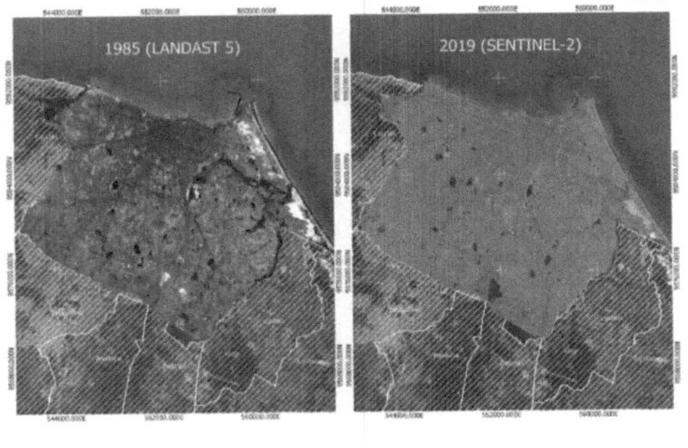

Figure 2. Process of urban expansion in Fortaleza-CE in the last 34 years. Produced by author using the QGIS Software on the images from Landsat 5, 1985, Sentinel-2 Satellite, 2019 [U.S. Geological Survey – USGS] and Google Satellite, 2019.

4 DEFENDING THE PARK

As expected, the intense process of urban expansion and environmental degradation galvanized environmentalist groups in the population of Fortaleza. The timeline of the ecological movement in the city (Figure 3) began in 1976, with the creation of the Cearense Society for the Defense of Culture and the Environment - Socema. Social pressures were initially motivated by the creation and delimitation of a park aiming to preserve the Cocó River and its ecosystem (Nottingham, 2006; Lopes, 2014). The demands and struggles for environmental issues continued in Fortaleza, being represented throughout its history by groups such as the SOS Cocó Movement, the Ceará Civil Society Forum on

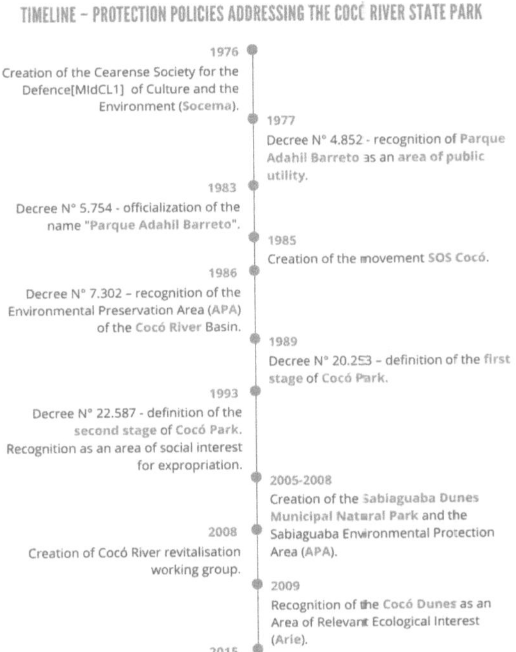

Figure 3. Timeline of the ecological movement in defence of the Rio Cocó State Park. Updated/modified by the authors from Frota (2018).

Environment and Development, SOS Litoral and SOS Lagoas (Costa, 2011).

Socema's first action was to block the construction of Banco do Nordeste (BNB) headquarters in the current Adahil Barreto Park (Figure 4), as this was a large architectural project. Lobbying prevented construction in the areas along the banks of the Cocó River and transformed these into a Conservation Unit (CU). Soon the Adahil Barreto Ecological Park became a symbol of the popular environmental struggle in the city. Mayor Evandro Ayres de Moura, through Decree No. 4.852, of March 29, 1977, made the area of public utility and made its name official in 1983 through Decree No. 5.754 (Frota, 2018).

In 1985, in response to society's growing involvement in environmental causes, the SOS Cocó movement emerged. With this, Socema was no longer the only popular organisation defending the ecological interests of a region that is constantly threatened by real-estate speculation. Under the administration of mayor Maria Luiza Fontenele, in 1986 the Cocó River basin region was recognised as an Environmental Preservation Area (APA). It took eight years of lobbying until the signing of Decree No. 7.302, of January 29.

The recognition of the APA, however, did not fully guarantee preservation of the region. On September 5, 1989, Governor Tasso Ribeiro Jereissati signed Decree No. 20.253 which delimited the first stage of Cocó Park, expanding the Adahil Barreto Park area to 1,046.22ha. The second stage was delimited by Decree No. 22.587, of June 8, 1993, recognising Cocó as an area of social interest for expropriation (LAPUR, 2017).

During city mayor Luizianne Lins's first term (2005-2008), the Sabiaguaba Dunes Municipal Natural Park and the Sabiaguaba Environmental Protection Area (APA) were created, both of enormous environmental value for the Cocó River APA. In 2009, together with councilor João Alfredo, the municipal government recognised the Cocó Dunes as an Area of Relevant Ecological Interest (Arie). The association of construction and real-estate entrepreneurs (Associação Cearense dos Empresários da Construção e Loteadores - Acecol) reacted immediately, claiming that Law 9.502/09 providing for the creation of the Arie was unconstitutional. In agreement with Acecol, Ernani Barreira, then president of the Court of Justice of Ceará, suspended the law (Coutinho, 2010). The order was soon overturned by the full court which, by 28 votes to 5, recognised the law's constitutional validity (Rebouças, 2011).

In 2008, by Decree of the Government of the State of Ceará, a working group was created to develop the Cocó River revitalisation program and to promote the formalisation of the Park Conservation Unit. On January 20th, 2015, the Federal Public Ministry created the Cocó Forum, whose objective was to discuss the issues related to Cocó Park in a multidisciplinary way. After a series of discussions, the Forum presented, on August 26 of the same year, a proposal for an Integral Protection Conservation Unit of 1,435 ha. The following year, in April, the polygon of the park was defined by the government at 1,050ha (Frota, 2018). According to Frente Cocó (2016), the exclusion of 385ha resulted from a lack of public funds for local expropriation.

On June 4th, 2017, the Cocó State Park was created. The final polygon, defined by Decree of Governor Camilo Santana, is 1,571.29 hectares (Figure 5). Although the polygon area defined is bigger than the original proposed by the Cocó Forum, the new extension took place in the southern part of Fortaleza, a region less valued by the real-estate market (Figure 6).

Despite the controversies and discussions involving the process of regulation of Cocó State Park, the

Figure 4. Adahil Barreto Park with image of Fortaleza in 1985. Produced by author using QGIS Software on the images from LANDSAT 5, 1985 [U.S. Geological Survey – USGS], and cartographic base for the polygonal of the Park of Fortaleza City Hall, 2019.

Figure 5. Cocó State Park and its polygons, before and after the 2017 regulation. Produced by author using the QGIS Software on images from Sentinel-2, 2019 [U.S. Geological Survey – USGS], Google Satellite, 2019 and cartographic base for the polygons of the Cocó Park Forum, 2017, and Government of the State of Ceará, 2017.

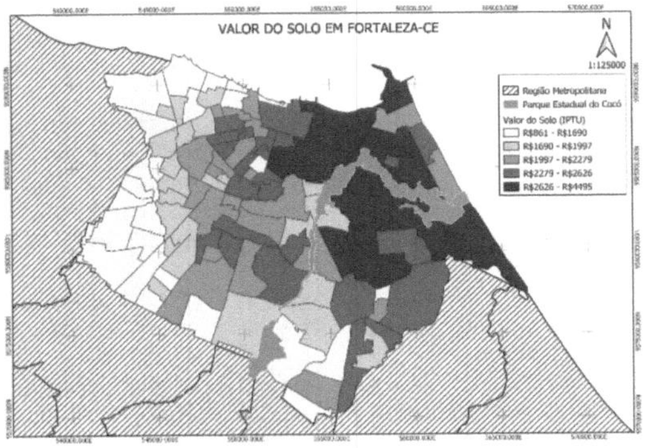

Figure 6. Cocó State Park (polygonal 2017) and the territorial classification of Fortaleza according to land values. Produced by author using the QGIS Software and cartographic base for the polygonal of the State Government Park, 2017, and land classification of Fortaleza City Hall, 2019.

government acted vigorously in safeguarding and maintaining the region. Actions included the construction of Cocó cinema, the promenade and accessibility ramps, the revitalisation of sports equipment, bridges, and trails, the identification of the species in the park (Aflorar Project), navigation on the Cocó River, the installation of a public outdoor gym and shared bicycle station in partnership with Unimed Fortaleza and the City Hall, the renovation of the Adahil Barreto Park, and creation of the "Viva o Parque" project.

5 CONCLUDING REMARKS

The study revealed that the efforts of socio-environmental movements were crucial in recognising and safeguarding the Cocó River. This case illustrates that urban green areas, a constant target of the fragmentation imposed by real-estate speculation, suffer directly from the government's slow legal responses. Government resistance to recognising the values of environmental heritage over the years is also noteworthy.

Although it has taken the initiative to execute programs and projects for the collective use of the Cocó Park, it still lacks an essential provision: to put into practice the correlation between green infrastructure and their respective ecosystem services in urban planning and management processes. Consolidating this measure by understanding the lessons of recent history will depend once again on the action of socio-environmental movements, summoning the will to act toward a common purpose under shared interpretation and meaning. To this end, it would be necessary to create a regulatory framework for an environmental policy designed to rectify external issues. This framework would then constitute the relevant element to achieve resilience in prevention, and responsibility in dependence.

REFERENCES

Alves, T. C. V. A. 2013. Parques urbanos de Fortaleza-CE: espaço vivido e qualidade de vida. 198. Rio Claro.

Baró, F. et al. 2015. Mismatches between ecosystem services supply and demand in urban areas: A quantitative assessment in five European cities. Ecological Indicators, [s.l.], v. 55, p.146–158. Elsevier BV. http://dx.doi.org/10.1016/j.ecolind.2015.03.013.

Bennett, G. et al. 2016. Natural infrastructure investment and implications for the nexus: A global overview. Ecosystem Services, [s.l.], v. 17, p.293–297. Elsevier BV. http://dx.doi.org/10.1016/j.ecoser.2015.05.006.

Brasil. Instituto Brasileiro de Geografia e Estatística – IBGE. População estimada 2018. Rio de Janeiro.

Brasil. Instituto Brasileiro de Geografia e Estatística – IBGE. 2010. Censo Demográfico 2010. Rio de Janeiro.

Brasil. Ministério do Meio Ambiente (MMA). Conselho Nacional do Meio Ambiente (CONAMA). Resolução Conama n. °369, de 28 de março de 2006.

Costa, A. S. 2011. Demandas do movimento ambiental por áreas verdes em Fortaleza. Fortaleza.

Costa, C. 2005. Fortaleza: expansão urbana e organização do espaço. In: Silva, Jose Borzacchiello da; Cavalcante, Tércia C.; DANTAS, Eustógio W. C. (Org.). Ceará: um novo olhar geográfico. Demócrito Rocha, v. 1, p. 51–100. Fortaleza.

Coutinho, T. 2010. Suspensa lei que impede novas construções no Cocó. O Povo, 16 jun. 2010 Política, p. 25. Fortaleza.

Ferreira, L. C. 2019. A Questão Ambiental: sustentabilidade e políticas públicas no Brasil. 2ª ed. Boitempo Editorial. São Paulo.

Fortaleza. Prefeitura Municipal 2003. Inventário ambiental de Fortaleza: diagnóstico versão final. Fortaleza.

Fortaleza. Prefeitura Municipal. Secretaria Municipal do Urbanismo e Meio Ambiente 2019. Fortaleza em Mapas. Fortaleza.

Frente Cocó 2016. Source: https://www.facebook.com/FrenteCoco/videos/1371871366173337/.

Freitas, C. F. S. 2019. Insurgent planning? City, [S.L.], v. 23, n. 3, p. 285–305. Informa UK Limited. http://dx.doi.org/10.1080/13604813.2019.1648030.

Frota, N. T. S. 2018. Planejamento urbano do institucional ao insurgente: Uma análise sobre a atuação dos movimentos socioambientais na proteção dos bens comuns urbanos. Fortaleza.

Kabisch, N. et al. 2016. Urban green space availability in European cities. Ecological Indicators, [s.l.], v. 70, p.586–596. Elsevier BV. http://dx.doi.org/10.1016/j.ecolind.2016.02.029.

Lapur. Laboratório de Planejamento Urbano e Regional 2017. Demarcação do Parque do Cocó: de 1977 até 2017. Fortaleza.

Lopes, G. 2014. Práticas sociais e cotidiano: o parque ecológico do cocó em análise. Fortaleza.

Madureira, H. et al. 2011. Green structure and planning Evolution. In Urban Forestry & Urban Greening, 10(2), 141–149.

Millennium Ecosystem Assessment 2005. World Resources Institute. Ecosystems and Human Well-Being:

Opportunities and Challenges for Business and Industry. 36 p. Washington, DC.

Moura, M. O. 2008. O clima urbano em Fortaleza na perspectiva do campo térmico. Fortaleza.

Nottingham, P. C. 2006. Tempos verdes em Fortaleza: um olhar sobre a experiência do movimento ambientalista (1976-1992). Fortaleza.

Pulighe, G. et al. 2016. Insights and opportunities from mapping ecosystem services of urban green spaces and potentials in planning. In Ecosystem Services, [s.l.], v. 22, p.1–10. Elsevier BV. http://dx.doi.org/10.1016/j.ecoser.2016.09.004.

Rebouças, H. 2011. Meio ambiente: justiça rejeita ação e mantém proteção às dunas do Cocó. In Jornal O Povo, 1 abr. 2011. Política, p.20. Fortaleza.

Rossatto, D. R. et al. 2008. Arborização urbana na cidade de Assis-SP: uma abordagem quantitativa. In Anais SBAU, v. 3, n. 3, p.1–16. Piracicaba.

Rufino, B. 2012. Incorporações da Metrópole: centralização do mercado imobiliário e nova produção do espaço de Fortaleza. São Paulo.

Sandifer, P. et al. 2015. Exploring connections among nature, biodiversity, ecosystem services, and human health and well-being: Opportunities to enhance health and biodiversity conservation. In Ecosystem Services, v.12, p.1–15. Elsevier BV. http://dx.doi.org/10.1016/j.eco-ser.2014.12.007.

Soares, J. (2005). Parque ecológico do cocó: a produção do espaço urbano no entorno de áreas de proteção ambiental. Fortaleza.

Stessens, P. et al. 2017. Analysing urban green space accessibility and quality: A GIS-based model as spatial decision support for urban ecosystem services in Brussels. In Ecosystem Services, [s.l.], v. 28, p.328–340. Elsevier BV. http://dx.doi.org/10.1016/j.ecoser.2017.10.016.

USGS. U.S. Geological Survey 2019. In Earth Explorer.

Author index